NANOSTRUCTURED SUPERCONDUCTORS

NANOSTRUCTURED SUPERCONDUCTORS

Victor V Moshchalkov
Joachim Fritzsche

Katholieke Universiteit Leuven, Belgium

World Scientific

NEW JERSEY · LONDON · SINGAPORE · BEIJING · SHANGHAI · HONG KONG · TAIPEI · CHENNAI

Published by

World Scientific Publishing Co. Pte. Ltd.

5 Toh Tuck Link, Singapore 596224

USA office: 27 Warren Street, Suite 401-402, Hackensack, NJ 07601

UK office: 57 Shelton Street, Covent Garden, London WC2H 9HE

British Library Cataloguing-in-Publication Data

A catalogue record for this book is available from the British Library.

Cover image: M.C. Escher's "Metamorphosis III" ©2010 The M.C. Escher Company-Holland.
All rights reserved. www.mcescher.com

ISBN-13 978-981-4343-91-6
ISBN-10 981-4343-91-9

Printed in Singapore by Mainland Press Pte Ltd.

A blank piece of paper is open for further structuring. It can be structured with words, symbols, patches of colour, etc., breaking the symmetry of a non-structured blank paper and producing a sensible or abstract pattern. This pattern tells us a story, depicts an object, captures a tune, invokes a hidden feeling.... In this sense, patterning can generate a variety of new properties. Breaking plane symmetry with patterns, like Escher's famous creations, can generate objects evolving, for example, from fishes to birds. In analogy to that, breaking the symmetry of superconducting films by nanopatterning can result in novel properties unseen in pristine materials. This opens new fascinating perspectives for superconductivity which celebrates in 2011 its centennial.

Preface

The book presents a variety of novel vortex phases and dynamic vortex states studied recently in nanostructured superconductors. In individual nanostructures a strong topology- and geometry-dependence of the critical fields and the formation of the symmetry induced anti-vortices is discussed. In superconductor/ferromagnet hybrids such new phenomena as field-induced superconductivity and the domain wall superconductivity are presented. The recent progress in the research of nanostructured superconductors, covered by the book, provides the basic principles for enhancing the superconducting critical parameters through nanostructuring and for developing a variety of novel fluxonics devices based on vortex manipulation. Nanostructuring can, in fact, create such conditions for the flux pinning by arrays of nanofabricated antidots or magnetic dots, which could maximize the second important superconducting critical parameter (critical current) up to its theoretical limit — the depairing current. The underlying physics of this enhancement has been discussed on the basis of the results obtained experimentally by using integrated response and local techniques (Lorentz microscopy, scanning probe microscopy) for the direct vortex visualization.

These studies have revealed novel static phases (multiquanta and composite vortex lattices, vortex-antivortex molecules, etc) and new dynamic effects (guided vortex motion and vortex rectification, multiple sign reversal of the vortex ratchet effect, etc). The concept of controlling the superconducting properties through nanostructuring, presented in this book, is now becoming the backbone for developing new elements and systems for various applications.

The book provides a balanced view over a very dynamic field where more problems have been formulated than the answers found. Interestingly, superconducting condensates have many similarities with superfluid helium and Bose-Einstein condensates. Nanoscale confinement effects studied in superconductors can be also projected on other quantum condensates in restricted geometries where similar effects could be also observed.

Alexander F. Andreev

Acknowledgments

The authors would like to acknowledge the support from the FWO-Vlaanderen, the Center of Excellence INPAC at the KULeuven, the Methusalem Funding by the Flemish Government, and the ESF Programme 'Nanoscience and Engineering in Superconductivity — NES'. The authors are thankful to the members of the NSM&PF group at the KULeuven for their contribution to our successful joint research of nanostructured superconductors, reflected in this book.

Discussions with the following colleagues are gratefully acknowledged (in alphabetic order): A. A. Abrikosov, J. A. Aguiar, A. Yu. Aladyshkin, A. F. Andreev, E. Babaev, A. Barone, S. J. Bending, V. Bruyndoncx, A. I. Buzdin, C. Carballeira, G. Carneiro, A. Ceulemans, Q. H. Chen, L. F. Chibotaru, V. H. Dao, L. De Long, C. C. De Souza Silva, J. T. Devreese, H. J. Fink, Ø. Fischer, V. M. Fomin, K. Fossheim, H. Fukuyama, W. Gillijns, V. N. Gladilin, A. A. Golubov, D. S. Golubovic, T. Ishida, B. Janko, V. V. Kabanov, K. Kadowaki, G. Karapetrov, M. Kato, P. H. Kes, J. Kolacek, D. Kölle, R. B. G. Kramer, W. Lang, M. Lange, I. F. Lyuksyutov, S. Maekawa, V. V. Metlushko, V. Misko, M. Morelle, T. Nishio, X. Obradors, N. F. Pedersen, F. M. Peeters, W. V. Pogosov, T. Puig, X. G. Qiu, J. G. Rodrigo, P. Samuely, A. V. Silhanek, A. Sudbø, H. Suderow, M. Tachiki, J. Tempere, G. Teniers, A. Tonomura, J. Vanacken, J. Van de Vondel, F. Vidal, S. Vieira, H. H. Wen, R. Wördenweber, Z. R. Yang, B. R. Zhao, Z. X. Zhao, and B. Y. Zhu.

Both authors would also like to thank their wives Nina and Annelies for their encouragement and appreciation of writing this book.

Contents

List of Figures

Chapter 1

Introduction

Nanostructured superconductors play a special role in nano science. These materials provide a unique opportunity to apply quantum mechanical principles to obtain specific superconducting properties needed for applications, by using nano structuring to modify and control the coherent quantum ensembles of correlated electrons or holes responsible for the appearance of superconductivity. Designing specific material properties through the application of quantum mechanical principles is 'quantum design' — a key idea in nano science. Superconductors, with their inherent quantum coherence over even macroscopic scale, not to mention nanoscopic scale, are in that respect superior to semiconducting, magnetic or normal metallic materials, where quantum coherence is much more difficult to achieve. In that respect, nanostructured superconductors is the best choice for the demonstration of applicability of quantum design to tailor specific properties of nano materials.

The possibilities of practical applications of superconducting materials are limited by their critical parameters: temperature T, field H and current j, which define the critical surface that separates the superconducting from the normal state in the space span by these parameters (Fig. 1.1). The state of zero resistance can only be achieved below the critical surface. Therefore, expanding the critical surface as far as possible is one of the main challenges in physics and material science of superconductors.

Remarkably, superconductors are materials where an artificial nanoscale modulation can substantially improve their critical parameters. Like for a particle in a box, properties of the confined condensate in nanostructured superconductors can be designed using quantum laws. In contrast to the classical approach, which relied upon the search for new bulk materials each time a specific combination of their physical properties was required, nano

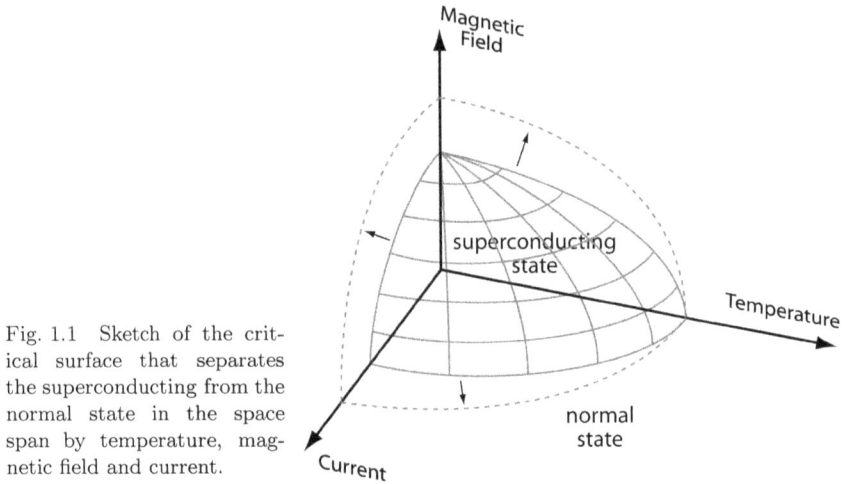

Fig. 1.1 Sketch of the critical surface that separates the superconducting from the normal state in the space span by temperature, magnetic field and current.

sciences rely upon the modification of the properties of the same material through its nanostructuring and the optimization of the confinement potential and topology. Vortices in superconductors carry a quantized flux, normally a single flux quantum or fluxon. Controlling the vortex behavior and creating a guided vortex motion via nanoengineering of arrays of pinning sites or channels in the superconductor, makes it possible to develop new devices like micronet superconducting transistors, vortex lenses, switches, pumps, etc. This paves the way for designing new generation of devices based on the controlled behavior of fluxons. The charge and the spin of electrons form the core of electronics and spintronics, respectively. In the same way, mastering fluxon behavior in nanostructured superconductors creates exciting new possibilities to develop the basics of fluxonics.

By optimizing condensate and flux confinement, the superconducting critical parameters can be enhanced through nanostructuring (Fig. 1.1). This book is focussed on the effects of nanostructuring on the nucleation of superconductivity, $T_c(H)$, and on the behavior of the vortex matter. We are not considering possible ways to increase the zero field superconducting critical temperature, which is currently a very challenging problem without any precise clues to solve it.

In this book we present a variety of nanoscale configurations used to confine flux and condensate in nanostructured superconductors, moving from single nanocells (loop, disc, triangle, square, etc.) via their clusters to their huge arrays (antidot lattices, etc.). In addition to that, nanoscale magnetic templates, where highly inhomogeneous local magnetic fields were generated by ferromagnetic nanodots and magnetic domains, are considered in superconductor–ferromagnet hybrid nanosystems.

Importantly, already in the introduction, the authors would like to note that the literature and the references used by the authors of this book by no means can be treated as a complete set. Due to the dynamic and rather complex character of the field of nanostructured superconductors, and also due to the limited space in this book, inevitably, a number of important and interesting contributions could have been unintentionally missed. Therefore, in a way, the references given in this book reflect a 'working list' of the publications the authors of this book have been dealing with.

1.1 Quantization and confinement in nano-materials

'Confinement' and 'quantization' are two closely related definitions: if a particle is 'confined' then its energy is 'quantized', and vice versa. According to the dictionary, to 'confine' means to 'restrict within limits', to 'enclose'. A typical example, illustrating the relation between confinement and quantization, is the restriction of the motion of a particle by enclosing it within an infinite potential well of size L_A. Due to the presence of an infinite potential $U(x)$ for $x < 0$ and $x > L_A$, see Fig. 1.2, the wave function $\psi(x)$ describing the particle is zero outside the well: $\psi = 0$ for $x < 0$ and $x > L_A$. In the region with $U(x) = 0$, i.e. for $0 \leq x \leq L_A$, the solutions of the one-dimensional Schrödinger equation correspond to standing waves with an integer number n of half wavelengths λ along L_A, i.e. $n\lambda_n/2 = L_A$.

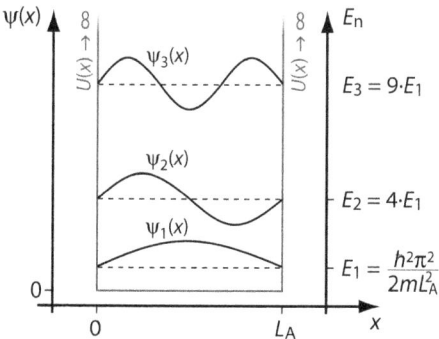

Fig. 1.2 Confinement and quantization of the motion of a particle by an infinite potential well applied on the length scale L_A for $n = 1$, 2 and 3.

This simple constraint results in the well-known quantized energy spectrum

$$E_n = \frac{\hbar^2 k_n^2}{2m} = \frac{\hbar^2 (2\pi/\lambda_n)^2}{2m} = \frac{\hbar^2 \pi^2}{2mL_A^2} n^2 \tag{1.1}$$

with the wave number k_n and the free electron mass m. To have an idea about the characteristic energy scales involved in this example and their dependence upon the confinement length L_A, the energies E_1 (Eq. 1.1) for electrons in an infinite potential well with the sizes 1 Å, 1 nm and 1 μm, are given in Table 1.1.

1.2 Nanostructuring

Recent impressive progress in nano fabrication has made it possible to realize the whole range of confinement lengths L_A: from 1 μm (photo- and e-beam lithography), via 1 nm to 1 Å (single atom manipulation) and, through that, to control the confinement energy (temperature) from a few mK higher up to far above room temperature (see Tab. 1.1).

This progress has strongly stimulated the experimental and theoretical studies of different nanostructured materials and individual nanostructures. The interest towards such structures arises from the remarkable principle of '*quantum design*', when quantum mechanics can be efficiently used to tailor the physical properties of nanostructured materials.

Nano structuring can also be considered as a sort of an artificially in-troduced nanoscale modulation. We can identify then the main classes of nanostructured materials using the idea of their modification along one, two or three axes, thus introducing one–dimensional (1D), 2D or 3D arti-ficial modulation, respectively (Fig. 1.3). The 1D or 'vertical' modulation represents then the class of superlattices or multilayers (upper left panel of Fig. 1.3) formed by alternating individual films of two ('A', 'B') or more different materials in a stack. Some examples of different types of multi-

Table 1.1   Confinement by the infinite potential well

Confinement length L_A	Energy E_1	Temperature T [a]
1 Å	38 eV	$\sim 4 \times 10^5$ K
1 nm	0.38 eV	$\sim 4 \times 10^3$ K
1 μm	0.38 μeV	~ 4 mK

[a]The corresponding temperature at which the kinetic energy of a classical 1D-particle is equal to E_1, given by Eq. (1.1).

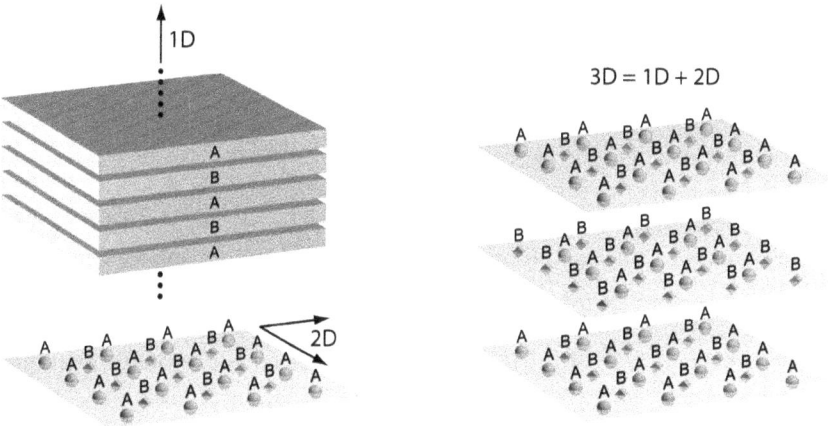

Fig. 1.3   Schematic presentation of the vertical modulation in superlattices or multilayers (upper left), of the horizontal modulation achieved by a lateral repetition of elements A and B (lower left) and of the 1D+2D=3D artificial modulation (right).

layers are superconductor/insulator (Pb/Ge, WGe/Ge, ...), superconductor/metal (V/Ag, ...), superconductor/ferromagnet (Nb/Fe, V/Fe, ...), ferromagnet/metal (Fe/Cr, Cu/Co, ...), etc. [1, 100, 123, 239, 241, 257, 310, 354, 392, 451, 486, 503].

The 'horizontal' (lateral) superlattices (lower left panel of Fig. 1.3) correspond to the 2D artificial modulation achieved by a lateral repetition of one ('A'), two ('A', 'B') or more elements ('elementary cells'). As examples, we should mention here antidot arrays or antidot lattices, when 'A' = micro hole ('antidot'), and arrays and lateral superlattices consisting of magnetic dots. If the 2D lateral modulation is applied to each individual layer of a multilayer or superlattice, then we deal with the 1D + 2D = 3D artificial modulation (right panel of Fig. 1.3). Finally, macroscopic nanostructured samples, with a huge number N of repeated elementary nanocells ('A', 'B', ...), are examples of very complicated systems if the confined charge carriers or flux lines are strongly interacting with each other and the relevant interaction is of a long range. In this case the essential physics of such systems can be understood much better if one first uses *clusters* of elements ($N \simeq 10$), instead of their *huge arrays* ($N \to \infty$), see Fig. 1.4. These clusters, occupying an intermediate place between individual nanostructures ($N = 1$) and nanostructured materials ($N \to \infty$), are very helpful model objects to study the interactions between flux lines or superconducting condensate confined by finite arrays (clusters) of elements 'A'. The growth of

clusters on the way from an individual object 'A' to a huge array of 'A's
can be done either in a 1D or 2D fashion (Fig. 1.4), thus realizing 1D chains
or 2D-like clusters of elements 'A'.

1.3 Confining the superconducting condensate

The nanostructured materials and individual nanostructures, introduced in
the previous section, can be prepared using modern nanofabrication facili-
ties. It is worth, however, first asking ourselves a few simple questions like:
why it is important to make such structures, what interesting new physics
can be expected, and why it is of interest to focus on superconducting (and
not, for example, normal metallic) nanostructured materials?

First of all, by making nanostructured materials, one creates an artificial
potential in which superconducting condensate or flux lines are confined.
The size L_A of an elementary 'cell' 'A', gives roughly the expected energy
scale in accordance with Table 1.1, while the positions and topology of
the elements 'A' determine the pattern of the potential modulation. The
concentration of charge carriers or flux lines can be controlled by varying
a gate voltage (in 2D electron gas systems) [135, 140, 427] or the applied
magnetic field (in superconductors) [377]. In this situation, different commensurability effects between the fixed number of elements 'A' in an array
and a tunable number of charge or flux carriers are observed.

Secondly, modifying the sample topology in nanostructured materials
creates a unique possibility to impose the desired boundary conditions,
and through that to change the properties of the sample. A Fermi liquid or
a superconducting condensate confined within such materials will be subjected to artificially introduced constraints and, as a result, the properties
of these materials will be strongly affected by the boundary conditions.

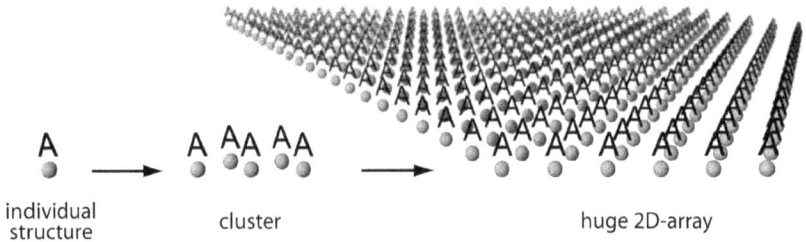

Fig. 1.4   Schematic presentation of an individual structure ('A' = 'nanoplaquette'), a
cluster of nanoplaquettes and a huge array of nanoplaquettes.

While a normal metallic system can be considered quantum-mechanically by solving the Schrödinger equation:

$$\frac{1}{2m}\left(-\imath\hbar\boldsymbol{\nabla} - e\boldsymbol{A}\right)^2 \psi + U\psi = E\psi \ , \tag{1.2}$$

a single component superconducting system is described by the two coupled Ginzburg–Landau (GL) equations:

$$\frac{1}{2m^\star}(-\imath\hbar\boldsymbol{\nabla} - e^\star\boldsymbol{A})^2\psi_s + \beta|\psi_s|^2\psi_s = -\alpha\psi_s \tag{1.3a}$$

$$\boldsymbol{j} = \boldsymbol{\nabla}\times\boldsymbol{h} = \frac{e^\star}{2m^\star}\left[\psi_s^\star(-\imath\hbar\boldsymbol{\nabla} - e^\star\boldsymbol{A})\psi_s + \psi_s(\imath\hbar\boldsymbol{\nabla} - e^\star\boldsymbol{A})\psi_s^\star\right] \ , \tag{1.3b}$$

with the vector potential $\boldsymbol{A}$ which corresponds to the microscopic field $\boldsymbol{h} = \boldsymbol{\nabla}\times\boldsymbol{A}/\mu_0$, the potential energy U, the total energy E, a temperature dependent parameter α changing sign from $\alpha > 0$ to $\alpha < 0$ as T is decreased through T_c, a positive temperature independent constant β, and the effective mass $m^\star$ which can be chosen arbitrarily and is generally taken as twice the free electron mass m.

In magnesium diboride MgB_2 [357] and other superconductors [227,266], the use of a single component order parameter ψ_s is not sufficient to obtain an adequate description of these materials. In these cases, instead of a single component, a more general Ginzburg–Landau formalism is applied with the two interacting components $\psi_{s,1}$ and $\psi_{s,2}$ of the order parameter $\psi_s = \psi_{s,1} + \psi_{s,2}$. In the case of MgB_2, $\psi_{s,1} = \psi_\pi$ and $\psi_{s,2} = \psi_\sigma$. A two-component Ginzburg–Landau free energy functional [19,20,22,67,515] is then the sum of two single band Ginzburg–Landau functionals with a Josephson coupling term corresponding to the interaction between the two bands:

$$F = \int dr^3 \left(F_\sigma + F_\pi + F_{\sigma\pi} + \frac{h^2}{2\mu_0}\right) \tag{1.4}$$

where F_α ($\alpha = \sigma, \pi$) is the free energy of each band and $h = |\boldsymbol{h}|$ is a magnetic field ($\mu_0\boldsymbol{h} = \boldsymbol{\nabla}\times\boldsymbol{A}$ with the vector potential $\boldsymbol{A}$) [465]:

$$F_\alpha = 2E_\alpha|\psi_\alpha^2| + |E_\alpha||\psi_\alpha|^4 + C\left|\left(-\imath\boldsymbol{\nabla} + \frac{2\pi}{\phi_0}\boldsymbol{A}\right)\psi_\alpha\right|^2 \ . \tag{1.5}$$

Here, E_α is the condensation energy [133], and $C = \frac{\phi_0}{2\pi}\sqrt{|E_\alpha|/\mu_0\kappa_\alpha^3}$ with the Ginzburg–Landau parameter κ_α for each band. $F_{\sigma\pi} = \frac{E_\gamma}{2}(\psi_\sigma^\star\psi_\pi + \psi_\pi^\star\psi_\sigma)$ is a Josephson coupling term, where E_γ is the coupling energy.

The two-component (or two-gap) superconductivity brings up a variety of new physics. To give a few examples, the first two-component superconductor MgB_2 [357], with two weakly coupled coexisting order parameters,

$\psi_{s,1} = \psi_\pi$ and $\psi_{s,2} = \psi_\sigma$, has opened remarkable new possibilities both for fundamental research and applications. Among the new research topics we find semi-Meissner state [22], the violation of the London law and Onsager-Feynman quantization [20], non-composite vortices [95], intrinsic Josephson effect [62], two-condensate Bose systems [21], superfluidity in liquid metallic hydrogen [23], etc. It should be also mentioned here that in mixtures of superfluids the entrainment of the two condensates leads to their interaction in the form of the Andreev–Bashkin effect [12]. The two-component character of MgB_2 [65, 185, 447, 453] is related to two different types of electronic bondings, π and σ, giving rise to two superconducting gaps with energies $\Delta_\pi(0) = 2.2\,\mathrm{meV}$ [138, 414] and $\Delta_\sigma(0) = 7.1\,\mathrm{meV}$ [98, 189, 230], respectively. Generally speaking, in the two-component superconductors (TCS), the two critical temperatures signalling the onset of the formation of the two supercomponents $\psi_{s,1}$ and $\psi_{s,2}$, may be different ($T_{c1} \neq T_{c2}$).

Interestingly, the first Ginzburg–Landau equation (Eq. (1.3a)), with the nonlinear term $\beta|\psi_s|^2\psi_s$ neglected

$$\frac{1}{2m^\star}(-\imath\hbar\nabla - e^\star \boldsymbol{A})^2\psi_s = -\alpha\psi_s \qquad (1.6)$$

is the analogue of the Schrödinger equation (Eq. (1.2)) with $U = 0$, when making a few substitutions: $\psi_s \leftrightarrow \psi$, $e^\star \leftrightarrow e$, $-\alpha \leftrightarrow E$ and $m^\star \leftrightarrow m$. The superconducting order parameter ψ_s corresponds in fact to the wave function ψ in Eq. (1.2). The effective charge $e^\star$ in the Ginzburg–Landau equations is $2e$, i.e. the charge of a Cooper pair, while the temperature dependent Ginzburg–Landau parameter α

$$-\alpha = \frac{\hbar^2}{2m^\star\,\xi^2(T)} \qquad (1.7)$$

plays the role of the eigenvalue E in the Schrödinger equation. Here $\xi(T)$ is the temperature dependent coherence length:

$$\xi(T) = \frac{\xi(0)}{\sqrt{1 - \frac{T}{T_{c0}}}}. \qquad (1.8)$$

The boundary conditions typically used for interfaces normal metal-vacuum and superconductor-vacuum or superconductor-normal metal are, however, different (Fig. 1.5):

$$\psi\psi^\star = 0 \qquad\qquad \text{normal-metal/vacuum} \qquad (1.9a)$$

$$(-\imath\hbar\nabla - e^\star\boldsymbol{A})\psi_s|_\perp = \frac{\imath\hbar}{b}\psi_s \qquad \begin{cases} \text{S/vacuum } (b=0) \\ \text{S/normal metal } (b>0) \end{cases} \qquad (1.9b)$$

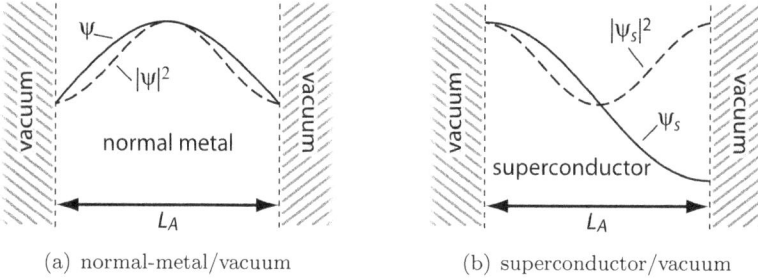

(a) normal-metal/vacuum (b) superconductor/vacuum

Fig. 1.5   Boundary conditions for (a) interfaces between a normal metal and vacuum (quantum particle in a box) and (b) interfaces between a superconductor and vacuum (confined superconducting condensate).

i.e. for normal metallic systems typically the *density* is zero (Dirichlet boundary condition), while for superconducting systems, the *normal component of the gradient* of ψ_s is zero (Neumann boundary condition). As a consequence, the supercurrent cannot flow through the boundary.

The nucleation of the superconducting condensate is favored at the superconductor/vacuum interfaces, thus leading to the appearance of superconductivity at the third critical field $H_{c3}(T)$, in a surface sheet with a thickness $\xi(T)$, see Fig. 1.6(a). For bulk superconductors the surface-to-volume ratio is negligible and, therefore, superconductivity in the bulk is not affected too much by a thin superconducting surface layer. For nanostructured superconductors with antidot arrays, however, the boundary conditions of Eqs. (4.6) and the surface superconductivity introduced through them become very important if $L_A \leq \xi(T)$. The advantage of superconducting materials in this case is that it is not even necessary to go to nanometer scale like for normal metals, since for L_A of the order of 0.1 – 1.0 µm, the temperature range in which $L_A \leq \xi(T)$ spreads over 0.01 – 0.1 K below T_c due to the divergence of $\xi(T)$ at $T \to T_{c0}$, see Eq. (1.8).

In principle, the mesoscopic regime $L_A \leq \xi(T)$ (and $L_A \leq \lambda(T)$, with λ the magnetic penetration depth) can formally be reached even in bulk superconducting samples with $L_A \sim 1\,\text{cm} - 1\,\text{m}$, since $\xi(T)$ and $\lambda(T)$ diverge at $T \to T_{c0}$. However, the temperature window where $L_A \leq \xi(T)$ is so narrow, not more than $\sim 1\,\text{nK}$ below T_{c0}, that one needs unrealistically perfect sample homogeneity and temperature stability.

In the mesoscopic regime $L_A \leq \xi(T)$, which is quite easily realized in nanostructured materials, the surface superconductivity can cover the whole space occupied by the material, thus spreading superconductivity all

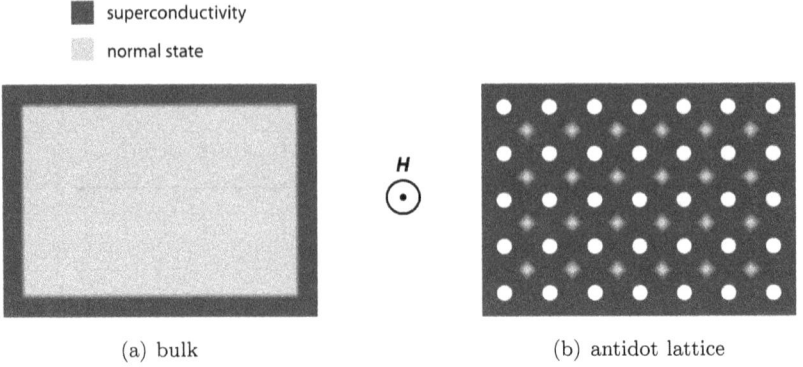

(a) bulk   (b) antidot lattice

Fig. 1.6   Schematic illustration of the surface superconductivity nucleated at the third critical field $H_{c3}(T)$. (a) In a bulk superconductor the surface superconducting sheet appears only at the external sample boundary, while in a laterally nanostructured superconductor with an antidot lattice (b) the surface superconductivity appears around each antidot, thus covering the whole sample interior. In the latter case the role of $H_{c2}(T)$ is played by $H_{c3}(T)$, since the area of surface superconductivity covers practically the entire sample.

over the sample. It is then evident that in this case the surface effects play the role of bulk effects, as illustrated in Fig. 1.6(b).

Using the similarity between the linearized Ginzburg–Landau equation (Eq. (1.6)) and the Schrödinger equation (Eq. (1.2)), the approach to determine $T_c(H)$ can be formalized as follows: since the parameter $-\alpha$ (Eqs. (1.6) and (1.7)) plays the role of the energy E in Eq. (1.2), the highest possible temperature $T_c(H)$ for the nucleation of the superconducting state in presence of the magnetic field H always corresponds to the lowest Landau level $E_{LLL}(H)$ found by solving the Schrödinger equation Eq. (1.2) with appropriate 'superconducting' boundary conditions (Eq. (1.9b)). Figure 1.7 illustrates the application of this basic rule to the calculation of the upper critical field $H_{c2}(T)$: indeed, if the well-known classical Landau solution for the lowest level in a bulk sample $E_{LLL}(H) = \hbar\omega/2$ is taken ($\omega = e^{\star}\mu_0 H/m^{\star}$ is the cyclotron frequency), one obtains from $-\alpha = E_{LLL}(H)$:

$$\frac{\hbar^2}{2m^{\star}\xi^2(T)} = \left.\frac{\hbar\omega}{2}\right|_{H=H_{c2}}. \tag{1.10}$$

With the help of Eq. (1.7) the above expression can be written as

$$\mu_0 H_{c2}(T) = \frac{\Phi_0}{2\pi\xi^2(T)}, \tag{1.11}$$

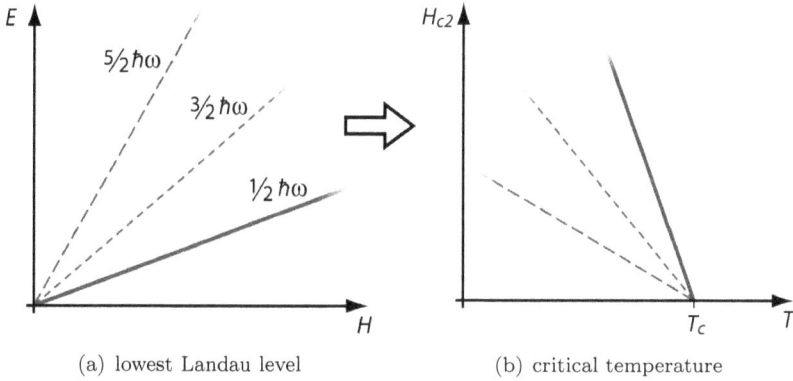

(a) lowest Landau level (b) critical temperature

Fig. 1.7 (a) The Landau level scheme for a particle in a magnetic field. From the lowest Landau level $E_{LLL}(H) = \hbar\omega/2$ the second critical field $H_{c2}(T)$ is derived, see the solid line in (b), by rotating plot (a) 90° counterclockwise (after [345]).

where $\Phi_0 = h/e^\star = h/2e$ is the superconducting flux quantum. In nanostructured superconductors, where the boundary condition of Eq. (1.9b) strongly influences the Landau level scheme, $E_{LLL}(H)$ is to be calculated for each specific confinement geometry. By measuring the shift of the critical temperature $T_c(H)$ in a magnetic field, one can compare the experimental phase boundary $T_c(H)$ with the calculated level $E_{LLL}(H)$ and thus explore the effect of the confinement topology on the superconducting phase boundary for a series of nanostructured superconducting samples. The transition between normal and superconducting states is usually very sharp and therefore the lowest Landau level can be easily traced as a function of applied magnetic field. The midpoint of the resistive transition from the superconducting to the normal state is usually taken as the criterion to determine $T_c(H)$.

1.4 Nucleation of superconductivity in presence of spatially modulated magnetic fields

So far we have been discussing the effect of the confinement on superconductivity in the presence of *homogeneous* magnetic fields. At the same time it is clear that nanoscale modulation of magnetic fields is possible in hybrids, such as the superconductor(S)/ferromagnet(F) systems. Alternatively, magnetic fields with strong non-uniformity can be produced by sending current through micro- or nano-loops and their arrays.

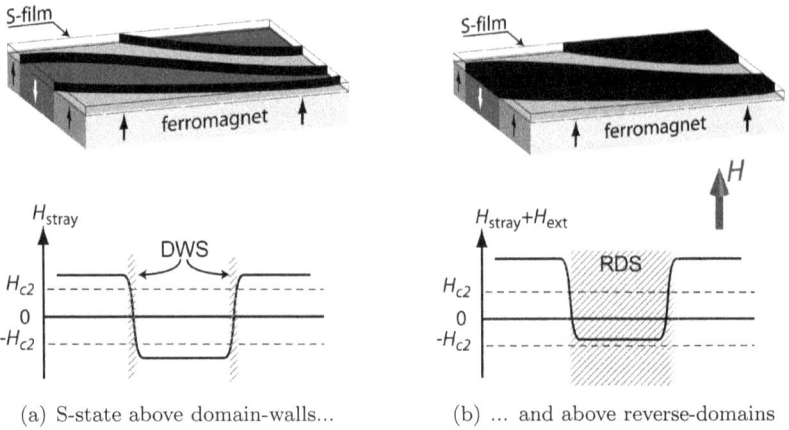

(a) S-state above domain-walls...   (b) ... and above reverse-domains

Fig. 1.8   Possible states of inhomogeneous superconductivity in a S/F-hybrid: (a) Super-conductivity can exist above domain-walls only (domain-wall superconductivity, DWS), and (b) the stray field of one type of magnetic domains is compensated by an external field (reverse-domain superconductivity, RDS).

The interaction between superconductivity and magnetism with direct charge transfer between the S- and F-systems has been intensively studied [76, 153].   Recently, hybrid superconductor–ferromagnet (S/F) systems have attracted considerable attention because it is believed that the interaction between superconducting- and magnetic order parameters at the mesoscopic length scale may lead to new physical phenomena [5, 77, 82, 137, 191, 192, 204, 215, 270, 274, 275, 297].

In the S/F hybrids with an insulating buffer layer between the S- and F-systems, an inhomogeneous distribution of magnetic field produced by the ferromagnet substantially modifies the nucleation of the order parameter and can lead to a significant change in the superconducting properties of the S-layer, including its critical temperature $T_c(H)$. Different ferromagnetic systems can be used in the S/F hybrids such as individual magnetic dots [191, 192] and their arrays [274] or non-patterned ferromagnetic thin films

Fig. 1.9   The principle of 'magnetic bias': the $T_c(H)$ phase boundary of a plain superconducting film (gray curve) is shifted by $\pm H_{\text{stray}}$ (black curve).

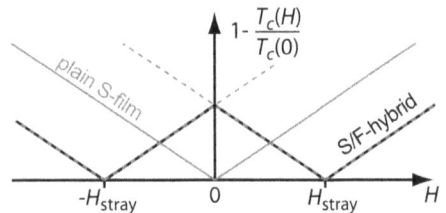

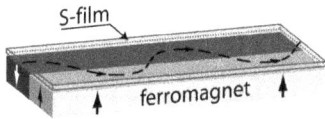

S-film

ferromagnet

Fig. 1.10 Schematics of the trajectory of an electron that is trapped at the magnetic domain wall of an underlying ferromagnet.

with bubble domains [275]. Depending on the domain structure of the ferromagnet, theory [5, 82] predicts a non-trivial T–H phase diagram for the nucleation of superconductivity in an external applied magnetic field H. Here, for simplicity, we shall consider planar geometries.

The basics of possible scenarios for the nucleation of superconductivity is illustrated in the upper panels of Fig. 1.8: A superconducting film (S) is placed on top of a ferromagnetic substrate (F) with the easy axis of the magnetization in the z-direction. In this case the stray fields in the thin S-film can be considered as almost homogeneous over the domain. With the decrease of temperature, superconductivity in zero applied field ($H = 0$) must first appear just above the domain wall, see the dark region in Fig. 1.8(a), thus realizing the so-called state of *domain-wall superconductivity* (DWS), because in that area the stray fields are the lowest [454]. Application of an external magnetic field results in a partial or complete compensation of the stray field above the reversed domains and favors the superconductivity to nucleate in the place corresponding to the minimum of the total magnetic field. In this case, superconductivity nucleating above domains of the opposite polarity will be strongly favored, see the dark areas in Fig. 1.8(b), leading to the state of *reverse-domain superconductivity* (RDS).

The field compensation effects — in simple terms — lead to a sort of '*magnetic bias*'. In this case, $T_c(H)$-curves plotted for both positive and negative field polarities are shifted by the stray magnetic fields $\pm H_{\text{stray}}$ generated by the magnetic domains. Accordingly, the resulting $T_c(H)$-curve has a pronounced 'W'-shape, see Fig. 1.9. Around $H = 0$ the nucleation of superconductivity is in the DWS regime. In that case, the electron can be trapped at the domain boundary (Fig. 1.10) and the corresponding Landau level goes down with increasing field.

1.5 Vortex matter in superconductors

The value of the Ginzburg–Landau parameter $\kappa = \lambda(T)/\xi(T)$ determines the behavior of a bulk superconductor in a magnetic field. Depending on κ being smaller or larger than $1/\sqrt{2}$, a distinction is made between type-1

($\kappa < 1/\sqrt{2}$) and type-2 ($\kappa > 1/\sqrt{2}$) single-component superconductors. A very special case can be found in close vicinity to the crossover point at $\kappa = 1/\sqrt{2}$, where Meissner (typical for type-1) and vortex states (typical for type-2) can co-exist in the same sample. This peculiar state in single-component superconductors was coined as 'intermediate–mixed state', see Ref. 225 for an overview and also the original publications [261, 418, 467].

Quite remarkably, in two-component superconductors, with $\psi = \psi_1 + \psi_2$, a *type-1.5 regime* can be realized when in the same material both type-1 ($\kappa_1 < 1/\sqrt{2}$) and type-2 ($\kappa_2 > 1/\sqrt{2}$) conditions are applied to the sub-components of the order parameter ψ_1 and ψ_2, respectively[1].

The two types of single-component superconductors are characterized by their different behavior in a magnetic field as a function of temperature. The H–T phase diagram for a bulk type-1 and a type-2 superconductor is presented in Figs. 1.11(a, b), respectively. A type-1 superconductor in a field lower than the thermodynamic critical field $H_c(T)$ is in the Meissner state, where the magnetic flux is completely expelled ($B = 0$). Above $H_c(T)$, superconductivity is destroyed and the sample is in the normal state.

The $H - T$ phase diagram of a type-2 superconductor is given in Fig. 1.11(b). At fields below the first critical field, i.e. for $H < H_{c1}(T)$, the superconductor is in the Meissner state. In the mixed state between the first and the second critical field, $H_{c1}(T) < H < H_{c2}(T)$, magnetic flux is able to partially penetrate the superconductor in quantized units, which are called flux lines or vortices [2]. In the field and temperature region $H_{c2}(T) < H < H_{c3}(T)$, superconductivity only exists in a thin sheet at the surface, see Fig. 1.6(a), while the rest of the material is in the normal state. Finally, above the third critical field at $H > H_{c3}(T)$, superconductivity is completely destroyed and the whole sample is in the normal state. The corresponding magnetization curves $-M(H)$ for ideal homogeneous type-1 and type-2 superconductors (without demagnetization effects) are shown in Fig. 1.12(a, b), respectively. The magnetization M can be written in terms of the applied field H and the flux density B in the sample:

$$\mu_0 M = B - \mu_0 H . \tag{1.12}$$

In type-1.5 superconductors, besides Meissner and mixed states, a new vortex state exists between them: the 'vortex cluster' or 'vortex clumps' phase, where vortex-rich areas ('vortex clusters' or 'vortex droplets') are

---

[1]For simplicity we shall omit the subscript 's' of the superconducting order parameter in the remainder of this book.

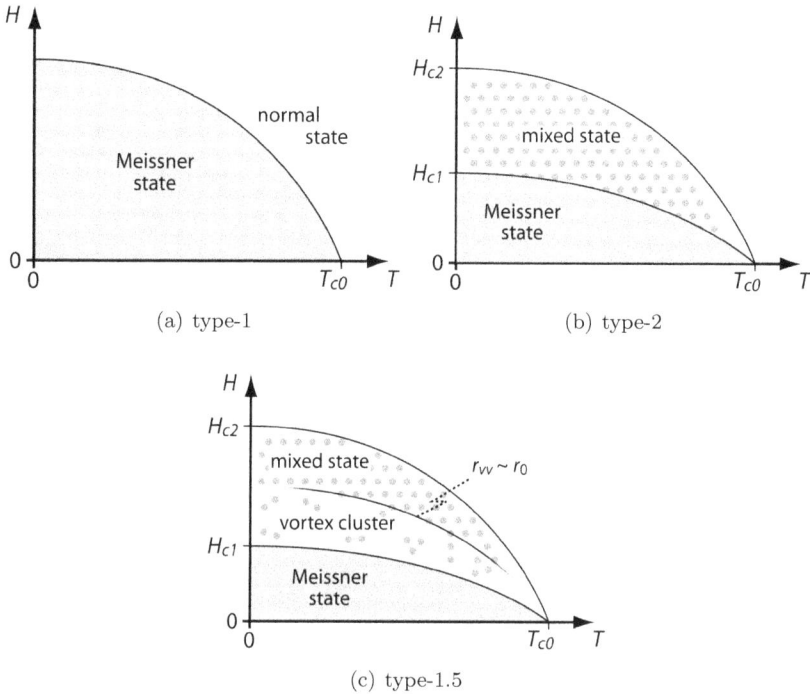

Fig. 1.11 The H–T phase diagram of a single-gap (a) type-1 and (b) type-2 super-conductor, and (c) a tentative schematic H–T phase diagram of a type-1.5 two-gap superconductor. Demagnetization effects are neglected and the area of surface superconductivity is not indicated.

separated by Meissner areas. In a simple model without coupling between ψ_1 and ψ_2 ($\eta = 0$), this phase has been introduced by Babaev [22] as a 'semi–Meissner' phase. In vortex clusters the order parameter is 'enriched' with the type-2 sub-component ψ_2, while in the surrounding Meissner state the order parameter is locally dominated by ψ_1. The transition into the inhomogeneous 'semi–Meissner' state with 'vortex droplets' represents a sort of a vortex matter analogy of the water droplets or a sublimation process in classical physics [22]. Remarkably, in two-gap superconductors in the type-1.5 regime, the co-existence of Meissner- and vortex states occurs in a much broader parameter space, compared to the close vicinity of the point $\kappa \approx 1/\sqrt{2}$ in single-gap superconductors in the intermediate-mixed state [16].

The $M(H)$ behavior in type-1, type-2 and type-1.5 superconductors is schematically presented in Fig 1.12. Without taking into account demag-

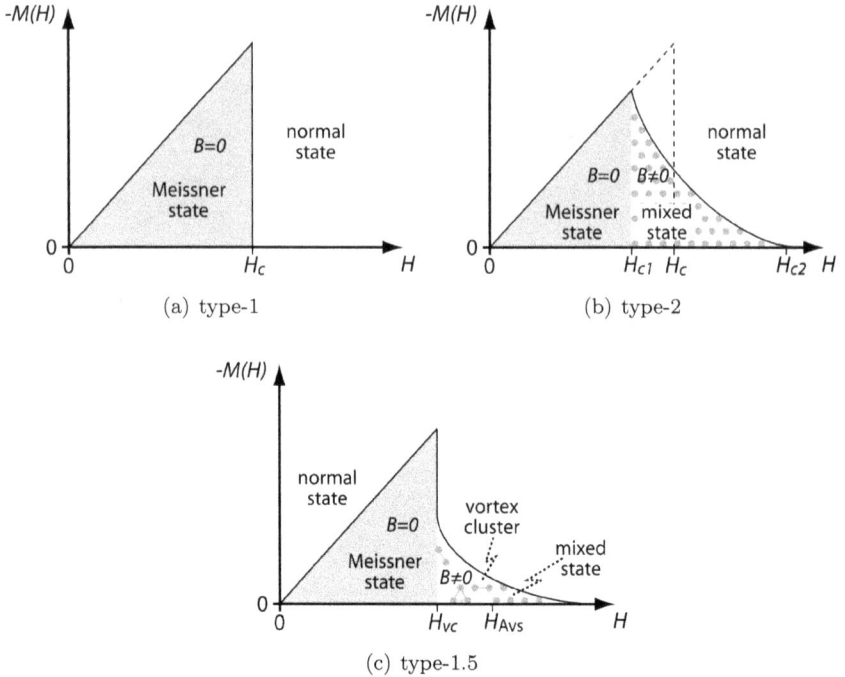

(a) type-1

(b) type-2

(c) type-1.5

Fig. 1.12   Reversible magnetization, without taking into account demagnetizing fields, as a function of the applied field $(-M(H))$ for (a) an ideal homogeneous type-1, (b) a type-2 superconductor and (c) a 1.5-superconductor. Note that in single-gap superconductors at the crossover point $\kappa \simeq 1/\sqrt{2}$, the abrupt $M(H)$ transition from the Meissner to the intermediate–mixed state takes also place [16].

netization fields, type-1 superconductors have an abrupt first-order phase transition from the Meissner to the normal state at $H = H_c$, accompanied by the $M(H)$ jump from a perfect diamagnetic response down to a very low M-value (practically to zero), corresponding to the normal state magnetization response. In type-2 superconductors, the transition from the Meissner to the mixed state leads to the appearance of a magnetization kink at $H = H_{c1}(T)$. The situation with the $M(H)$ behavior in type-1.5 materials is predicted [22] to be just half way between type-1 and type-2: formation of the vortex clusters surrounded by the Meissner phase results in the $M(H)$ jump (but *not* down to zero as in the type-1 case) at $H = H_{vc}$ ('vc' stands for 'vortex cluster'). As the field increases above H_{vc}, vortex clusters grow, but the vortex–vortex distance r_{vv} in the clusters remains practically unchanged: $r_{vv} \approx r_{opt}$, where r_{opt} is the optimal vortex–vortex separation in the clusters caused by the competition of the vortex–vortex repulsion and vortex–vortex attraction in 1.5-materials [22, 364]. Such behavior of the vortex matter in the inhomogeneous vortex-cluster state is in sharp contrast with the conventional Abrikosov vortex lattice in type-2 superconductors, where $r_{vv} \sim \sqrt{\Phi_0/B}$ and the vortex–vortex distance is uniformly shrinking with increasing magnetic field. In higher fields $H \gtrsim H_{\mathrm{Avs}}$ ('Avs' stands for 'Abrikosov vortex state'), when r_{opt} approaches $r_{vv} \sim \sqrt{\Phi_0/B}$, the conventional Abrikosov vortex state is recovered in type-1.5 materials and growing vortex clusters are merging with each other to form a homogeneous vortex array, with the vortex–vortex distance steadily decreasing with increasing field.

Interestingly, the unusual properties of vortex matter in type-1.5 superconductors have similarities with properties of single-component superconductors with the special κ, namely $\kappa = 1/\sqrt{2}$ (intermediate–mixed state, see Ref. 225). Such single component superconductors, just around the transition point from the type-1 to the type-2 regime, demonstrate the magnetization jump at 'H_{c1}' and also co-existence of vortex-clusters with the Meissner phase [68, 261].

The various critical fields of a type-2 superconductor can be expressed in terms of the characteristic length scales. $H_c(T)$ is the thermodynamic critical field, given by:

$$\mu_0 H_c(T) = \frac{\Phi_0}{2\sqrt{2}\,\pi\lambda(T)\,\xi(T)} \ . \tag{1.13}$$

The first and second critical fields of a type-2 superconductor can be written as:

$$\mu_0 H_{c1}(T) = \frac{\ln(\kappa)}{\sqrt{2}\,\kappa}\,\mu_0 H_c(T) = \frac{\ln(\kappa)\,\Phi_0}{4\pi\lambda^2(T)} \tag{1.14}$$

and (see also Eq. (1.11))

$$\mu_0 H_{c2}(T) = \frac{\Phi_0}{2\pi\xi^2(T)} = \sqrt{2}\,\kappa\,\mu_0 H_c(T) \ . \tag{1.15}$$

For thin type-2 superconducting films in a perpendicular field, the first critical field can be extremely small as a result of strong demagnetizing effects [408].

In two-component superconductors, including type-1.5 materials, the following form of the Ginzburg–Landau energy density is applicable:

$$F = \frac{\hbar^2}{4m_1}\left|\left(\boldsymbol{\nabla} + \imath\frac{2e}{\hbar c}\boldsymbol{A}\right)\psi_1\right|^2 + \frac{\hbar^2}{4m_2}\left|\left(\boldsymbol{\nabla} + \imath\frac{2e}{\hbar c}\boldsymbol{A}\right)\psi_2\right|^2 + \dots$$
$$\dots + V(|\psi_{1,2}^2|) + \eta[\psi_1^*\psi_2 + \psi_2^*\psi_1] + \frac{\boldsymbol{H}}{8\pi} \ , \tag{1.16}$$

where $\psi_{1,2} = |\psi_{1,2}|e^{i\varphi_{1,2}}$ and $V(|\psi_{1,2}^2|) = \sum_{n=1,2} -b_n|\psi_n^2| + c_n|\psi_n^4|/2$. In Eq. (1.16), vortices with phase winding in only one field have logarithmically divergent energy per unit length if $\eta = 0$, and linearly divergent if $\eta \neq 0$ [19]. Neglecting the effects of thermal fluctuations, we first restrict our attention to vortices with constant phase difference $\varphi_1 - \varphi_2$ in the case of $\eta = 0$, which have finite energy per unit length. Moving η from zero will be considered below in this section.

There are three characteristic length scales in Eq. (1.16): two partial coherence lengths $\xi_{1,2} = \hbar/\sqrt{4m_{1,2}b_{1,2}}$ and the penetration depth $\lambda = (c/\sqrt{8\pi}e)[|\bar{\psi}_1|^2/m_1 + |\bar{\psi}_2|^2/m_2]^{-1/2}$. It is also convenient to introduce formally characteristic critical fields for the two sub-components of the order parameter

$$H_{c(1,2)} = \frac{\Phi_0}{2\sqrt{2}\pi\xi_{1,2}\lambda_{1,2}} \tag{1.17a}$$

$$H_{c1(1,2)} = \frac{\Phi_0}{4\pi\lambda_{1,2}^2} \tag{1.17b}$$

$$H_{c2(1,2)} = \frac{\Phi_0}{2\pi\xi_{1,2}^2} \tag{1.17c}$$

Note, however, the formal character of the definition of partial fields for each sub-component of the order parameter. Depending on the $\xi_{1,2}$ and $\lambda_{1,2}$ relative values, several different regimes can be realized. If $\xi_1 \approx \xi_2 \gg \lambda$ or $\xi_1 \approx \xi_2 \ll \lambda$, the two-component superconductor just shows a typical type-2 or type-1 behavior [22]. The most interesting situation holds for the type-1.5 materials with $\lambda_1/\xi_1 < 1/\sqrt{2}$ and $\lambda_2/\xi_2 > 1/\sqrt{2}$. If, additionally, $\xi_1 > \lambda$, then a vortex solution for the subcomponent ψ_1 will have the core size exceeding the penetration depth.

1.5.1   *The structure of a single vortex*

The amount of flux carried by a single vortex in single-component superconductors is quantized and it equals usually to one flux quantum Φ_0. A single vortex consists of a normal core of radius ξ, around which shielding currents are circulating. In Fig. 1.13, a schematic presentation of a vortex is given, showing the local field $h = |\boldsymbol{h}|$ and the superconducting electron density n_{sc} as a function of the distance r from the vortex center. The superconducting order parameter vanishes in the center of the vortex core. The local field h is highest in the normal core and decays, due to the screening currents, over a distance given by the penetration depth λ.

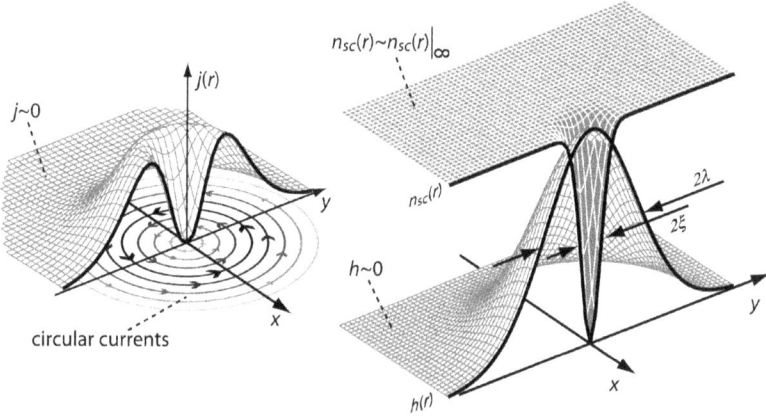

Fig. 1.13 Structure of a single vortex, showing the radial distribution of the local field $h(r)$, the circulating supercurrents $j(r)$ and the density of superconducting electrons $n_{sc}(r)$.

The local field and current distribution in the mixed state ($H_{c1} < H < H_{c2}$) of a type-2 superconductor can be calculated in the framework of the London model [292], which is valid provided that $\kappa \gg 1$ and that the order parameter (ψ or n_{sc}) is nearly constant in space. In the London limit, the free energy per unit length (the line energy) can be expressed as [118]:

$$E_l = \frac{1}{2\mu_0} \int \left(h^2 + \lambda_l^2 \left| \boldsymbol{\nabla} \times \boldsymbol{h} \right|^2 \right) d\boldsymbol{r} , \qquad (1.18)$$

where the first and second terms are the magnetic and the kinetic energy, respectively, λ_L is the London penetration depth, and the integration volume does not include the hard cores of the vortices. Minimization of E_l with respect to the local field $\boldsymbol{h}$ yields the London equation for the local field distribution outside the vortex core:

$$\boldsymbol{h} + \lambda_L^2 \boldsymbol{\nabla} \times \boldsymbol{\nabla} \times \boldsymbol{h} = 0 . \qquad (1.19)$$

This equation can also be derived from the Ginzburg–Landau theory by setting $\boldsymbol{\nabla}\psi(\boldsymbol{r})$ to zero. In order to include also the vortex core, Eq. (1.19) may be adjusted to:

$$\mu_0 \left(\boldsymbol{h} + \lambda_L^2 \boldsymbol{\nabla} \times \boldsymbol{\nabla} \times \boldsymbol{h} \right) = \boldsymbol{\Phi}_0 \delta_2(\boldsymbol{r}) \qquad (1.20)$$

where the small core is represented by a two-dimensional delta-function and with $\boldsymbol{\Phi}_0$ along the local field direction. By combining this equation with the Maxwell equations, the local field and current distribution of a vortex can be calculated.

The local field distribution around a vortex can be expressed as [118]:

$$\mu_0 h(r) = \frac{\Phi_0}{2\pi\lambda^2} K_0\left(\frac{r}{\lambda}\right) , \qquad (1.21)$$

where K_0 is the zero-order modified Bessel function. The asymptotic expressions are:

$$\mu_0 h(r) = \frac{\Phi_0}{2\pi\lambda^2} \ln\left(\frac{\lambda}{r}\right) \qquad \xi < r \ll \lambda \text{ and} \qquad (1.22a)$$

$$\mu_0 h(r) = \frac{\Phi_0}{2\pi\lambda^2} \sqrt{\frac{\pi\lambda}{2r}} \exp\left(-\frac{r}{\lambda}\right) \qquad \text{for } r \gg \lambda. \qquad (1.22b)$$

From the radial dependence of the local field, the current distribution of a single vortex can be calculated [387]:

$$j(r) = \frac{\Phi_0}{2\pi\mu_0\lambda^2} K_1\left(\frac{r}{\lambda}\right) . \qquad (1.23)$$

K_1 is the first order modified Bessel function of which the asymptotic behavior is used to obtain expressions for the current very close to the normal core and at large distance:

$$j(r) = \frac{\Phi_0}{2\pi\mu_0\lambda^2 r} \qquad \text{for } \xi < r \ll \lambda \text{ and} \qquad (1.24a)$$

$$j(r) = \frac{\Phi_0}{2\pi\mu_0\lambda^2} \sqrt{\frac{\pi\lambda}{2r}} \exp\left(-\frac{r}{\lambda}\right) \qquad \text{for } r \gg \lambda. \qquad (1.24b)$$

The line energy is the sum of the field and the kinetic energy of the currents and a small core contribution, and reads [118]:

$$E_l = \frac{1}{4\pi\mu_0} \left(\frac{\Phi_0}{\lambda}\right)^2 \left[\ln\left(\frac{\lambda}{\xi}\right) + \epsilon\right] . \qquad (1.25)$$

The numerical constant $\epsilon \simeq 0.12$ describes the contribution of the normal core. Since E_l is a quadratic function of the magnetic flux, it is energetically unfavorable in a homogeneous superconductor to form multiquanta vortices, carrying more than one flux quantum. For example, when comparing E_l for a $2\Phi_0$-vortex and two Φ_0-vortices, on gets $(2\Phi_0)^2 > 2\Phi_0^2$, and therefore $E_l(2\Phi_0) > 2E_l(\Phi_0)$. In this simple consideration, however, the vortex-vortex interaction has been completely ignored. The line energy becomes definitely larger if one takes the lattice consisting of $2\Phi_0$-vortices instead of single Φ_0-vortices. But at the same time the vortex-vortex interaction energy will be reduced if $2\Phi_0$-vortices are formed, since the distance

a_v between them increases: $a_v(2\Phi_0) > a_v(\Phi_0)$. Under certain conditions, see Ref. 341, multiquanta vortex lattices have a lower energy than a conventional Φ_0-flux line lattice. The formation of the multiquanta vortices can be induced by fabricating arrays of relatively large pinning centers, as we shall discuss in Chap. 4. In individual symmetric superconducting nano- and microstructures, symmetry-induced multiquanta vortices may also appear [90].

1.5.2 *The vortex lattice — general considerations*

Due to the repulsive electromagnetic interaction between vortices in single-component superconductors, they tend to assume positions as far away from each other as possible, resulting in the well-known Abrikosov vortex lattice [2]. The repulsive interaction energy per unit length between two vortices 'i' and 'j' at a mutual distance r_{ij} is given by [118]:

$$U_{ij}(r_{ij}) = \frac{\Phi_0^2}{2\pi\mu_0\lambda^2} K_0\left(\frac{r_{ij}}{\lambda}\right) , \qquad (1.26)$$

which decreases exponentially at large distances, $r_{ij} \gg \lambda$, as:

$$U_{ij}(r_{ij}) = \frac{\Phi_0^2}{2\pi\mu_0\lambda^2} \sqrt{\frac{\pi\lambda}{2r_{ij}}} \exp\left(-\frac{r_{ij}}{\lambda}\right) , \qquad (1.27)$$

and diverges at short distances ($\xi < r_{ij} \ll \lambda$) as:

$$U_{ij}(r_{ij}) = \frac{\Phi_0^2}{2\pi\mu_0\lambda^2} \ln\left(\frac{\lambda}{r_{ij}}\right) . \qquad (1.28)$$

The repulsive force between two vortices can be calculated from the energy U_{ij}:

$$f_{ij}(r_{ij}) = -\frac{\partial U_{ij}}{\partial r_{ij}} = \frac{\Phi_0^2}{2\pi\mu_0\lambda^3} K_1\left(\frac{r_{ij}}{\lambda}\right) . \qquad (1.29)$$

For a given vortex density, the maximum mutual distance between the vortices is obtained for a triangular arrangement, which is the most favorable vortex configuration, see Fig. 1.14(a). The density of vortices n_v increases with increasing field. The distance a_v between nearest neighbor vortices in the triangular lattice is related to the induction B through the relation:

$$B(\triangle) = \Phi_0 n_v = \frac{2}{\sqrt{3}} \frac{\Phi_0}{a_v^2} . \qquad (1.30)$$

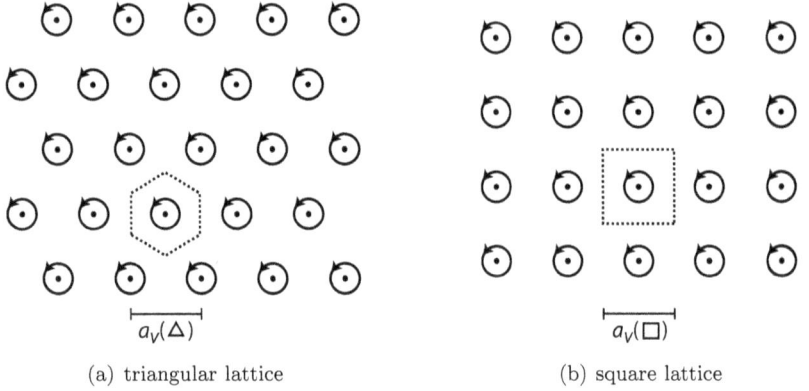

(a) triangular lattice   (b) square lattice

Fig. 1.14   Schematic diagram of (a) a triangular and (b) a square vortex lattice with period a_v. The dashed lines indicate the unit cells.

The second most favorable configuration, with only a slightly higher energy, is a square vortex lattice as illustrated in Fig. 1.14(b). The distance between nearest neighbor vortices in a square lattice depends on the induction B as follows:

$$B(\square) = \Phi_0 n_v = \frac{\Phi_0}{a_v^2} \ . \tag{1.31}$$

Vortices in superconductors consist of a normal core with size ξ (coherence length) and supercurrents flowing over a distance λ (penetration depth). If two vortices are generated in a type-1 superconductor, the normal cores would overlap first, due to the larger value of ξ with respect to λ, thus leading to a gain in the condensation energy and, consequently, to vortex-vortex attraction [66, 262], see Fig. 1.15(a). Although vortices are thermodynamically unstable in a type-1 superconductor, they still can be considered as zeros appearing in the Ginzburg–Landau equations applied to a type-1 material. Two vortices in a type-2 material, however, would have first their supercurrents overlapping, in view of the bigger λ, leading to vortex-vortex repulsion as illustrated in Fig. 1.15(b). An attractive vortex-vortex interaction results in the formation of macroscopic normal domains in the intermediate state [226], while vortex-vortex repulsion leads to the appearance of the Abrikosov lattice [2].

For the case of the type-1.5 superconductor MgB$_2$, the BCS expression $\xi(0) = \hbar v_F / \pi \Delta(0)$ with the Fermi velocity $v_F = 5.35 \cdot 10^5 \, m/s$ for the π-band and $v_F = 4.40 \cdot 10^5 \, m/s$ for the σ-band [71], leads to the two coherence lengths $\xi_\pi(0) = 51 \, \text{nm}$ and $\xi_\sigma(0) = 13 \, \text{nm}$. The calculated $\xi_\pi(0)$ value

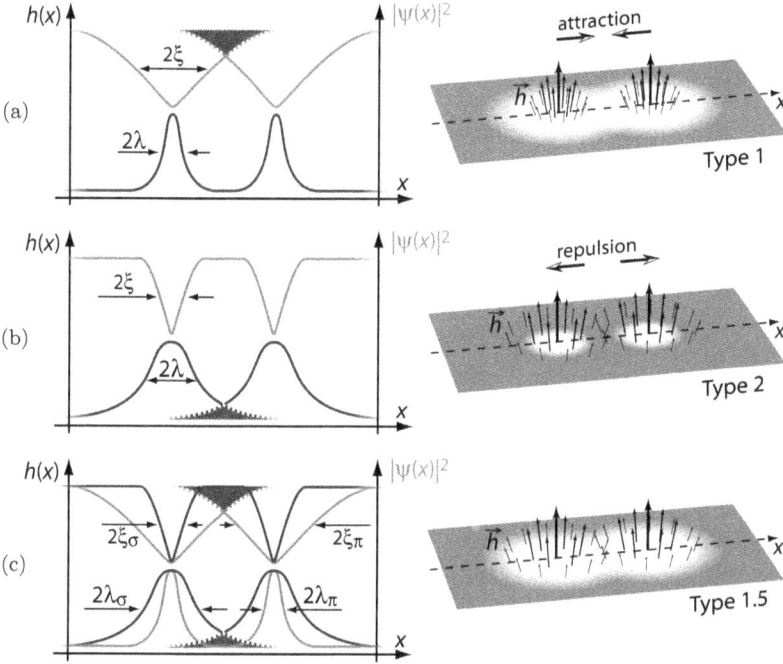

Fig. 1.15 Schematics of the vortex-vortex interactions: in type-1 superconductors, vortex cores are overlapping first, thus causing an attraction between vortices (a). In type-2 materials, the first to overlap are the local fields $h(x)$, which leads to a vortex-vortex repulsion (b). Type-1.5 superconductors combine both the vortex-vortex attraction and repulsion, since type-1 and type-2 conditions are formally fulfilled for the two subcomponents of the order parameter simultaneously (c). Note also that this is only a tentative, schematic and simplified picture for non-interacting subcomponents ψ_1 and ψ_2. In reality, both subcomponents will always generate a common λ_{eff} ($\lambda_{\mathrm{eff}}^{-2} \sim \lambda_\sigma^{-2} + \lambda_\pi^{-2}$).

is in agreement with $\xi_\pi(0) = 49.6 \pm 0.9\,\mathrm{nm}$ obtained from the fit of the vortex profile measured by scanning tunneling spectroscopy [138]. The London penetration depths λ_π and λ_σ can be estimated from the respective calculated plasma frequencies $\omega_{p\pi}$ and $\omega_{p\sigma}$ [311]: $\lambda_\pi(0) = 33.6\,\mathrm{nm}$ and $\lambda_\sigma(0) = 47.8\,\mathrm{nm}$. As a result, at least in the clean limit, the π- and σ-components of the order parameter in $\mathrm{MgB_2}$ are expected to be in different regimes: $\kappa_\pi = \lambda_\pi(0)/\xi_\pi(0) = 0.66 \pm 0.02 < 1/\sqrt{2}$ (type-1) and $\kappa_\sigma = \lambda_\sigma(0)/\xi_\sigma(0) = 3.68 > 1/\sqrt{2}$ (type-2). Therefore, clean $\mathrm{MgB_2}$ is an excellent candidate to search for type-1.5 superconductivity, where the vortex-vortex interaction is the result of the competing interaction between short-range repulsion and long-range attraction, see Fig. 1.15(c).

1.5.3 *Vortex lattices in thin films*

In continuous and non-perforated thin films where the thickness $\tau \ll \lambda(T)$ and the external field is perpendicular to the film plane, the vortex lattice is quasi two-dimensional. The small film thickness has important consequences for the vortex structure:

First of all, the screening currents are restricted to the thickness τ of the film, resulting in a larger effective penetration depth $\Lambda(T)$ [118]:

$$\Lambda(T) = \frac{\lambda(T)^2}{\tau} \ . \tag{1.32}$$

Secondly, the field around a single vortex in a thin film is less effectively screened at larger distances. The radial distribution of the local field around a vortex is given by:

$$\mu_0 h(r) = \frac{\Phi_0}{4\pi\Lambda(T)r} \qquad \text{for } \xi < r \ll \Lambda \text{ and} \tag{1.33a}$$

$$\mu_0 h(r) = \frac{2\Phi_0\Lambda(T)}{\pi r^3} \qquad \text{for } r \gg \Lambda. \tag{1.33b}$$

Instead of an exponential screening at large distances, given by Eq. (1.22b), the field only falls off as $1/r^3$ in a thin film.

Thirdly, the current decreases within Λ as $1/r$, but even at large distances it decays only as $1/r^2$:

$$j(r) = \frac{\Phi_0}{2\pi\Lambda(T)r} \qquad \text{for } x < r \ll \Lambda \text{ and} \tag{1.34a}$$

$$j(r) = \frac{\Phi_0}{\pi r^2} \qquad \text{for } r \gg L. \tag{1.34b}$$

Finally, the repulsive energy between two vortices decreases like $1/r$, similar to a Coulomb repulsion between two electric charges [118]. This implies that the repulsive interaction between vortices in a thin film is of longer range than in a bulk superconductor.

1.5.4 *Vortex lattices in type-1.5 superconductors*

In this paragraph, we shall analyze the vortex patterns at low vortex densities in high quality MgB$_2$ single crystals, which can be seen as a possible model system for type-1.5 superconductors with $\psi = \psi_\pi + \psi_\sigma$ (see Sec. 1.5.2). The vortex patterns are compared with the results of molecular dynamics simulations based on a two-gap Ginzburg–Landau theory which produces peculiar equilibrium vortex structures such as gossamer-like vortex patterns and vortex stripes.

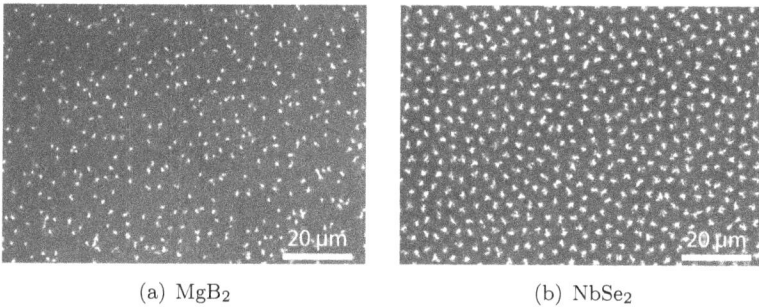

(a) MgB$_2$ (b) NbSe$_2$

Fig. 1.16 Images of the vortex structure obtained by Bitter decoration at $T = 4.2$ K and $H = 1$ Oe in (a) MgB$_2$ and (b) NbSe$_2$ single crystals. Notice that the density of vortices in the decoration experiments represents the internal field B rather than the applied field H. Due to differences in thicknesses of the MgB$_2$ and NbSe$_2$ samples and, therefore, differences in demagnetization factors, this leads to different numbers of decorated vortices for NbSe$_2$ and MgB$_2$, even at the same applied field (after [351]).

The MgB$_2$ single crystals were grown using a high pressure method [243] and their basic superconducting properties were analyzed according to Tab. 1.2. From the extrapolated value $H_{c2}(0) = 5.1$ T one obtains $\xi_{ab}(0) = 8.0$ nm close to the value found from the BCS theory for $\xi_\sigma(0)$. In order to estimate the penetration depth of the σ-band, the theoretical expression for the field can be used at which the first vortex will penetrate a two component superconductor, see Eq. (4) in Ref. 22. Considering $\lambda_\pi(0) = 33.6$ nm [311] for the type-1 component and taking $\xi_{ab}(0) = \xi_\sigma(0) = 8.0$ nm and $H_{c1}(0) = 0.241$ T from the measurements, one obtains $\lambda_{ab} = \lambda_\sigma = 38.2$ nm for these samples.

For two coexisting interpenetrating coupled order parameters, the vortex-vortex interaction can be derived from the Ginzburg–Landau theory by numerically minimizing the free energy of two vortices with a variational procedure [237]. The corresponding two-band Ginzburg–Landau

Table 1.2 Characteristics of the considered MgB$_2$-single crystals

Superconducting Property	Method of Investigation
$T_{c0} = 38.6$ K	ac-susceptibility
$H_{c1}(T)$	magnetization measurements [350, 385]
$H_{c2}(T)$	ac-susceptibility

functional is:

$$F_{GL}[\psi_\sigma, \psi_\pi, \boldsymbol{A}] = \int d^3 r \left[\frac{E_\gamma}{2}(\psi_\sigma^* \psi_\pi + \psi_\pi^* \psi_\sigma) + \frac{(\boldsymbol{\nabla} \times \boldsymbol{A})^2}{2\mu_0} \right] + \dots$$
$$\dots + F_\sigma + F_\pi \qquad (1.35)$$

where F_α with $\alpha = \sigma, \pi$ is given by

$$F_\alpha = \int d^3 r \left[2E_{c\alpha}|\psi_\alpha|^2 + |E_{c\alpha}||\psi_\alpha|^4 + \dots \right.$$
$$\left. \dots + \frac{\Phi_0}{2\pi}\sqrt{\frac{|E_{c\alpha}|}{\mu_0 \kappa_\alpha^3}} \left| \left(-\imath \boldsymbol{\nabla} + \frac{2\pi}{\Phi_0}\boldsymbol{A}\right)\Psi_\alpha \right|^2 \right]. \qquad (1.36)$$

Values from Ref. 133 were used for the estimations of the intrinsic condensation energies $E_{c\alpha}$ and the coupling energy E_γ, together with the values of κ_σ and κ_π obtained from the investigated samples. The result of the minimization shows that the interaction between vortices is short-range repulsive and weakly long-range attractive, similarly to Ref. 22. For modelling a system of overdamped vortices by molecular dynamics simulations, the equation of motion for a vortex 'i' is

$$\boldsymbol{F}_i = \boldsymbol{F}_i^{vv} + \boldsymbol{F}_i^T = \eta \boldsymbol{v}_i , \qquad (1.37)$$

where $\boldsymbol{F}_i^{vv}$ represents the vortex-vortex interaction and $\boldsymbol{F}_i^T$ the thermal stochastic force satisfying

$$\langle \boldsymbol{F}_i^T(t) \boldsymbol{F}_j^T(t') \rangle = 2\eta \delta_{i,j} \delta(t - t') k_B T \qquad \text{and} \qquad (1.38a)$$
$$\langle \boldsymbol{F}_i^T(t) \rangle = 0 . \qquad (1.38b)$$

Here, $\eta = \Phi_0 H_{c2}/\rho_n$ is the viscosity, $H_{c2} = 5.1\,\text{T}$ is the upper critical field and $\rho_n = 0.7\,\mu\Omega\cdot\text{cm}$ [133] is the normal state resistivity. The system size used in in the simulations [351] is $2000\times2000\,\lambda^2(0)$. Two systems with $N_v = 150$ and 400 vortices are initially prepared in a high temperature molten state and then slowly annealed to $T = 4.2\,\text{K}$ with 1000 temperature steps, letting the system stabilize during 2000 time steps in each step of temperature. Bitter decoration experiments on MgB_2 single crystals were performed at $4.2\,\text{K}$ after cooling down in the presence of an applied field perpendicular to the sample surface (field cooling). In this way, a homogeneous vortex distribution all over the sample is expected. A Bitter decoration image at $H = 1\,\text{Oe}$ shows clear evidence of an inhomogeneous distribution of vortices, see Fig. 1.16(a), reminding gossamer patterns: local groups of vortices with intervortex distances shorter than the average vortex distance $(2\Phi_0/\sqrt{3}B)^{1/2} \sim 5\,\mu\text{m}$ are separated by randomly distributed vortex voids

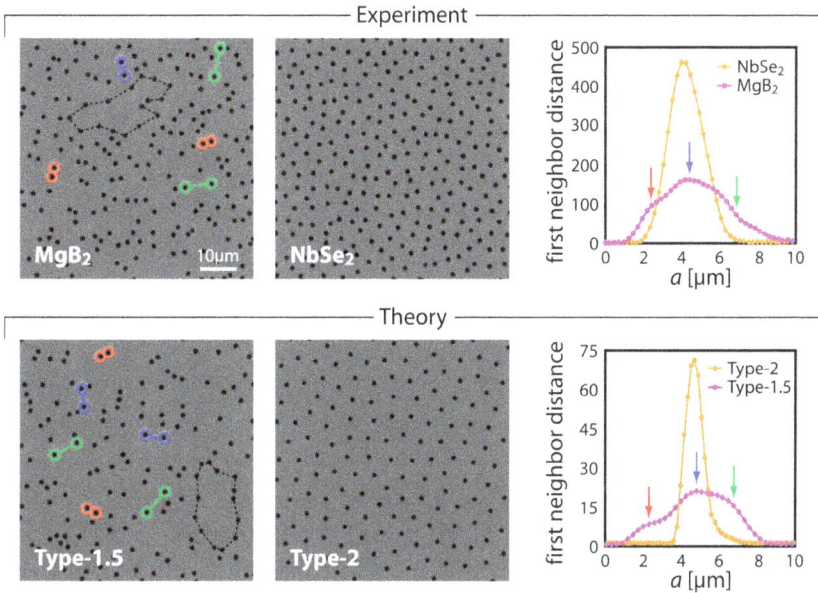

Fig. 1.17 The upper left image shows the experimentally determined vortex locations in a selected part of the image of Fig. 1.16(a). The vortex configuration resulting from the numerical simulations in a two-component superconductor at low density is shown in the lower left image, giving evidence for an inhomogeneous spatial distribution of vortices. In both cases, the regions enclosed by the dashed black line indicate voids of vortices, or Meissner areas, caused by the inhomogeneous distribution. In the upper middle image the vortex pattern obtained by magnetic decoration of a $NbSe_2$ crystal at 1 Oe is shown and corresponds to the vortex pattern obtained by a numerical simulation of a type-2 superconductor (lower middle image). The two graphs on the right display the distribution of the first neighbor distance, P_a, of the experimental and theoretical vortex structures. In the case of MgB_2, P_a shows additional peaks at distances shorter and longer (see the red and green arrows) than the most probable separation. Pairs of vortices that are separated at those distances where the additional peaks are located are highlighted in the left images by red and green circles, while blue circles correspond to pairs of vortices separated by the most probable distance (after [351]).

with size of a few micrometers. This is in striking contrast with the conventional homogenous vortex pattern formed in reference $NbSe_2$ single crystals, shown in Fig. 1.16(b). The observed clustering of vortices in MgB_2 samples is consistent with the theoretical modeling [22] for a two-component superconductor in the semi-Meissner state.

In the upper left panel of Fig. 1.17 the vortex positions in a selected region of the image shown in Fig. 1.16(a) are indicated as black dots, while in the lower left panel of Fig. 1.17 the results from the numerical simulations for a two-component superconductor are shown. Similarly, the lower middle panel of Fig. 1.17 shows the calculated vortex configuration for a conventional type-2 superconductor. The obtained vortex structure, considering $\lambda = 69$ nm and $\xi = 7.7$ nm [216], is similar to the one observed in NbSe$_2$ samples, which is shown in the upper middle panel of Fig. 1.17.

A possibility to characterize the inhomogeneity of the vortex structure is to calculate the distribution of the first neighbor distance P_a, resulting in the diagrams shown at the right in Fig. 1.17. To do so, the first neighbor distance a can be calculated by means of the Delaunay triangulation of the vortex structure. For the vortex structure of NbSe$_2$, the distribution is Gaussian with a relative standard deviation $\delta = \mathcal{SD}/ < a >= 0.224$, where $\mathcal{SD}$ is the standard deviation of the Gaussian fit to the experimental data and $< a >$ is the average first neighbor distance. On the other hand, in MgB$_2$ samples, P_a is quite broad as a consequence of the inhomogeneous arrangement of vortices and has additional peaks at distances shorter and longer than the most probable separation, see the red and green arrows in the diagrams of Fig. 1.17. The distribution of the first neighbor separation of the vortex structure obtained by simulations of a two-component material has also a three-peak structure which is fairly broader when compared to the one obtained in the case of a one-component conventional type-2 superconductor. The peaks at short distances in the diagrams of Fig. 1.17 correspond to an average minimum separation between vortices.

The upper left image of Fig. 1.18 shows a Bitter decoration image of the vortex structure in a MgB$_2$ crystal at $H = 5$ Oe and $T = 4.2$ K. The vortex distribution appears also to be inhomogeneous at this field but in a rather different manner than in the experiment described above. In some regions of the sample, vortices agglomerate forming stripes, while voids are formed on considerably larger areas.[2] The present experimental results seem to be in contradiction with previous reports of the vortex state at low fields in MgB$_2$ samples [489]. However, it is important to note that in Ref. 489 the authors show an image of the vortex state at 4 Oe in a very small region of the sample (approximately $10 \times 10 \ \mu m^2$). Therefore, it is not possible to determine whether the vortex distribution in the samples studied in Ref. 489 is uniform all over the sample or not.

---

[2]It is important to note that this agglomeration in the form of vortex stripes is not related to the presence of steps or other surface defects in the crystal. In fact, there are regions where steps can be observed (not shown) but at those locations the vortex configuration revealed by Bitter decoration is unaffected by them.

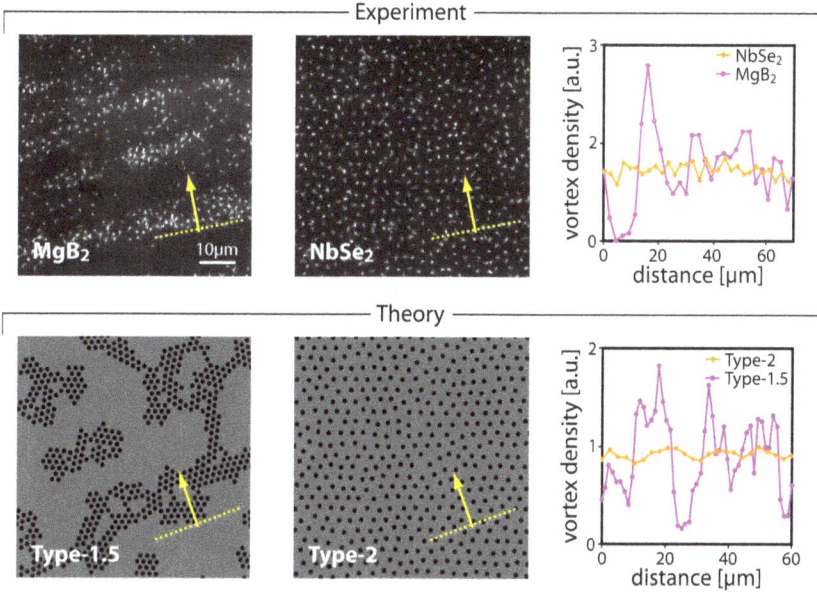

Fig. 1.18 The two upper images show the magnetic decoration of a MgB_2 single crystal (left) and of a $NbSe_2$ crystal (right) at $H = 5\,Oe$. The formation of vortex stripes, see in MgB_2, is also reproduced in numerical simulations of a two-component type-1.5 superconductor (lower left image) in contrast to a homogeneous vortex distribution in a type-2 superconductor at the same vortex density (lower middle image). The diagrams on the right show the vortex density along lines parallel to the vortex stripe direction (yellow dashed lines in the images). The variation of the vortex density is calculated as a function of the distance measured along the direction perpendicular to the stripes, marked by yellow arrows in the images. The curves are normalized by their respective average density (after [351]).

Although the vortex stripe pattern is rather disordered, it is still possible to determine an average direction of vortex stripes as the one defined by the dashed yellow lines in Fig. 1.18, and to calculate the vortex density in lines parallel to the vortex stripes as a function of the distance measured along the direction of the yellow arrows in Fig. 1.18. The upper diagram of Fig. 1.18 shows the linear vortex density normalized by the average value for both the MgB_2 and $NbSe_2$ vortex state at 5 Oe. In the case of MgB_2, fluctuations of the vortex density of the order of 50% are observed. A similar calculation along lines perpendicular to the stripes shows that the standard deviation of the mean value is of the order of 30%. The large fluctuations of the vortex density in MgB_2 are in contrast to what is observed in $NbSe_2$ crystals where the standard deviation of the vortex density is approximately 1% of the average value. A remarkable similarity is found between the experiments and simulations at still low density but higher than the one shown in Fig. 1.17. Disordered vortex stripes are formed in the two-component superconductor while a homogeneous distribution is apparent in the case of a conventional type-2 material, see the two lower images of Fig. 1.18. Consistently, the vortex density is seen to fluctuate in the direction perpendicular to the vortex stripes in the type-1.5 material, as shown in the lower diagram of Fig. 1.18.

Composition analysis with an electron microprobe in a field emission scanning electron microscope, performed in an area across the stripes, i.e. in the direction of the yellow arrows in Fig. 1.18, shows no significant variations in magnesium or boron content, thus ruling out the possibility to attribute the stripe formation to inhomogeneous surface pinning distribution. There is also no observed correlation between the vortex positions and locations of microdefects. At $H = 10\,\text{Oe}$ the vortex structure in MgB_2 samples is similar to the one in $NbSe_2$ crystals, indicating that the superconducting phase in two-component type-1.5 superconductors is only accessible at very low applied fields, as predicted in Ref. 22.

Interestingly, the vortex patterns in type-1.5 superconductors (Fig. 1.17) have many similarities with the co-existing vortex-clusters and Meissner areas in the intermediate–mixed state in superconductors with κ very close to the crossover point $\kappa = 1/\sqrt{2}$ between type-1 and type-2 regimes [67, 68]. Superconductors with $\kappa \approx 1/\sqrt{2}$ spontaneously decay into an inhomogeneous mixture of type-1 and type-2 areas, which is responsible for the appearance of the type-1.5-like vortex patterns. Contrary to the $\kappa \approx 1/\sqrt{2}$ superconductors with type-1/type-2 phase separation, in type-1.5 superconductors the coexistence of the two different regimes is driven by the simultaneous presence of the two sub-components of the order parameter $\psi = \psi_\pi + \psi_\sigma$, which, from a formal point of view, are individually in two different regimes: type-1 (ψ_π) and type-2 (ψ_σ).

Note also that vortex stripes (or vortex chains) can appear in layered type-2 superconductors when the magnetic field is applied at certain angles to the layers [83].

The competing magnetic responses in the two components of the order parameter in a type-1.5 superconductor can also result in a vortex-vortex interaction that generates giant vortices and unusual vortex rings in the absence of any extrinsic artificial pinning or confinement [114]. The normalized Ginzburg–Landau functional $\overline{F}$ for a two-gap superconductor

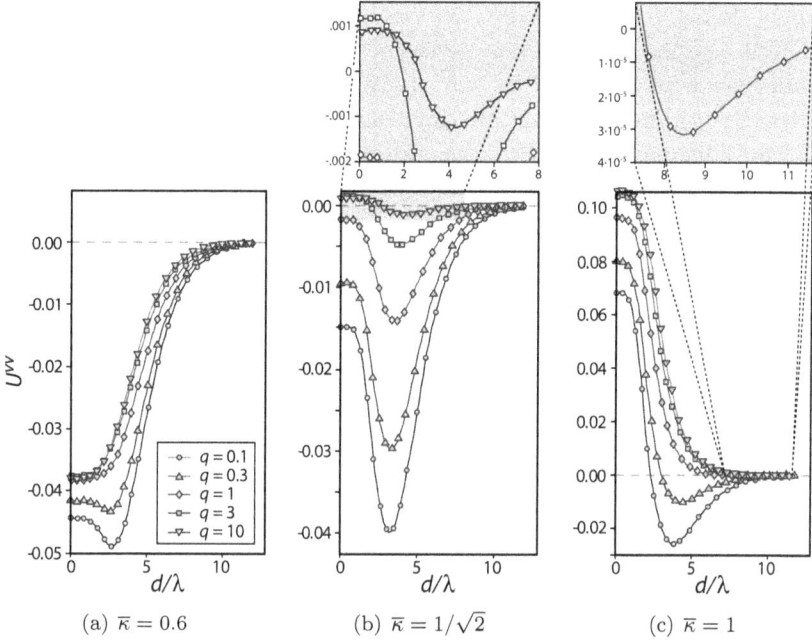

Fig. 1.19 The potential of the vortex interaction $U^{vv}(d)$ as a function of vortex separation d for different values of the average Ginzburg–Landau parameter $\overline{\kappa}$ and the normalized Josephson coupling q (after [114]).

is [114]

$$\overline{F} = \frac{1}{2\pi} \left\{ \overline{\kappa}^2 \left[p_\sigma(|\psi_\sigma^2 - 1|)^2 + p_\pi(|\psi_\pi^2 - 1|)^2 + q|\psi_\sigma - \psi_\pi| \right] \right. $$
$$\left. + r_\sigma |D\psi_\sigma|^2 + r_\pi |D\psi_\pi|^2 + \frac{1}{2}(\nabla \times \mathbf{A})^2 \right\}, \tag{1.39}$$

where $D = (-\imath\nabla + \mathbf{A})$, $\overline{\kappa}^{-1} = \sqrt{p_\sigma}\kappa_\sigma^{-1} + \sqrt{p_\pi}\kappa_\pi^{-1}$ ($p_{\sigma,\pi}$ are the condensation energy fractions), q is the normalized Josephson coupling factor and $r_{\sigma,\pi}$ are the superfluid-density fractions.

In contrast to single-component superconductivity where the vortex-vortex interaction depends only on κ, in two-component superconductors this interaction depends on four parameters $(\overline{\kappa}, \kappa_\pi, \kappa_\sigma, q)$. It is important to note here that the Josephson term actually locks ψ_σ and ψ_π when $q \to \infty$ at the critical temperature, where $\psi_\sigma = \psi_\pi$ and $\kappa = \overline{\kappa}$ are recovered in this limit, if $\overline{\kappa} \neq 1/\sqrt{2}$.

Interestingly, not exactly at T_c but still relatively close to it (where the Ginzburg–Landau approach remains valid), a well defined type-1.5 regime can be found (see Fig. 1.19). As one moves closer to T_c and the normalized Josephson coupling g increases, the vortex-vortex interaction potential looses its type-1.5 character (with a well-defined U^{vv} minimum at $d/\lambda \simeq 2 - 4$), and it is continuously converted into a conventional type-1 ($\overline{\kappa} = 0.6$) or type-2 ($\overline{\kappa} = 1$) regime with pure attractive or repulsive vortex-vortex interaction. This can drive, still within the limits of the Ginzburg–Landau theory, the sequence of transitions, like type-1 → type-1.5 → type-2 → type-1.5, as temperature decreases [114].

Finally, since the two-gap superconductivity is a relatively new but rapidly growing field, a fully consistent picture of the vortex matter in two-gap materials is currently only emerging. For a broader parameter space, especially for the type-1.5 regime, the application of the Usadel and Eilenberger equations, in addition to the two-component Ginzburg–Landau approach, is needed [258, 358, 457].

1.6 Flux pinning

A superconductor can carry a *dc*-current without losses if the current density is smaller than the superconducting critical current density j_c of the material [118]. High values for the critical current density in an applied field can be obtained if the flux lines are prevented from moving, since a moving vortex induces an electric field parallel to the current density j and hence dissipates energy. The origin of the flux motion and the associated dissipation is the Lorentz force from the current density j acting on the flux lines, which is per unit length and for one vortex given by:

$$\boldsymbol{f}_L = \boldsymbol{j} \times \boldsymbol{\Phi}_0 . \tag{1.40}$$

The Lorentz force tends to move the vortices transverse to the current. In order to suppress the motion of vortices, the Lorentz force should be counteracted by a pinning force $\boldsymbol{f}_P$. The total force acting on a flux line per unit length is the sum of several contributions [250],

$$\boldsymbol{f} = \boldsymbol{f}_L - \eta \boldsymbol{v}_v - \boldsymbol{f}_M \tag{1.41}$$

with $\eta \boldsymbol{v}_v$ a small friction-like contribution proportional to the vortex velocity $\boldsymbol{v}_v$, and $\boldsymbol{f}_M$ the Magnus force which usually is negligible. A general formula for $\boldsymbol{f}_P$ can not be given, since it strongly depends on the specific type of the pinning center. The average macroscopic pinning force per unit volume $\langle f_P \rangle$ is linked to the critical current density by the expression:

$$f_P = j_c B . \tag{1.42}$$

The theoretical upper limit for the critical current is determined by the depairing critical current $I_c^{GL}(T)$ at which the superconducting Cooper pairs are destroyed, and is given by the following expression [465]:

$$I_c^{GL}(T) = \frac{4}{3\sqrt{6}} \frac{H_c(T)}{\lambda(T)} .$$ (1.43)

1.6.1 *Structural pinning*

A vortex can be trapped or pinned due to the presence of spatial variations in the free energy of the flux line lattice. Almost any kind of defect can create local minima in the free energy landscape, e.g. crystalline imperfections, columnar defects, grain boundaries, thickness variations of a film, etc. Depending on the interaction between the vortex and the pinning center, a distinction can be made between two contributions to the pinning mechanism: core pinning and electromagnetic pinning [112]. *Core pinning* is the origin of flux pinning at most point defects and nano engineered pinning sites, and refers to the local variation of T_c or κ which reduces the free energy if the vortex core is located at the position of the (normal) defect or pinning center. *Electromagnetic pinning* is due to the kinetic energy of the confined screening currents around the defect and the perturbation of the magnetic field of a vortex. Theoretically, it can be described by assuming the presence of an antivortex image, in analogy with electrostatics. Thickness variations of a film can lead to the admixture of this type of pinning. The vortices are pinned at locations of smallest thickness where their energy is lowest, see Eqs. (1.25) and (1.32). The typical length scale for this kind of pinning is the penetration depth λ. Both, the core and the electromagnetic contributions define the pinning, which eventually controls the critical current density j_c.

1.6.2 *Magnetic pinning*

In a model system, consisting of a superconductor and a magnetic dipole at a fixed distance and with fixed magnetization, the interaction energy between the dipole and the screening currents is constant. In this specific case, any variation of the free energy of the system in the presence of vortices, either induced by the dipole or by an external field, can be attributed to rearrangements of the vortex pattern. The part of the interaction energy that is proportional to the magnetization of the dipole and sensitive to the vortex positions is usually referred as to the magnetic pinning energy E_p.

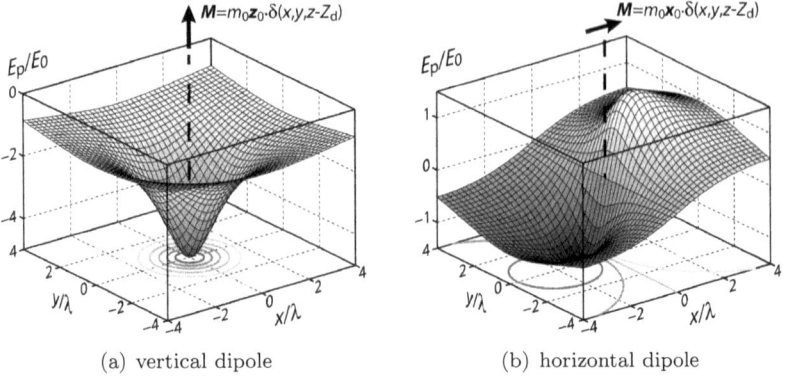

(a) vertical dipole   (b) horizontal dipole

Fig. 1.20   The spatial variation of the energy of a single vortex in a film of thickness D_s in the field of a point magnetic dipole, $M = m_0\delta(x, y, z - Z_d)$, calculated according to Eq. (1.44) for the height $Z_d/\lambda = 1$ and the dipole strength $m_0/(\Phi_0\lambda) = 10$. In (a) and (b), the dipole is magnetized vertically ($\mathbf{m}_0 = m_0\mathbf{z}_0$) and horizontally ($\mathbf{m}_0 = m_0\mathbf{x}_0$), respectively. The constant E_0 is equal to $\Phi_0^2 D_s/2\pi\lambda^2$.

The angular dependence of the interaction between a point magnetic dipole of fixed magnetization and a vortex can be given as [85, 325, 327]:

$$
E_p \approx \frac{\Phi_0 D_s}{2\pi\lambda} \left(-\frac{m_{0,z}}{2\Phi_0\lambda} \frac{1}{\sqrt{r_\perp^2 + Z_d^2}} \right) + \dots
$$

$$
\dots + \frac{m_{0,x}}{2\Phi_0\lambda} \frac{r_\perp \cos(\phi)}{\sqrt{r_\perp^2 + Z_d^2}(Z_d + \sqrt{r_\perp^2 + Z_d^2})} ,
$$

(1.44)

assuming a thin superconducting film with thickness $D_s \ll \lambda$ and for vortex-dipole separations smaller than the effective penetration depth $\lambda_{2D} = \lambda^2/D_s$. Here $m_{0,x}$ and $m_{0,z}$ are the in-plane and out-of-plane components of the dipolar moment, $r_\perp = \sqrt{x^2 + y^2}$, $\mathbf{R}_d = (0, 0, Z_d)$ is the position of the dipole, and ϕ is the angle between the x-axis and the vector position of the vortex in the plane of the superconductor $\mathbf{r}_\perp$. The resulting pinning potentials for an in-plane and an out-of-plane magnetized dipole, derived from Eq. (1.44), are shown in Fig. 1.20.

This magnetic pinning potential has a depth of $m_0/(4\pi\lambda)$ and penetrates a distance λ into the film [85], whereas its range, parallel to the film surface, is a few times λ. Since typical λ values are similar for a quite broad spectrum of superconducting materials, magnetic pinning represents a promising way of increasing the critical current not only in conventional but also in high-T_c superconductors. However, the fact that the magnetic pinning range is determined by λ sets a limit for the minimum distance between magnetic particles, beyond which the vortex lines cannot resolve the field modulation and, therefore, the pinning efficiency decreases. Having in mind that many practical applications typically involve high-T_c superconductors with $\xi \ll \lambda$, it is clear that this maximum density of magnetic pinning sites for trapping individual vortices is much lower than the one typical for core pinning.

It is important to note that real nanoengineered ferromagnetic elements are quite far from point dipoles but rather consist of extended volumes of a magnetic material, such as magnetic dots and stripes. In this case, the magnetic pinning energy should be integrated over the volume of the ferromagnet V_F to determine the interaction between a vortex line and a finite size permanent magnet

$$E_p(\boldsymbol{r}) = - \int_{V_F} \boldsymbol{M}(\boldsymbol{r}') \cdot \boldsymbol{B}_v(\boldsymbol{r} - \boldsymbol{r}') \, d^3r'. \tag{1.45}$$

Equation (1.45) shows that the pinning potential does not only depend on the size and the shape of the ferromagnetic elements but also on the particular distribution of the magnetization, i.e. on their exact magnetic state. It can be expected that the average pinning energy is less efficient for magnetic dots in a multi-domain state, whereas a maximum magnetic pinning is reached for single domain ferromagnetic nanostructures. In other words, if the size of the magnetic domains is small in comparison with λ or with the separating distance Z_d, a vortex line is subjected to the interaction of the average field emanating from the domains and, therefore, the magnetic pinning will be strongly suppressed. This flexibility of magnetic pinning centers makes it possible to tune the effective pinning strength by inducing a crossover from multi- to single domain states of magnetic pinning centers.

Chapter 2

Individual Nanostructures

In this chapter, we present experimental data on the $T_c(H)$ phase boundary[1] of individual superconducting mesoscopic structures of different topology/geometries, that is a line, a loop, a dot, a triangle, a square and a rectangle. In order to relate the differences in $T_c(H)$ directly to topological and geometrical effects, it is important to keep other parameters of these samples constant, like the material from which they are made (aluminium) or the width w of the lines.

Topological effects can be also continuously tuned in nanofabricated structures, providing a smooth cross-over from loop to dot or from an elongated rectangular sample to a square. Moreover, the presence of discrete symmetry of nanostructures induces the formation of antivortices close to the $T_c(H)$ boundary.

Antivortices turned out to be a very convenient tool for complying with the imposed boundary conditions. The vortex-antivortex patterns in an equilateral triangle and in a square will be analyzed in the framework of the Ginzburg–Landau theory. Experimentally, the effects of discrete symmetry on the nucleation line are presented by using the $T_c(H)$ data obtained on an equilateral triangle, square, rectangle and disc. Additionally, the $T_c(H)$ phase boundary in individual nanostructures is tuned by creating inhomogeneous local magnetic fields by combining a ferromagnetic dot with loops, discs, squares etc. Finally, dynamic effects (vortex ratchets) are briefly introduced for individual nanostructures with a broken symmetry.

---

[1] The magnetic field H is always applied perpendicular to the structures.

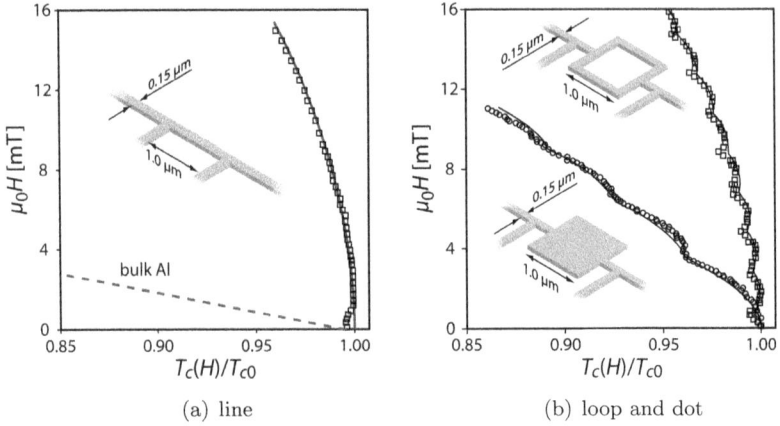

(a) line

(b) loop and dot

Fig. 2.1   The measured superconducting/normal-state phase boundary as a function of the reduced temperature $T(H)/T_{c0}$ for (a) a line and (b) a loop and a dot. The solid line in (a) is calculated using Eq. (2.1) with $\xi(0) = 110\,\mathrm{nm}$ as a fitting parameter. The dashed line represents $T_c(H)$ for bulk aluminium. Comparing $T_c(H)$ for these three different mesoscopic structures, made of the same material, one clearly sees the effect of the confinement topology of the superconducting condensate on $T_c(H)$ (after [345]).

2.1   Line

In Fig. 2.1(a) the phase boundary $T_c(H)$ of a mesoscopic line [349] is shown. The solid line gives the $T_c(H)$ calculated from the well-known formula [465]

$$T_c(H) = T_{c0}\left[1 - \frac{\pi^2}{3}\left(\frac{w\xi(0)\mu_0 H}{\Phi_0}\right)^2\right] \tag{2.1}$$

which, in fact, describes the parabolic shape of $T_c(H)$ for a thin film of thickness w in a parallel magnetic field. Since the cross-section, exposed to the applied magnetic field, is very similar for a film of thickness w in a parallel magnetic field and for a mesoscopic line of width w in a perpendicular field, the same formula can be used for both [349]. Indeed, the solid line in Fig 2.1 is a parabolic fit of the experimental data with Eq. (2.1), where $\xi(0) = 110\,\mathrm{nm}$ was obtained as a fitting parameter. The coherence length found with this method, coincides reasonably well with the dirty limit value $\xi(0) = 0.85(\xi_0 l)^{1/2} = 132\,\mathrm{nm}$, calculated from the known BCS coherence length $\xi_0 = 1600\,\mathrm{nm}$ for bulk aluminium [118] and the mean free path $l = 15\,\mathrm{nm}$, estimated from the normal state resistivity at 4.2 K [407].

Another simple argument can be used as well to explain the parabolic relation $T_c(H) \propto H^2$: the expansion of the energy $E(H)$ in powers of H, as given by the perturbation theory, is [499]:

$$E(H) = E_0 + A_1 L H + A_2 S_e H^2 + \dots \,, \tag{2.2}$$

where A_1 and A_2 are constant coefficients, the first term E_0 represents the energy levels in zero field, the second term is the linear field splitting with the orbital quantum number L and the third term is the diamagnetic shift, with S_e the area exposed to the applied magnetic field. When this area is very strongly reduced, like, for example, in superconducting tips for scanning tunneling microscopy [406], $E(H)$ is lowered and the critical fields can be enhanced quite dramatically.

Now, for the topology of the line with a width w much smaller than the Larmor radius $r_H \gg w$, any orbital motion is impossible due to the constraints imposed by the boundaries onto the electrons inside the line. Therefore, in this particular case $L = 0$ and $E(H) = E_0 + A_2 S_e H^2$, which immediately leads to the parabolic relation $T_c(H) \propto H^2$. This diamagnetic shift of $T_c(H)$ can be understood in terms of a partial screening of the magnetic field H due to the non-zero width of the line [465].

2.2 Loop

The $T_c(H)$ of the mesoscopic loop [349] shown in Fig. 2.1(b) demonstrates very distinct Little–Parks oscillations [380] superimposed on a monotonic background. A closer investigation leads to the conclusion that this background is very well described by the same parabolic dependence as the one which we just discussed for the mesoscopic line [349], see the solid line in Fig. 2.1(a). As long as the width of the strips w, forming the loop, is much smaller than the loop size, the total shift of $T_c(H)$ can be written as the sum of an oscillatory part, and the monotonic background given by Eq. (2.1) [200]:

$$T_c(H) = T_{c0} \left[1 - \frac{\pi^2}{3} \left(\frac{w\xi(0)\mu_0 H}{\Phi_0} \right)^2 - \frac{\xi^2(0)}{R^2} \left(n - \frac{\Phi}{\Phi_0} \right)^2 \right], \tag{2.3}$$

where $R^2 = R_1 R_2$ is the product of inner and outer loop radius, and the magnetic flux threading the loop $\Phi = \pi R^2 \mu_0 H$. The integer n has to be chosen so as to maximize $T_c(H)$ or, in other words, selecting the lowest Landau level $E_{LLL}(H)$.

The Little–Parks oscillations originate from the fluxoid quantization requirement, which states that the complex order parameter $\psi = |\psi| \exp(\imath\varphi)$

should be a single-valued function when integrating along a closed contour

$$\oint \varphi \cdot dl = n \cdot 2\pi \qquad n = \ldots - 2, -1, 0, 1, 2, \ldots . \qquad (2.4)$$

The fluxoid quantization gives rise to a circulating supercurrent in the loop when $\Phi \neq n\Phi_0$, which is periodic with the applied flux Φ/Φ_0.

Using the sample dimensions and the value for $\xi(0)$ obtained above for the mesoscopic line (with the same width $w = 0.15\,\mu\text{m}$), the $T_c(H)$ for the loop can be calculated from Eq. (2.4) without any free adjustable parameter. The solid line in Fig. 2.1(b) shows indeed a very good agreement with the experimental data [349]. It is worth noting here that the amplitude of the Little–Parks oscillations is about a few mK — in qualitative agreement with the simple estimate given in Table 1.1 for $L_A = 1\,\mu\text{m}$.

Interestingly, for very small radii in Eq. (2.3), the amplitude of the Little-Parks oscillations becomes bigger than T_{c0} itself. In this case, even at $T = 0$, the $T_c(H)$ phase boundary splits the superconducting area into several separate S-areas, thus corresponding to clear reentrant superconductivity as the field is ramped up even at $T = 0$. This reentrant superconducting behavior was observed in small ultrathin aluminum and $Au_{0.7}In_{0.3}$ cylinders by Liu *et al.* [291].

The susceptibility of a single mesoscopic aluminium ring, showing Little–Parks oscillations, has been studied by Zhang and Price [511] who found an excellent agreement with the Ginzburg–Landau theory for the susceptibility below T_c.

The analogue of the lower critical field of a loop can be found from the condition that half a flux quantum is applied, thus giving [45]:

$$\mu_0 H_{c1}^{\text{loop}} = \frac{1}{2} \frac{\Phi_0}{\pi R^2} \qquad (2.5)$$

which is totally different from the H_{c1} value of a bulk superconductor (see Eq. (1.14)).

The superconducting loop (Fig. 2.1(b)) nicely demonstrates a classical Little–Parks oscillatory $T_c(H)$ phase boundary [290, 380], related to the quantization of the total flux threading the loop area. Moreover, in this simple homogeneous superconducting loop, the two supercurrents I_1 and I_2 flowing in the different branches of the loop gain *opposite* phases $\pm\pi\Phi/\Phi_0$ from the perpendicular applied field [149, 152].

$$I_1 \propto \sin\left(\beta - \alpha - \pi\frac{\Phi}{\Phi_0}\right) \qquad (2.6\text{a})$$

$$I_2 \propto \sin\left(\beta - \alpha + \pi\frac{\Phi}{\Phi_0}\right) . \qquad (2.6\text{b})$$

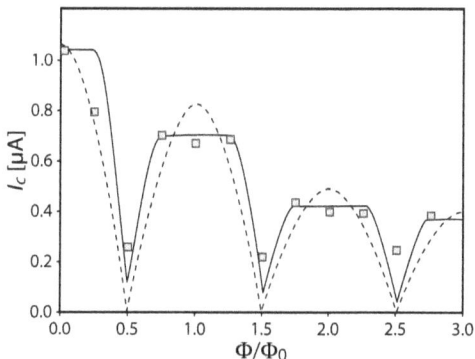

Fig. 2.2 Oscillations of the critical current I_c of a $1 \times 1\,\mu m^2$ square aluminium loop as a function of normalized flux Φ/Φ_0. The solid line corresponds to the calculations based on the theory of Ref. 149, while the dashed line is the sinusoidal variation of I_c in a conventional SQUID with two Josephson junctions (after [348]).

The interference of these currents produces an oscillatory dependence of the total current $I = I_1 + I_2$

$$I \propto \sin(\beta - \alpha) \cos\left(\pi \frac{\Phi}{\Phi_0}\right). \tag{2.7}$$

As in the first Josephson equation $I \propto \sin(\Delta\varphi)$ [3, 118], the current amplitude in Eq. (2.7) is determined by the phase difference $\Delta\varphi = \beta - \alpha$. The critical current I_c for the homogeneous loop is obtained by taking $\sin(\Delta\varphi) = 1$ in Eq. (2.7) [152]:

$$I_c \propto \left| \cos\left(\pi \frac{\Phi}{\Phi_0}\right) \right|. \tag{2.8}$$

Equation 2.8 implies that the critical current of a mesoscopic superconducting loop without artificial Josephson weak links oscillates with the applied magnetic field in the same way as it does in a classical SQUID with extrinsic weak links. The existence of these oscillations, predicted in Refs. 152 and 149, has been confirmed experimentally by Moshchalkov *et al.* [348] and is shown in Fig. 2.2.

Another interesting feature of a mesoscopic loop and other mesoscopic structures is the unique possibility they offer for studying nonlocal effects [450]. In fact, a single loop can be considered as a 2D artificial quantum orbit with a *fixed radius*, in contrast to Bohr's description of atomic orbitals. In the latter the stable radii are found from the quasi-classical quantization rule, stating that only an integer number of wavelengths can be set along the circumference of the allowed orbits. For a superconducting loop with an arbitrary fixed circumference, however, supercurrents must flow in order to fulfill the fluxoid quantization requirement of Eq. (2.4), thus causing oscillations of the critical temperature T_c versus magnetic field H.

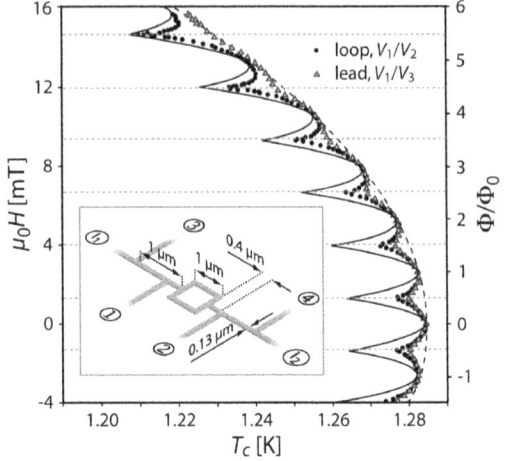

Fig. 2.3 Local (V_1/V_2) and non-local (V_1/V_3 or V_2/V_4) phase boundaries $T_c(H)$. The measuring current is sent through the contacts $I_{1,2}$. The solid and dashed lines correspond to the theoretical $T_c(H)$ of an isolated loop and a one-dimensional line, respectively, both made of strips of width $w = 0.13\,\mu m$. The inset shows a schematics of the structure, where the distance $P = 0.4\,\mu m$ (after [450]).

In order to measure the resistance of a mesoscopic loop, electrical contacts have, of course, to be attached to it, and as a consequence, the confinement geometry is changed. A loop with attached contacts and the same loop without any contacts are, strictly speaking, different mesoscopic systems. This 'disturbing' or 'invasive' aspect ('Schrödinger cat') of probing a quantum object can now be exploited for the study of nonlocal effects [450]. Due to the divergence of the coherence length $\xi(T)$ at $T = T_{c0}$ (Eq. (1.8)) the coupling of the loop with the attached leads is expected to be very strong for $T \to T_{c0}$.

Figure 2.3 shows the results of such measurements [450]. For both cases, 'local' (potential probes across the loop V_1/V_2) and 'nonlocal' (potential probes aside of the loop V_1/V_3 or V_2/V_4), Little–Parks oscillations are clearly observed. For the 'local' probes there is an unexpected and pronounced increase of the oscillation amplitude with increasing field, in disagreement with previous measurements on aluminium micro-cylinders [200]. In contrast to this, for the 'nonlocal' Little–Parks effect, the oscillations rapidly vanish when the magnetic field is increased.

When increasing the field, the background suppression of T_c (Eq. (2.1)) results in a decrease of $\xi(T)$. Hence, the change of the oscillation amplitude with H is directly related to the temperature-dependent coherence length. As long as the coherence of the superconducting condensate protrudes over the nonlocal voltage probes, the nonlocal Little–Parks oscillations can be observed. On the other hand, the importance of an 'arm' attached to a mesoscopic loop, was already demonstrated theoretically by [119]. For a

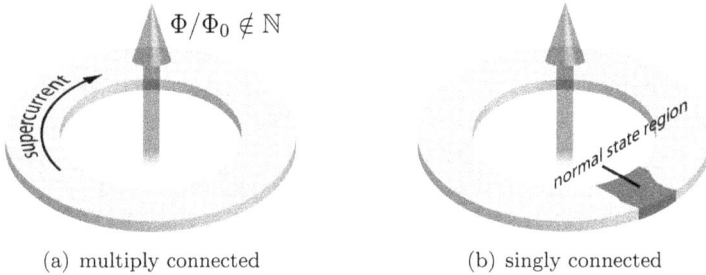

(a) multiply connected   (b) singly connected

Fig. 2.4 Schematic view of a mesoscopic loop in the multiply connected state (a) and the singly connected state (b), where a normal spot is spontaneously created and consequently no circular supercurrent, like in (a), can flow.

perfect one-dimensional loop with vanishing width of the strips, adding an 'arm' will result in a decrease of the Little–Parks oscillation amplitude, which was indeed observed at low magnetic fields where $\xi(T)$ is still large. With these experiments, it has been proven that adding probes to a structure considerably changes both the confinement topology and the phase boundary $T_c(H)$.

The effect of topology on $T_c(H)$, related to the presence of the sharp corners in a square loop, has been considered by Fomin *et al.* [157]. In the vicinity of the corners the superconducting condensate sustains a higher applied magnetic field, since at these locations the superfluid velocity is reduced in comparison with the ring. Consequently, in a field-cooled experiment, superconductivity will nucleate first around the corners [157]. Eventually, for a square loop, the introduction of a local superconducting transition temperature seems to be needed. As a result of the presence of a corner, the $H_{c3}(T)$ of a wedge with an angle θ will be strongly enhanced at the corner resulting in the ratio $H_{c3}/H_{c2} \approx 3.79$ for $\theta \approx 0.44\pi$ [155].

Another interesting possibility for a superconducting ring has been analyzed by Horane *et al.* [221] and by Berger *et al.* [48–51]: under certain conditions superconductivity spontaneously breaks at some spot along the perimeter of the ring, so that the topology of the *superconducting area changes from multiple to single connectivity*. The physics behind this interesting theoretical prediction is the following. The oscillatory $T_c(\Phi)$ phase boundary is caused by a periodic variation of a circular supercurrent which is induced in a ring if the applied flux Φ/Φ_0 is not integer, see Fig. 2.4(a), in order to fulfill fluxoid quantization. The highest current, and therefore the strongest reduction of $T_c(\Phi)$, is realized for half integer flux when

$\Phi/\Phi_0 - n = 1/2$, see Figs. 2.1(b) and 2.3. In this situation it may turn out, however, that somewhere in the ring the order parameter is spontaneously suppressed and a sort of 'normal core' is created along the ring circumference, as sketched in Fig. 2.4(b). The energy of this normal state core, below the $T_c(\Phi)$ line, is of course higher than the energy corresponding to a superconducting state everywhere in the ring, but at the expense of that the circular supercurrent is interrupted, thus effectively opening the ring for entrance and removal of flux. While Horane *et al.* [221] predicted the existence of the singly connected state for rings made of 'one-dimensional' strips, Berger *et al.* [49] showed that the temperature region where the singly connected state exists, can be enhanced by proper tuning the nonuniform 'strip width' profiles along the ring.

2.3 Disc

Among singly connected nanostructures such as discs, triangles, squares and rectangles, the cylindrical symmetry of the disc makes this system the evident choice for calculating the relevant Landau levels, including the lowest Landau level E_{LLL} which determines the $T_c(H)$ line. On the other hand, the discrete symmetry of an equilateral triangle or a square makes it possible to observe in these systems symmetry-induced antivortices. For the theoretical analysis of the latter, instead of the conventional gauge $\boldsymbol{A} = 1/2 H r \boldsymbol{e}_\varphi$, a specially developed analytical gauge transformation [92] will be used.

We will start our analysis of the $E_{LLL}(H)$ behavior from cylindrically symmetric structures. For a dot of cylindrical symmetry, the choice of the coordinates (r, φ, z) and the gauge $\boldsymbol{A} = 1/2 H r \boldsymbol{e}_\varphi$, where $\boldsymbol{e}_\varphi$ is the tangential unit vector, is well suited. The solution of the Hamiltonian in Eq. (1.6) in cylindrical coordinates has the following form [126]:

$$\psi_s(r, \varphi) = e^{\pm \imath L \varphi}\, r^L\, \gamma^{(L+1)/2}\, e^{-\gamma r^2/2} \cdot M(-N, L+1, \gamma r^2). \qquad (2.9)$$

Here $\gamma = e^\star \mu_0 H/\hbar$ and the energy $E_\perp$ of the motion in the plane perpendicular to $\boldsymbol{H}$ is determined by the orbital quantum number L and parameter N, which is not necessarily an integer number, as we shall see later:

$$E_\perp = \frac{e^\star \hbar \mu_0 H}{2\, m^\star} \left(2N \pm L + L + 1\right). \qquad (2.10)$$

The function M is the Kummer function defined as:

$$M(a, c, y) = 1 + \frac{a}{c}y + \frac{a(a+1)}{c(c+1)} \frac{y^2}{2!} + \frac{a(a+1)(a+2)}{c(c+1)(c+2)} \frac{y^3}{3!} + \cdots, \qquad (2.11)$$

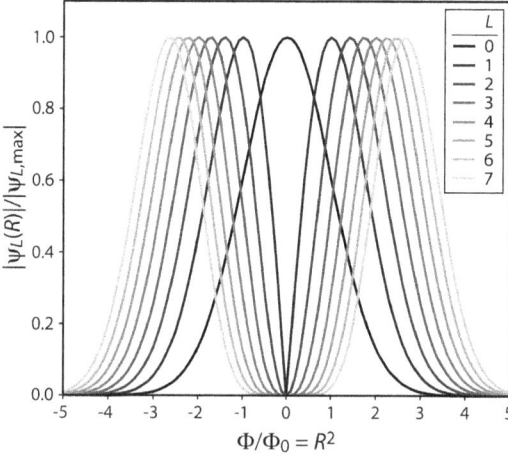

Fig. 2.5 The set of functions ψ_L of Eq. (2.12) for L values ranging from 0 to 7, normalized to the same maximum value at $R = \sqrt{L}$ (after [346]).

where $a = -N$, $c = L + 1$, $y = \gamma r^2$. Introducing the dimensionless radius $R = \sqrt{\gamma r^2} = \Phi/\Phi_0$, the superconducting order parameter (see the insets of Fig. 2.8) can be written in the form

$$\psi_L(R, \varphi) = e^{\pm \iota L \varphi} R^L e^{-R^2/2} \cdot M(-N, L + 1, R^2) . \qquad (2.12)$$

The representation of the order parameter $\psi = \sum_L c_L \psi_L$ as an expansion over states with different L for infinite samples has been analyzed by Moshchalkov *et al.* [346], where $M(0, L+1, R^2) = 1$ has been taken. Under these conditions, the functions $|\psi_L|$ have their maxima at $R^2 = L$, i.e., the area enclosed by the circle with the radius corresponding to the $|\psi_L|$ maximum is always penetrated by an integer number L of the flux quanta: $\Phi/\Phi_0 = L$ (see Fig. 2.5).

Here, we shall analyze the case of *finite* samples, where the N value has to be found from the boundary condition Eq. (1.9b). It is important to note that in the general form, there are no limitations on the parameter N in Eq. (2.11) and Eq. (2.10), that is N is not necessarily an integer number. The only argument, which is usually given in favor of taking integer N is the possibility to get a cut off in the summation of Eq. (2.11). Indeed, if one inserts an integer N into the summation, then by adding 1 to $-N$ in each new term one inevitably comes to the situation where $-N + N = 0$ and all subsequent terms in the summation will be equal to zero. Thus, by using the cutoff, we just keep a finite number of terms in the summation and, accordingly, M is finite in this case. But we should keep in mind that any *converging, but infinite* row also gives a finite solution for M. Therefore, not only positive integer N in Eq. (2.10), but also noninteger

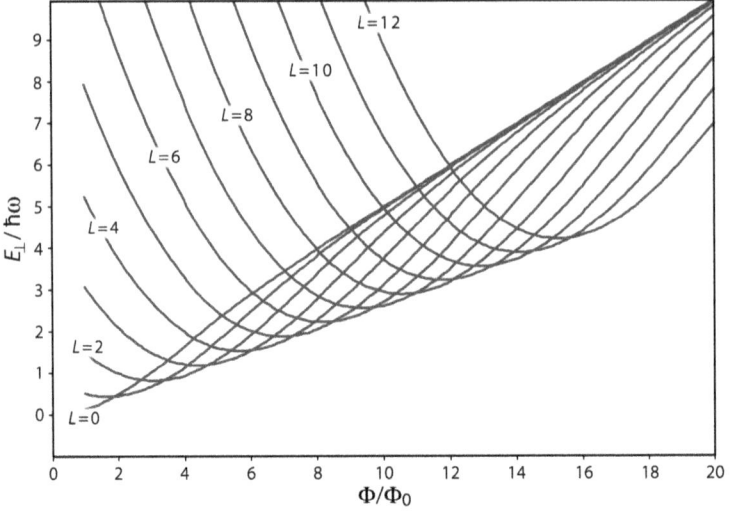

Fig. 2.6   Energy levels versus normalized flux Φ/Φ_0 for a superconducting cylinder in a parallel magnetic field. The lowest cusp-like $H_{c3}(T)$ line is formed due to the change of the orbital quantum number L (after [352]).

and even negative N values are possible. In finite size samples the N value, which we further denote as $N(L, R_0)$, has to be found from the boundary condition Eq. (1.9b) at $R = R_0$, where R_0 is the normalized disk radius:

$$\frac{\partial |\psi(R)|}{\partial R}\bigg|_{R=R_0} = 0 \ . \tag{2.13}$$

Since we are looking for the lowest possible energy state, we should take the minus sign in the argument of the exponent $\exp(-\imath L\varphi)$ in the solution given by Eq. (2.10). In this case $-L$ and $+L$ in Eq. (2.10) cancel and for any L the energy levels are given by:

$$E_\perp = \hbar\omega \left(N + \frac{1}{2} \right) , \tag{2.14}$$

where $\omega = e^\star \mu_0 H/m^\star$ is the cyclotron frequency. This result coincides with the well-known Landau quantization, but now N is *any real number, including negative real numbers*, which is to be calculated from Eq. (2.13). Using the expression

$$\frac{dM(a, c, y)}{dy} = \frac{a}{c}M(a+1, c+1, y) \tag{2.15}$$

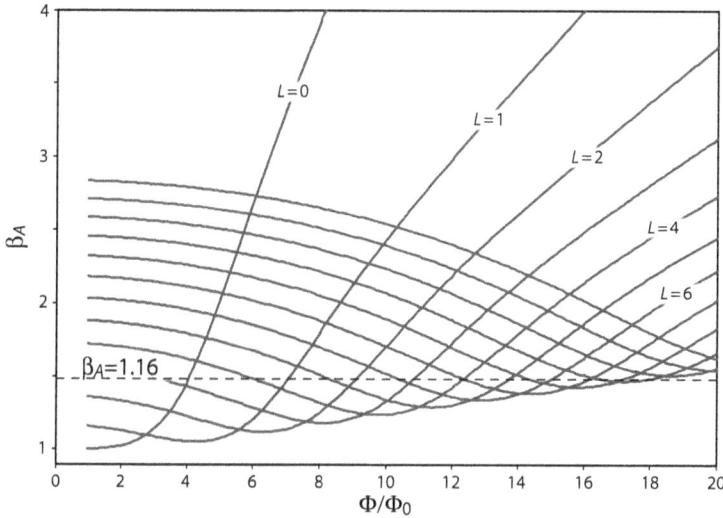

Fig. 2.7   Field dependence of the Abrikosov parameter β_A for different values of the angular momenta L (after [352]).

for the derivative of the Kummer function, we can find the $N(L, R_0)$ value, which obeys the boundary condition Eq. (2.13):

$$(L - R_0^2)M(-N, L+1, R_0^2) - \frac{2NR_0^2}{L+1}M(-N+1, L+2, R_0^2) = 0 . \quad (2.16)$$

Remarkably, negative values for $N(L, R_0)$ are found from the solutions of Eq. (2.13), meaning that the energy $E_\perp$ in Eq. (2.14) is lower than $1/2\,\hbar\omega$. Therefore, *as a result of the confinement by 'superconducting' boundary conditions, the energy levels in finite samples lie below the classical value $1/2\,\hbar\omega$ for infinite samples* [75, 415]. The whole energy level scheme (Fig. 2.6), found by Saint–James [415], can be reconstructed by calculating $E_\perp$ versus R_0^2 for different L values. From this diagram one can easily go over to a plot of field versus temperature, using the relation $E_\perp = -\alpha$. The corresponding values of the Abrikosov parameter $\beta_A = \overline{|\psi|^4}/(\overline{|\psi|^2})^2$, giving an idea about the 'flatness' of $|\psi|$ [3], are plotted in Fig. 2.7. It should be noted that β_A for certain values of L and H (see the levels below the dashed line in Fig. 2.7) is smaller than the well-known minimum possible value $\beta_A = 1.16$ for the triangular Abrikosov vortex lattice in bulk samples.

To conclude this discussion, we note that N is a 'bad quantum number' in finite samples, but rather is a parameter which has to be found from the appropriate boundary condition. By contrast, a 'good quantum number'

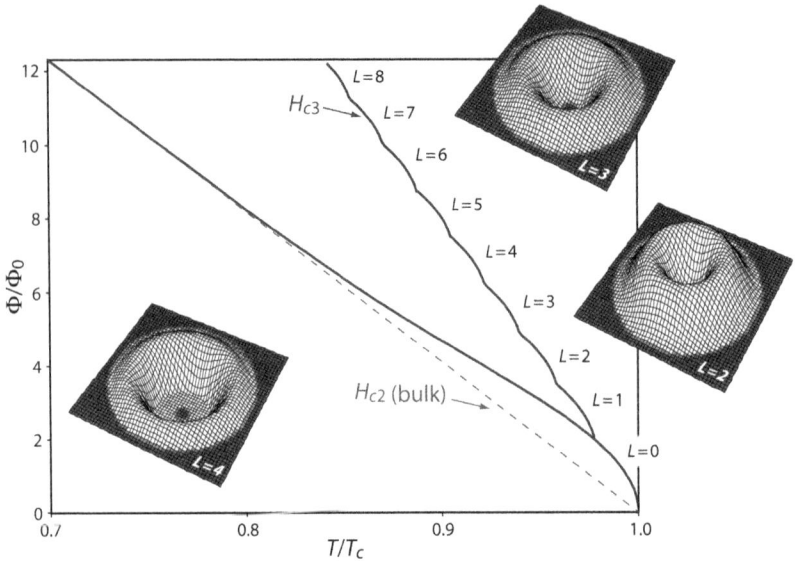

Fig. 2.8   Third critical field H_{c3} (theory) and the bulk upper critical field H_{c2} (dashed line) versus T/T_c for a superconducting cylinder. The cusp-like $H_{c3}(T)$ line is formed due to the change of the orbital quantum number L. In the three-dimensional plots, $|\psi|$ is shown as a function of the spatial coordinates for several values of L (after [352]).

of the problem is L. When forming a superconducting condensate with a proper finite L and $N(L, R_0) < 0$, we conserve the rotational momentum and at the same time reduce the energy below $1/2\hbar\omega$ [352].

As shown above, due to the onset of surface superconductivity at $H_{c3}(T)$ corresponding to negative N in Eq. (2.14), superconductivity can appear at magnetic fields well above the $H_{c2}(T)$ line found for $N = 0$. By changing the variable $E_\perp$ in Fig. 2.6 into T, we obtain the cusplike phase boundary $H_{c3}(T)$ as shown in Fig. 2.8, which is due to switching between different orbital momenta L. The phase boundary of the superconducting disk (Fig. 2.8) has been observed experimentally by Buisson et al. [75, 415] and by Moshchalkov et al. [349]. It should be emphasized that the presence of the oscillations in the $H_{c3}(T)$ curve is crucially dependent on the imposed Neumann boundary conditions [45]. Contrary to that, the equivalent eigenvalue spectrum with Dirichlet boundary conditions does not show any oscillations, if one follows the lowest Landau level $E_{LLL}(H)$ [359].

The Landau level scheme for a cylindrical dot with the 'superconducting' boundary conditions of Eq. (1.9b) is presented in Fig. 2.6. Each level is characterized by a certain orbital quantum number L where $\psi = |\psi| \exp(\mp \imath L \varphi)$ [352]. The levels corresponding to the plus sign in the argument of the exponent are not shown since they are situated at energies higher than the ones with minus sign. The lowest Landau level in Fig. 2.6 represents, in fact, a cusp-like envelope, obtained from the switching between different L values with changing magnetic field. Following our main guideline that $E_{LLL}(H)$ determines $T_c(H)$, a cusp-like superconducting phase boundary with nearly perfect linear background is expected for the dot. The measured phase boundary $T_c(H)$, shown in Fig. 2.1(b), can be nicely fitted by the calculated one of Fig. 2.6, thus proving that $T_c(H)$ of a superconducting dot consists indeed of cusps with different L's [415]. Each fixed L describes a giant vortex state that carries L flux quanta Φ_0. The linear background of the $T_c(H)$ dependence is very close to the third critical field $H_{c3}(T) \simeq 1.69\, H_{c2}(T)$, which is obtained in the limit of $L \to \infty$ [118, 415]. Contrary to the very thin loop where the Little–Parks oscillations are perfectly periodic, the dot demonstrates a certain aperiodicity [347] in very good agreement with the theoretical calculations [45, 75, 415]. The lower critical field of a cylindrical dot H_{c1}^{dot} corresponds to the change of the orbital quantum number from $L = 0$ to $L = 1$, i.e. to the penetration of the first flux line [45]:

$$\mu_0 H_{c1}^{\mathrm{dot}} = 1.924\, \frac{\Phi_0}{\pi R^2} \ . \tag{2.17}$$

For the long mesoscopic cylinder described above, demagnetization effects can be neglected. On the contrary, for a thin superconducting disk, these effects are quite essential [124, 428]. For a mesoscopic disk, made of a Type-1 superconductor, the phase transition between the superconducting and the normal state is of the second order if the expulsion of the magnetic field from the disk can be neglected, i.e. when the disk thickness is comparable to ξ and λ. However, when the disk thickness is larger than a certain critical value, first order phase transitions should occur. The latter has been confirmed in ballistic Hall magnetometry experiments on individual Al disks [166, 168, 169]. A series of first order transitions between states with different orbital quantum numbers L has been seen in magnetization curves $M(H)$ in the field range corresponding to the crossover between the Meissner and the normal states [168]. Besides the cusp-like $H_{c3}(T)$ line found earlier in transport measurements [75, 349, 415], transitions between the $L = 2$ and $L = 1$ states have been observed by probing the superconducting state below the $T_c(H)$ line with Hall micromagnetometry [168]. Even deeper in the superconducting area, the recovery of the normal Φ_0-

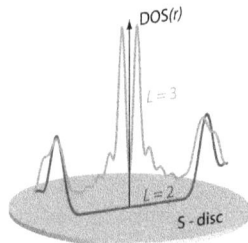

Fig. 2.9   Sketch of the spatial distribution of the thermodynamically averaged local density of states (DOS) in a superconducting disc for multi-quanta vortices $L = 2$ and $L = 3$.
DOS(r) is directly related to a zero-bias local tunnelling conductance between two reservoirs. (Adapted with permission
from Macmillan Publishers Ltd: Nature, [313], copyright
2002.)

vortices and the decay of the giant vortex state might be expected [346].
The former has been considered in Ref. 81 in the London limit by using the
image method. Magnetization and stable vortex configurations have been
also analyzed in mesoscopic disks in Refs. [124, 428].

Interestingly, the electronic structure of the giant vortices can be probed
with scanning tunneling measurements. A spacial map of the tunneling
density of states for multiquanta giant vortex states has been calculated
by Tanaka *et al.* [458] and by Mel'nikov and Vinokur [313]. In particular,
the zero-bias tunneling spectrum taken at the center of the disc is very
different for even and odd vorticities, see Fig. 2.9. Up to now, however,
no direct experimental observation of such STM spectra in the core of the
giant vortices has been reported.

2.4   The cross-over from loop to dot

In this section, we will discuss the influence of an opening in the interior
of a superconducting disk on the $T_c(H)$ phase boundary by varying the
diameter of the hole. The case of loops with finite width was studied within
the London theory by Bardeen [35], who calculated that in cylinders of very
small diameters and with a wall thickness of the order of the penetration
depth, the flux is quantized in units of $\nu\Phi_0$ (with $\nu < 1$). Later on, it
was shown that the flux through an area $S = \pi r_m^2$ is quantized in units
of Φ_0, with the arithmetic mean $r_m = 1/2 (r_i + r_o)$ of the inner and outer
radius r_i and r_o, respectively [200]. Arutunian and Zharkov [14] found in
the London limit an effective radius of $r_{\text{eff}} = \sqrt{r_i r_o}$, such that inside this
ring the flux was exactly quantized.[2] Baelus *et al.* [26] could show that the
value of $r_i < r_{\text{eff}} < r_o$ is dependent on the vorticity L, and they found it to
be an oscillating function of the magnetic field.

---

[2]Note that these two different values r_m and r_{eff} are nearly identical for the narrow
ring.

A self-consistent treatment of the full nonlinear Ginzburg–Landau equations for a square loop has been carried out by Fomin *et al.* [157, 158], who could demonstrate that the distribution of the order parameter $|\psi|$ is strongly inhomogeneous due to the presence of sharp corners. The precise shape of the $T_c(H)$ curve crucially depends on the area fraction for which $T_c(H) \neq 0$.

Bruyndoncx *et al.* [73] investigated very thin loops of finite width using the linearized Ginzburg–Landau equation, where the additional magnetic field, induced by supercurrents, can be neglected.[3] They limited their investigations to cylindrically symmetric solutions, hence studying the giant vortex state only. Berger and Rubinstein [52] studied thin loops of finite width using the nonlinear Ginzburg–Landau theory, neglecting the induced fields. Baelus *et al.* [26] considered flat disks of nonzero thickness with a circular (but not necessarily centered) hole, studying the superconducting properties also deep in the superconducting state. For small disks with a centered hole, they only found the giant vortex state. However, for larger perforated superconducting disks, a re-entrant behavior was found from the giant vortex state to a state with separated Φ_0-vortices, and back to the giant vortex state.

In the following, we will discuss the evolution of the superconducting state for the transition from a disk geometry to a thin ring, as it is illustrated in the insets in Fig. 2.11(a – e). The superconducting properties of the different considered structures (see Tab. 2.1 for some basic characteristics) were investigated by transport measurements, carried out close to the transition line between superconductivity and the normal state.

In order to minimize the influence of the current and voltage leads on the superconducting properties of the mesoscopic structures, wedge-shaped contacts with an opening angle of $15°$ were used [337]. The importance of the specific design of the contacts lies in the fact that in a superconducting

Table 2.1 Characteristics of the considered disc and loops

Inner Radius r_i	Coherence Length[a] $\xi(0)$	Film Thickness[b]
0.0, 0.1, 0.3 and 0.5 μm	156 nm	39 nm
0.7 μm	120 nm	54 nm

[a] determined with a co-evaporated reference film
[b] determined with X-ray diffraction and atomic force microscopy

---

[3]This assumption is valid near the phase boundary where $\psi \to 0$.

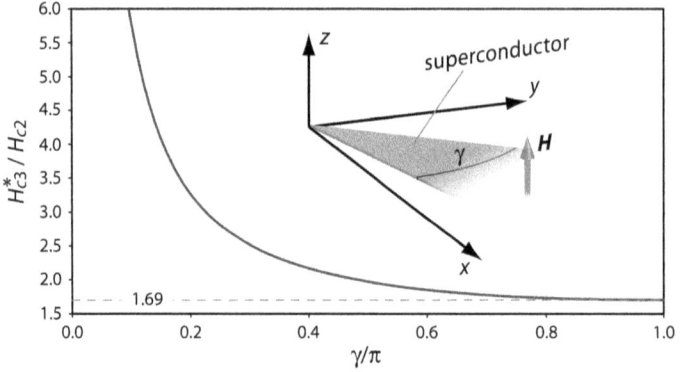

Fig. 2.10 Nucleation field enhancement factor $H_{c3}^*(T)/H_{c2}(T)$ (solid line) for an infinitely long superconducting wedge with a corner angle γ. The nucleation field asymptotically diverges as $\gamma \to 0$. The usual enhancement factor of 1.69 is naturally recovered for $\gamma = \pi$ (dashed line). The inset shows a schematics of the geometry.

wedge with an opening γ, the nucleation field $H_{c3}^*(T)$ is enhanced compared to the case of bulk material, since sharp corners reduce the superfluid velocity $\boldsymbol{v}_s$. The problem of a wedge has been addressed by several authors independently [155, 156, 429, 481], all leading to similar conclusions. The linear Ginzburg–Landau equation was solved either numerically [429] on a finite space grid, or with a variational approach using trial for the case of $\gamma < \pi$ functions [155, 156, 481]. To very good accuracy, the nucleation field for a wedge can be fitted to [429]:

$$\frac{H_{c3}^*(T)}{H_{c2}(T)} = \frac{\sqrt{3}}{\gamma}\left(1 + 0.14804\,\gamma^2 + \frac{0.746\,\gamma^2}{\gamma^2 + 1.8794}\right). \qquad (2.18)$$

This function is shown as a solid line in Fig. 2.10. The ratio H_{c3}^*/H_{c2} is a monotonically decreasing function of γ, which tends to the value 1.69 (horizontal dashed line) as γ increases to π. For very small opening angles ($\gamma \ll 1$), an analytical treatment of the linear Ginzburg–Landau equation becomes possible. In such case, ψ varies only weakly with the angle φ. A rigorous proof of the validity of this assumption was given by Klimin *et al.* [254]. They have used an 'adiabatic approach' and calculated the contribution of the non-adiabaticy operator on the lowest energy eigenvalue of the linearized Ginzburg–Landau equation, which is negligible for wedges with angles $\gamma \lesssim 0.1\,\pi$. In the low angle limit, the solutions of the linear Ginzburg–Landau equation are known and can be expressed in terms of Legendre polynomials [254]. The ground state was shown to have a vortex in the corner, meaning that a confined circulating superconducting current appears in a wedge in the vicinity of the corner. This vortex can exist for sufficiently sharp wedges with angles $\gamma \lesssim 0.2\,\pi$.

In contrast to a disc, loops have two superconducting/vacuum interfaces, one at the outer and one at the inner radius. Therefore, the appropriate boundary condition (Eq. 1.9b) has to be fulfilled at both r_o and r_i:

$$\left.\frac{\partial |\psi(r)|}{\partial r}\right|_{r=r_{o,i}} = 0 . \tag{2.19}$$

The experimentally determined $T_c(H)$ phase boundary of the disc is shown in Fig. 2.11(a), using a resistance criterium of $2/3\,R_n$. In order to fit the theoretical curve (solid line) to the experimental data, using the formalism described above, a slightly different coherence length of $\xi(0) = 135$ nm was used instead of the value determined with the reference sample (Tab. 2.1). With this minor correction, an excellent agreement between theory and experiment is found for both the position and the amplitude of the cusps. Similarly, the phase boundary of the ring with $r_i = 0.1\,r_o$ (Figure 2.11(b)) can be fitted very well with the theory, using an outer radius of $1.02\,\mu$m in order to achieve good matching between the position of the cusps in the experimental and in the theoretical curves.[4] Having a linear background superimposed with oscillations, the H–T diagram of the ring with the smallest hole resembles closely the $T_c(H)$ line of the disk.

Figure 2.11(c) shows the phase boundary obtained from the ring with $r_i = 0.3\,r_o$. In this case, a linear dependence is only seen for vorticity $L > 4$, while at lower magnetic fields a parabolic background suppression of T_c is observed. The crossover from the linear to the parabolic regime occurs at $\pi r_o^2/\xi^2(T) \approx 20$, corresponding to a value $r_o - r_i \approx 1.8\,\xi(T)$. This result is in good agreement with the thickness $\tau = 1.84\,\xi(T)$ of the studied structure, at which a crossover from a two-dimensional to a three-dimensional regime has been predicted for a thin film in a parallel magnetic field [145, 425]. Although the positions of the cusps are in good agreement with those of the theoretical curve, the amplitude of the experimentally determined oscillations deviates slightly from the calculated ones, which are more pronounced at $L = 1$ and weaker at higher vorticity. The first vortex penetrates the ring at a lower magnetic field value compared to the ring with the smallest hole, while the transitions $L = 1 \leftrightarrow 2$ to $L = 5 \leftrightarrow 6$ occur at a higher magnetic field. The fact that the transitions occur at lower magnetic fields, if the ring has thinner lines, can be expected, since the transition between L and $L+1$ occurs at $\Phi/\Phi_0 = L+1/2$ for an infinitely

---

[4]The flux Φ on the x-axis denotes the flux $\Phi = \mu_0 H \pi r_o^2$ through the ring and the hole.

thin loop or a cylinder. At higher magnetic fields, a giant vortex state is formed [73] and the disk with a small hole in the center behaves like a disk without hole. However, this consideration cannot fully explain why the introduction of a small hole in a disk delays changes in the vorticity at higher magnetic fields.

The measured $T_c(H)$ phase boundary of the ring with ratio $r_i = 0.5\,r_o$ is shown in Figure 2.11(d). In the temperature range accessible with the used experimental setup [334], only a parabolic background dependence of the critical temperature on the magnetic field could be observed. Comparison of the experimental results with the calculations demonstrates a similar behavior as in the previous case with $r_i = 0.3\,r_o$: the positions of the cusps match for experiment and theory, but a poor agreement is found for the amplitude of the oscillations. For vorticities of $L = 1$ and $L = 2$, the amplitude is lower in the experimental curve, while for $L > 3$ the reverse is observed. At low vorticities, the transition between states of different L occurs at lower magnetic field than for the disk, while the transitions $L = 3 \leftrightarrow 4$, $L = 4 \leftrightarrow 5$ and $L = 5 \leftrightarrow 6$ take place at a higher magnetic fields, similar to the case of $r_i = 0.3\,r_o$.

Using the nonlinear Ginzburg–Landau equations, Zhao *et al.* [513] found that a vortex state with non-uniform vorticity can exist, but it is unclear whether these solutions are thermodynamically stable or metastable. If these solutions have a lower energy than those with single vorticity in the ring, the phase boundary found from the linearized Ginzburg–Landau equation would be slightly altered. Such solution — or another solution with vortex molecules — could be responsible for the observed difference between the theoretical and experimental curves. In order to analyze the solution with non-uniform vorticity using the linearized Ginzburg–Landau equation, the ring was devided into two regions of different vorticity, and the order parameter was imposed to be zero along a certain radius. However, the critical field of the considered sample, obtained with these new boundary conditions, does not exceed H_{c2} in the temperature range of our measurements. Therefore, a state with non-uniform vorticity will probably not exist at the phase boundary, although it might become stable deeper in the superconducting state.

Finally, the H–T diagram for the ring with the thinnest line with $r_i = 0.7\,r_o$ is shown for a resistance criterium of $^4/_5\,R_n$ in Figure 2.11(e). Again, a quasi-parabolic background suppression of T_c is observed. However, in this case, the amplitude of the oscillations is larger than in the other samples and the transition between states of different vorticity is almost periodic in field.

The correspondence between the theoretical curve and our experimental data is reasonably good.

Comparing the phase boundaries of the five considered samples directly with each other, see Fig. 2.11(f), it becomes clear that these $T_c(H)$ curves overlap for $L = 0$. This means that an opening in the disk does not affect the phase boundary, as long as no vortex is trapped inside the superconductor. Only the magnetic field range, over which the state of $L = 0$ exists at the phase boundary, is lowered by introducing a hole in the disk. The $T_c(H)$ line of the disk with the smallest hole does not deviate strongly from the phase boundary of the disk without any opening — only small changes in the positions of the cusps are observed at low vorticity. For larger holes ($r_i = 0.3\, r_o$ and $0.5\, r_o$), a crossover from a parabolic to a linear behavior is clearly seen in the $T_c(H)$ curves. However, the sample with the thinnest line ($r_i = 0.7\, r_o$) does not show the linear regime in the studied interval, and only a parabolic dependence is seen.

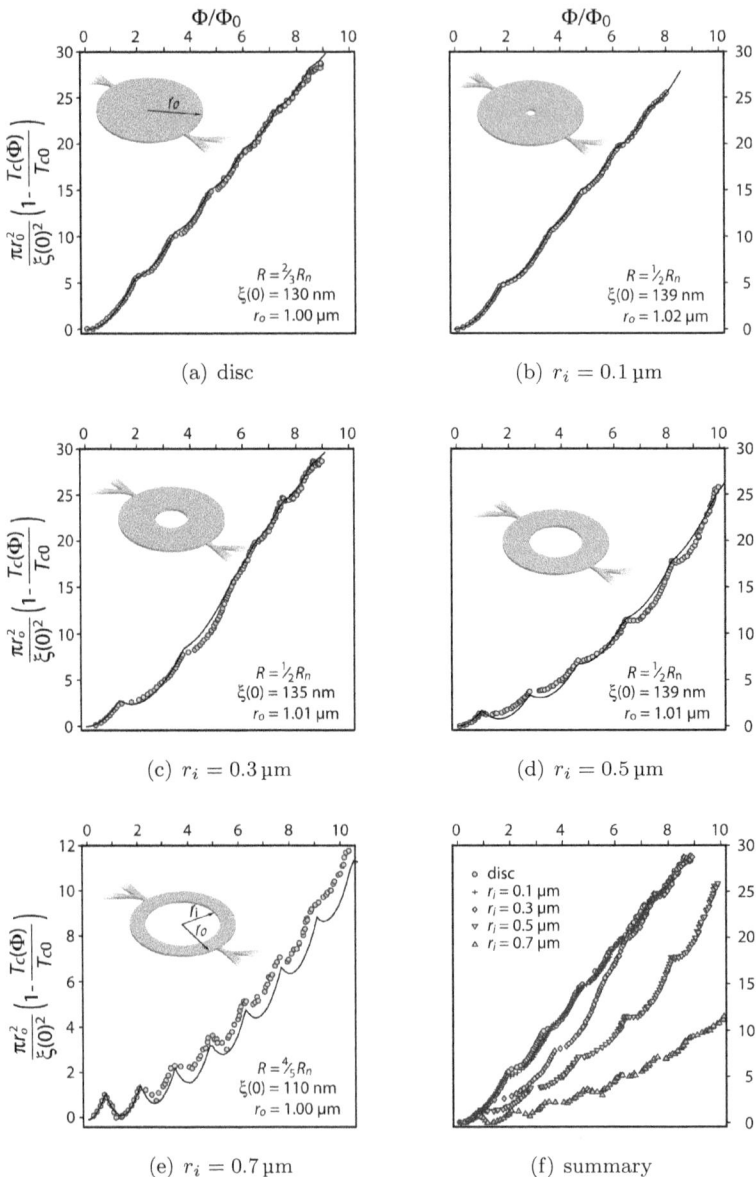

(a) disc

(b) $r_i = 0.1\,\mu m$

(c) $r_i = 0.3\,\mu m$

(d) $r_i = 0.5\,\mu m$

(e) $r_i = 0.7\,\mu m$

(f) summary

Fig. 2.11   Experimental $T_c(H)$ phase boundaries of (a) a disc with radius $r_o = 1.0\,\mu m$ and (b – e) loops of the same size as the disc but with inner radii $r_i = 0.1$, 0.3, 0.5 and 0.7 μm. Solid lines show theoretical $T_c(H)$ curves, calculated with the indicated values of $\xi(0)$ and r_o. For comparison, the five experimental curves are shown together in panel (f) (after [334]).

2.5 Symmetry-induced antivortices in mesoscopic superconductors

Recently, a new important step in our understanding of the flux confinement phenomena in individual nanostructures has been made by using a Ginzburg–Landau approach combined with experiments. An answer to a puzzling question of the symmetry conservation of the vortex patterns close to the phase boundary has been found for equilateral triangles, squares and other polygons [25,90,91,94,312,317]. As already discussed in Secs. 2.1–2.4, superconducting critical parameters of nanostructures are very sensitive to their geometry and topology. In the presence of a magnetic field, an important issue about the stability of the symmetry-compatible vortex pattern should be addressed. A very interesting problem arises here: *how to put, for instance, two vortices into a triangle or three vortices into a square still keeping their threefold and fourfold symmetries?*

In disks and cylinders, for example, the giant vortex state — a big single whirl in the center — is compatible with the symmetry of the boundary conditions. However, in superconducting equilateral triangles, squares etc, a very different situation arises [90–92]. For example, it is quite easy to keep the imposed symmetry when trying to put three vortices into a triangle or four into a square. But what about 'less lucky' numbers, such as two vortices in a triangle or three vortices in a square? The latter can be compared to "...pushing a triangular peg into a square hole" [131]. Seemingly, the only possible solution of this puzzle is to 'merge' single quantum vortices into a single 'giant' vortex and to place it in the center.

In the language of numbers, it is like presenting the number of applied flux quanta through the sample $2 = 2+0$ instead of $2 = 1+1$ for two vortices in triangle, or similarly $3 = 3+0$ instead of $3 = 1+1+1$ for three vortices in a square. But such a 'merger' of all vortices into a single giant vortex costs too much energy in triangles and squares, although its occurrence is possible in a disc or a cylinder. Excluding the formation of the giant vortex state, it looks as if there is no solution for the puzzle. Remarkably, spontaneous formation of a vortex-antivortex pair saves the whole scheme. Indeed, using the same symbolic language of numbers, we can claim that in the equilateral triangle, not only $2 = 2+0$ but also $2 = 1+1+1-1$, meaning three vortices in three corners of a triangle and one antivortex in the center (which is fully symmetry consistent), or for a square, instead of $3 = 3+0$, we can use $3 = 1+1+1+1-1$, thus making it possible to put comfortably four vortices into four corners and one antivortex into the center. This looks like a 'discovery' of a usefulness of negative numbers. antivortices in this case provide the necessary means for the vortex system to comply with the imposed symmetric boundary conditions [25, 90–92, 94, 312, 317].[5]

The experimental evidence in favor of this symmetry-conserving solution and formation of antivortices is obtained from the experimentally measured phase boundary $T_c(H)$ which is in a very good agreement with the calculations [90–92, 335]. Antivortices in vortex matter — like anti-particles in 'ordinary' matter — seem to play an essential role in conserving the symmetry close to $T_c(H)$. The symmetry-induced nucleation of antivortices is not restricted to superconductors, but may also be applied to symmetrically confined superfluids and Bose–Einstein condensates. The vortex-antivortex patterns are very compact, which makes a direct visualization of antivortices quite difficult. At the same time these patterns can be strongly expanded by using a magnetic dot with perpendicular magnetization in the center (see Sec. 2.8.1) of the symmetric structures (triangle, square) [84], or a 2×2 antidot cluster in a square (see Sec. 3.2). Therefore, the chances to resolve experimentally separate vortices and antivortices are better for these structures.

2.5.1   *Square*

The nucleation of superconductivity is normally analyzed in terms of the linearized Ginzburg–Landau equation (Eq. (1.6)) with the applied magnetic fields $\boldsymbol{H} = \nabla \times \boldsymbol{A}$. However, the superconducting boundary conditions considerably complicate the solution of equation Eq. (1.6) and, therefore, a recently developed analytical gauge transformation can be used, such that the vector potential $\boldsymbol{A}$ has no component normal to the sample boundaries of arbitrary regular polygons [92].

The solutions of the linearized Ginzburg–Landau equation for the square are then characterized by irreducible representations (irreps) A, B, E_- and E_+ with the characters $\exp(\imath\, n\, \pi/2)$, where $n = 0$, 2, -1 and 1 under the fourfold rotation, respectively [271]. The ground Landau level shows an oscillatory cusp-like behavior as a function of flux (defined as $\Phi = a^2 H$ with the side length a of the square), corresponding to a crossover of states belonging to different irreps, see Fig. 2.12(a). This resembles the situation in a superconducting disk discussed in Sec. 2.3, with the important

---

[5]Note, that these symmetry-induced antivortices are qualitatively different from the ones appearing due to the Kosterlitz–Thouless transition in two-dimensional superconductors.

Table 2.2   Characteristics of the considered aluminium square

Superconducting Property	Method of Investigation
$\xi(0) = 156\,\text{nm}$	slope of $T_c(H)$ of reference film
$\tau = 39\,\text{nm}$	X-ray diffraction
$T_{c0} = 1.32\,\text{K}$	resistive measurement of reference film

difference that the latter has the symmetry C_∞ of a two-dimensional rotator with an infinite number of irreps ('rotational quantum numbers') [415]. As a consequence, there is no 'repulsion' of levels in the disk, whereas for the square solutions of the linearized Ginzburg–Landau equation show a regular pattern of avoided crossings between levels belonging to the same irreps.

Figure 2.12(b) compares the calculated and the measured phase boundary for a mesoscopic aluminium square. The theoretical curve is obtained from Fig. 2.12(a), where the ground state level is selected for all flux values. In the experiment, the $T_c(H)$ boundary of a $2 \times 2\,\mu\text{m}^2$ aluminium square was measured resistively, using an electronic feedback circuit [72] (see Table 2.2 for some sample characteristics). The agreement between the calculated lowest Landau level and the measured $T_c(\Phi)$ is quite good (especially for the positions of the cusps), even without use of fitting parameters.

At the cusp positions on the phase boundary, the vorticity L changes by one, starting from zero (no fluxoids) at low magnetic fields. In the case of a disk (C_∞) the vorticity equals the orbital quantum number L, defining the flux $L\,\Phi_0$ that is carried by the giant vortex [415]. By contrast, for the square, the rotational axis is of *finite* order (C_4) and, therefore, the distribution of vortices in symmetry-consistent solutions is not *a priori* evident. The seven insets in Fig. 2.12(b) show schematically the distribution of vortices, which are clearly different from the giant vortex states in the $|\psi|^2$ pattern of the disc. In general, the density of $|\psi|^2$ increases near the corners of the square, which is in line with the general trend of nucleation of surface superconductivity [415].

In the case of small L's, vortices can occupy one central and four diagonal positions. In contrast to the diagonal vortices which always enclose a single quantum Φ_0, the central vortex can have different winding numbers in order to conserve the total vorticity of a given state. The contribution of the two kinds of vortices (central 'n' and four diagonal 'm') to the total winding number of the states, shown in Fig. 2.12(b), is given by $L = n + 4m$, where $n = 0, 1, 2, -1$ and $m = 0, 1$, see Tab. 2.3.

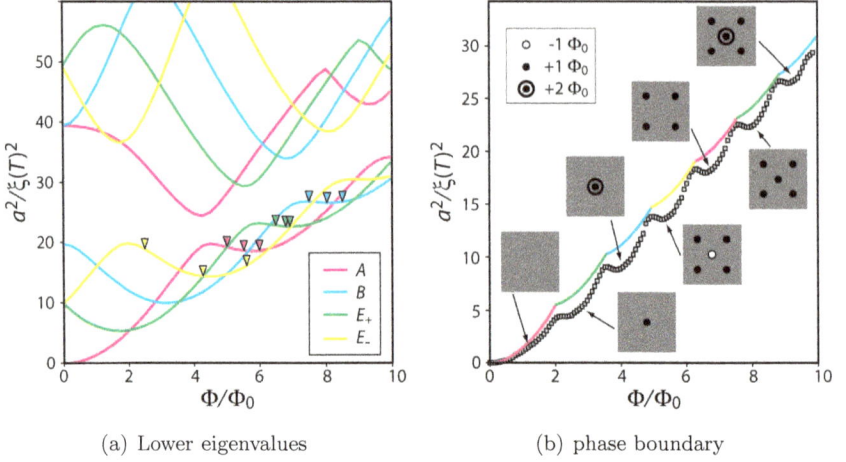

(a) Lower eigenvalues   (b) phase boundary

Fig. 2.12   (a) Lower eigenvalues of the linearized Ginzburg–Landau equation for a mesoscopic square as a function of magnetic flux, with superconductor/vacuum boundary
conditions. The different colors correspond to the four irreducible representations A
(magenta), B (cyan), E_+ (green) and E_- (yellow). The triangular markers on the lower
lines indicate the values of Φ for which vortex patterns are shown in Fig. 2.13. (b) Comparison between the calculated (solid curve) and the measured $T_c(\Phi)$ phase boundary
(squares). The experimental data have been corrected for the presence of the measuring
leads. Insets show schematically the vortex structure in different regions of the phase
diagram. In the range of $5.0 < \Phi/\Phi_0 < 6.3$, an antivortex is formed spontaneously at
the center of the square, coexisting with four Φ_0-vortices along the diagonals (after [90]).

    The nature of the central vortex changes whenever vorticity is changed
by one. Thus the central vortex is absent in the first state $(n = 0)$, it is
a Φ_0-vortex in the second state $(n = 1)$, it is a giant vortex in the third
state $(n = 2)$ and it is an *antivortex* in the fourth state $(n = -1)$, as
illustrated in Fig. 2.12(b). Moreover, the sequence of winding numbers

Table 2.3   Composition of the total winding number

Central Position		Diagonal Position		Vorticity
0	+	$4 \cdot 0$	=	0
1	+	$4 \cdot 0$	=	1
2	+	$4 \cdot 0$	=	2
-1	+	$4 \cdot 1$	=	3
0	+	$4 \cdot 1$	=	4
1	+	$4 \cdot 1$	=	5
2	+	$4 \cdot 1$	=	6

of the central vortex (-1, 0, 1, 2) is periodically repeated when going to the right of the phase diagram[6]. Since the kinetic energy of a vortex is proportional to L^2, the system prefers to split the giant vortex into a sum of Φ_0-vortices [3] if there are no special symmetry restrictions. This explains why only four numbers mentioned above appear as winding numbers for the central vortex. On the other hand, the formation of antivortices is controlled completely by the discrete symmetry. Indeed, in the state with $L = 3$, one cannot distribute three Φ_0-vortices on the square keeping the symmetry, which is solved by distributing four Φ_0-vortices and adding one antivortex in the center.

The flux penetration into mesoscopic superconductors turns out to be influenced as well by the spontaneous generation of antivortices. The theoretical analysis shows that in regular polygons with N edges the flux enters always by N singly quantized vortices through the edge centers, because these are the symmetry points with the lowest values of $|\psi|^2$ on the borders. When increasing the field further, it is energetically favorable for these vortices to reorient towards the corners of the polygon as shown in Fig. 2.12(b), thus paving the way for the entrance of the next set of N Φ_0-vortices. However, such a reorientation *cannot* be performed by a *continuous rotation* of the vortex patterns, since that would violate the imposed symmetry. Therefore the formation of additional antivortices and vortices turns out to be indispensable.

This is further illustrated in Fig. 2.13 for the case of a square, where three stages of the evolution of the vortex patterns — entrance of vortices (initial), the transient state (intermediate) and the formation of diagonal vortices (final) — are shown for the four irreps. The dynamics of this transformation differs dramatically for states of different symmetry. In the case of irreps E_+, the central region changes into a chess-like pattern of vortices and antivortices in the intermediate state. In a subsequent step (which is not shown here) the vortices form four pairs of lateral vortices, which then approach each other and 'rotate' by 45°. As the flux further increases, the antivortices and the 'rotated' Φ_0-vortices move towards the corners and eventually merge to form single diagonal Φ_0-vortices (final state).

For the case of irrep B, the intermediate state is associated with a change of the winding number of the central vortex: The reorientation mechanism of this symmetry involves the transformation of the central $2\,\Phi_0$ giant vortex

---

[6]The same counts for the irreps (E_-, A, E_+, B) of the corresponding states and, therefore, the states with winding numbers differing by multiples of four correspond to the same irrep.

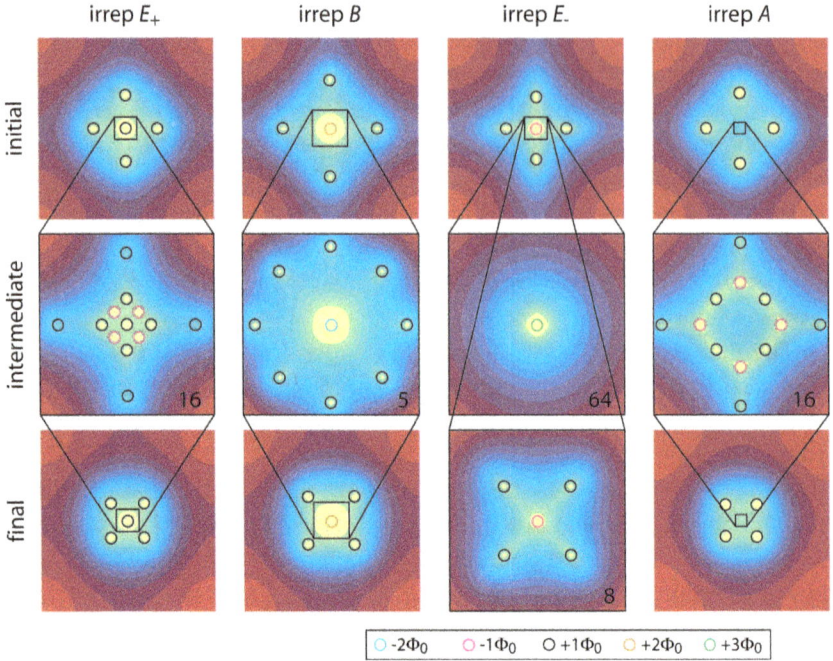

Fig. 2.13   Entrance of vortices from the edges of the square for the states of symmetry E_+, B, E_- and A, showing the density of $|\psi|^2$ for the initial, intermediate and final stages, corresponding to the Φ values marked (from left to right) in Fig. 2.12(a). The highest and lowest values of the density of $|\psi|^2$ are colored red and yellow, respectively, the latter thus indicating the positions of vortices and antivortices. Note the different magnification factors indicated in some of the panels (after [90]).

(initial state) into a giant antivortex ($-2\,\Phi_0$-vortex) and four diagonal Φ_0-vortices (intermediate state). Under this transformation the total vorticity remains invariant, i.e. $8 \times \Phi_0 - 1 \times 2\,\Phi_0 = 6\,\Phi_0$. While the diagonal vortices move to the corners, the four lateral Φ_0-vortices further approach the center and are absorbed by the central $2\,\Phi_0$-antivortex, restoring in this way the initial $2\,\Phi_0$ giant vortex in the center that is characteristic for the irrep B.

Similarly, for the E_- mode, the lateral Φ_0-vortices approach and merge with the central antivortex to form a giant $3\,\Phi_0$-vortex in the middle. Upon increasing field, the $3\,\Phi_0$-vortex will split into four diagonal Φ_0-vortices and the initial single antivortex (final state). Note that the $4\,\Phi_0$ vortices, surrounding the central antivortex in this pattern, are rotated by $45°$ if the Bogoliubov–de Gennes approach is used instead of the Ginzburg–Landau equations [245]. In order to be able to see the four separate Φ_0-vortices, the panel corresponding to the final state had to be magnified — this is due to the strong attraction of the Φ_0-vortices to the central antivortex. For the same reason, the spacing between the three markers on the E_- curve in Fig. 2.12(a) is large, as compared to the other irreps.

Finally, as already in the case of irrep E_+, four vortices and four antivortices arise in the intermediate state of symmetry A. In the initial stage, there is no central vortex in this specific case. The approach of the lateral Φ_0-vortices induces the creation of a central vortex–antivortex pattern in the intermediate state, which is however rotated by $45°$, as compared to the case of irreps E_+. At higher flux (final state), the lateral vortices and antivortices will annihilate, leaving four diagonal Φ_0-vortices which move towards the corners. The dynamics of the vortex–antivortex annihilation process in a square was analyzed by Sardella *et al.* [416].

The appearance of symmetry-induced antivortices remains valid for all other superconducting polygons (triangle, pentagon, etc.), and is a very fundamental property of symmetrically confined vortex matter in general. These findings are applicable not only to superconductors, but also to superfluids (^{4}He and ^{3}He) and Bose–Einstein condensates. Superfluids, rotated in a triangular or square vessel, may as well generate antivortices in order to comply with the imposed symmetry, and by proper arrangement of the laser fields, the vortex patterns in Bose–Einstein condensates confined by triangular or square traps could also reveal symmetry-induced antivortices. Vortex patterns in superconducting squares in case of the d-wave superconductivity ('d-dot') were analyzed theoretically by Kato *et al.* [246] and Koyama *et al.* [260].

2.5.2 *Triangle*

As a second example of symmetry-induced antivortices in mesoscopic superconductors, we shall now consider an equilateral triangle and use the linearized Ginzburg–Landau theory to investigate its phase boundary $T_c(H)$, the properties of different vortex states and the evolution of vortex patterns as a function of the applied field. As before in Sec. 2.5.1, the validity of the symmetry-consistent solutions will be checked by comparing the calculations with the measured phase boundary of a mesoscopic aluminium triangle. Again, the presence of a C_N symmetry of polygons (in this case $N = 3$) leads to a spontaneous formation of antivortices, which appear to conserve the discrete symmetry imposed by the boundary conditions.

As we pointed out in the previous section, the superconducting boundary condition of Eq. (1.9b) can be reduced to the Neumann boundary condition

$$\nabla \psi|_n = 0 , \tag{2.20}$$

if the magnetic vector potential $\boldsymbol{A}$ is chosen in a form with zero normal
component at the boundary of the sample. With this choice of gauge, the
solution of the linearized Ginzburg–Landau equation reduces to an eigenvalue problem in a basis set of functions obeying Neumann boundary conditions. It is convenient to take for such a basis set the eigenfunctions of
the zero field problem. Given the threefold rotational symmetry of the considered triangle, the solutions of Eq. (1.6) are characterized by irreducible
representations of the cyclic group C_3. This group contains the three irreps
A, E_-, E_+ with the characters $\exp(i n \pi/2)$ with $n = 0$, -1 and 1 under the
threefold rotation, respectively [271]. The unnormalized eigenfunctions of
the zero field problem, i.e. of the particle in the equilateral triangular box
obeying Neumann boundary conditions, can be obtained in the following
form for the irrep A [91]:

$$\psi_{pq}^A(x, y) = \cos \frac{\pi y(2p + q)}{h} \cdot \cos \frac{\sqrt{3}\pi x q}{h} + \cos \frac{\pi y(q - p)}{h} \cdot \cos \frac{\sqrt{3}\pi x(p + q)}{h}$$

$$+ \cos \frac{\pi y(p + 2q)}{h} \cdot \cos \frac{\sqrt{3}\pi x p}{h} , \qquad p \geq q = 0, 1, 2, \ldots, \tag{2.21a}$$

$$\psi_{pq}^A(x, y) = - \cos \frac{\pi y(2p + q)}{h} \cdot \sin \frac{\sqrt{3}\pi x q}{h} + \cos \frac{\pi y(q - p)}{h} \cdot \sin \frac{\sqrt{3}\pi x(p + q)}{h}$$

$$- \cos \frac{\pi y(p + 2q)}{h} \cdot \sin \frac{\sqrt{3}\pi x p}{h} , \qquad p \geq q = 1, 2, 3, \ldots . \tag{2.21b}$$

The first (Eq. (2.21a)) and second (Eq. (2.21b)) set of eigenfunctions are
symmetric and antisymmetric with respect to the vertical symmetry planes,
respectively. For the irreps E_+ and E_- with $\psi^{E_-} = (\psi^{E_+})^*$, the eigenfunc-

tions of the zero field problem are [91]:

$$\psi_{pq}^{E_+}(x,y) = \exp\left(\frac{\pi[(2p+q)y - \sqrt{3}xq]\imath}{h}\right) + \exp\left(\frac{\pi[(p+2q)y - \sqrt{3}xp]\imath}{h}\right)$$

$$+ \exp\left(\frac{\pi[(-p+q)y - \sqrt{3}(p+q)x]\imath}{h} - 2\pi q\imath \pm \frac{2\pi\imath}{3}\right)$$

$$+ \exp\left(\frac{\pi[-(p+2q)y - \sqrt{3}xp]\imath}{h} - 2\pi\imath p \mp \frac{2\pi\imath}{3}\right)$$

$$+ \exp\left(\frac{\pi[(p-q)y + \sqrt{3}(p+q)x]\imath}{h} - 2\pi\imath p \pm \frac{2\pi\imath}{3}\right)$$

$$+ \exp\left(\frac{\pi[-(2p+q)y - \sqrt{3}qx]\imath}{h} + 2\pi\imath q \mp \frac{2\pi\imath}{3}\right),$$

$$(2.22)$$

where the two signs correspond to the following quantum numbers with $n = 0, 1, 2, \ldots$:

$$q = n + 1/3 \qquad p = q, q + 1, q + 2, \ldots \quad \text{upper sign, and}$$
$$q = n + 2/3 \qquad p = q, q + 1, q + 2, \ldots \quad \text{lower sign.}$$

In Eqs. (2.21) and (2.22), $h = \sqrt{3/4}\,a$ is the height and a the edge of the triangle. The above solutions are close by their structure to the solutions of the corresponding problem of a particle in a box with the $\psi = 0$ boundary condition [286]. In the presented calculations, a basis of only 40 lowest eigenstates was used for each irrep.

Figure. 2.14(a) shows the results of the calculations of the lowest eigenstates. The lowest Landau level shows an oscillatory cusp-like behavior as a function of Φ/Φ_0, corresponding to a crossover of states belonging to different irreps. The measured phase boundary for the mesoscopic aluminium triangle is given in Fig. 2.14(b). In order to obtain a good correspondence for the experimental and the theoretical positions of the transitions $L \to L + 1$, an effective sample area $S_{\text{eff}} = 1.76\,\mu m^2$ has been used. The $T_c(H)$ boundary was measured resistively, with the transport current fed through contacts I^+ & I^-, and the voltage measured across the contacts V^+ & V^- (see the inset in Fig. 2.14). The theoretical phase boundary was obtained from Fig. 2.14(a), as the ground state level, selected for all flux values. The agreement between the calculated lowest Landau level and the measured $T_c(\Phi)$ is good (especially the cusp positions coincide very well),

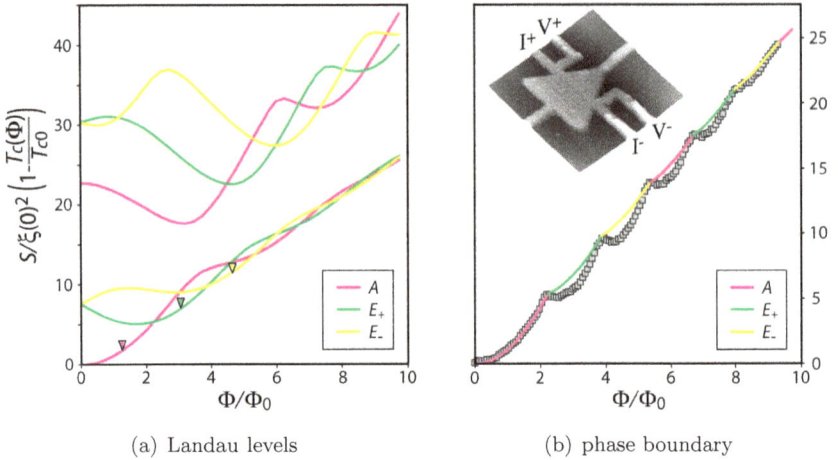

(a) Landau levels   (b) phase boundary

Fig. 2.14   (a) Calculated Landau levels for an equilateral triangle with superconducting boundary conditions. Magenta, green and yellow lines correspond to states of symmetry A, E_+ and E_-, respectively. Triangular markers indicate the values of Φ for which vortex patterns are shown in Fig. 2.15. (b) Comparison between the calculated (solid curve) and the experimental (squares) $T_c(H)$ phase boundary of an equilateral mesoscopic triangle of side length $a = 1.9\,\mu\text{m}$. The experimental data have been corrected for the presence of the measuring leads by subtracting a parabolic term that arises due to the $w = 0.33\,\mu\text{m}$ wide contact lines. The inset shows an AFM micrograph of the studied triangle (after [91]).

which justifies both the applicability of the linearized Ginzburg–Landau theory and the truncation of the basis set used in the calculations. The asymptotic behavior of the phase boundary line, at large flux, is in a good agreement with the ratio of critical fields $H_{c2} = 0.41\,H_{c3}$ found for a $1/3\,\pi$-wedge [222].

The density distribution of the order parameter $|\psi|^2$, shown in Fig. 2.15, is always in agreement with the C_{3v} symmetry of the equilateral triangle, which is higher than the C_3 symmetry of the linearized Ginzburg–Landau equation. This density is indeed invariant under the mirror reflections in vertical planes, contained in the group C_{3v}, since these induce complex conjugation of the order parameter. In the case of small L's, vortices can occupy one central position and three positions on the lines from the center to the corners of the triangle. Integration of the gradient of phase of the order parameter along the contours encircling the vortices has shown that the vortices in corner positions are always Φ_0-vortices. In contrast to that, the central vortex can have different winding numbers in order to adjust the total vorticity of a given state. The contribution of the two kinds of

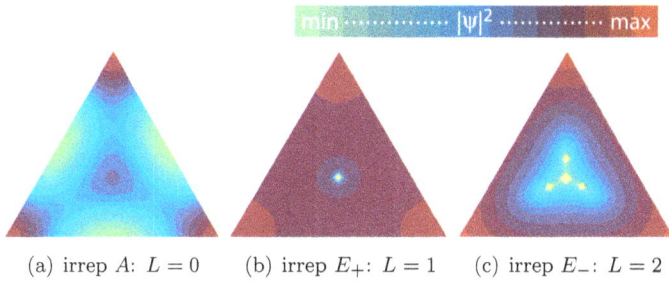

(a) irrep A: $L = 0$ (b) irrep E_+: $L = 1$ (c) irrep E_-: $L = 2$

Fig. 2.15 Density of the order parameter at the middle of the lowest three cusps in the $T_c(H)$ phase diagram of Fig. 2.14. The highest $|\psi|^2$ density is shown in red, followed by blue and green. The lowest values, drawn in yellow, indicate the positions of vortices and antivortices. The panel for the E_--state (16 times magnified) shows the presence of an Φ_0-antivortex in the center (after [91]).

vortices (central + three corner) to the total winding number of the states, shown in Fig. 2.15, is given by:

$$L = n + 3m , \qquad n = 0, 1, -1 , \qquad m = 0, 1 . \qquad (2.23)$$

The nature of the central vortex changes, whenever vorticity is changed by one. Thus the central vortex is absent in the first state, it is a Φ_0-vortex in the second state and it is an antivortex in the third state.

The flux penetration into the triangle, following one of the three continuous lines in the ground state, is schematically shown in Fig. 2.16. In analogy to the case of the square (Sec. 2.5.1), for a field sweep along eigenstates of the same symmetry, the flux enters by three Φ_0-vortices through the edges centers. This flux entry obeys the C_{3v} symmetry constraints of the $|\psi|^2$ distribution, and moreover is physically justified because the density is the lowest just at these three middle points on the sides of triangle. The entrance itself takes place exactly when the system passes over the top of an avoided crossing barrier (see Fig. 2.14(a)). The vortices, which entered laterally, approach the center and then, at some specific value of the applied field, are transformed into three corner-directed vortices which are dispatched towards the corners, thus paving the way for the entrance of the next vortex triade.

The stages of the vortex pattern evolution with increasing field are shown schematically in Fig. 2.16 for the three irreps. In the E_+ state, each of the entered vortices transforms into two Φ_0-vortices and one Φ_0-antivortex when approaching the center of the triangle. While the Φ_0-vortices move in pairs towards the corners, the three antivortices annihilate

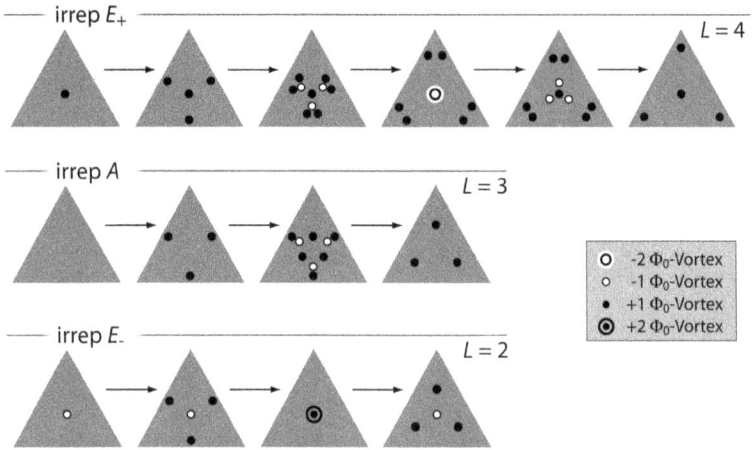

Fig. 2.16   Schematic plots showing different stages of the vortex entry and vortex pattern rotation in the states with different irreps and vorticity (after [91]).

with the central Φ_0-vortex to form a $2\Phi_0$ giant antivortex, which then explodes into the pattern consisting of the Φ_0-vortex in the center and three Φ_0-antivortices dispatched to the corners. Later on, each of these antivortices meets two Φ_0-vortices and recombines with them to form a single Φ_0-vortex. In the A state, when the entered vortices approach the center of the triangle, the central region 'explodes' into three Φ_0-vortices and three Φ_0-antivortices. While the antivortices annihilate in pairs with the entered vortices, the three Φ_0-vortices from the central region move towards the corners of the triangle. Finally, the simplest scenario of the vortex reorientation and vortex pattern rotation is achieved in the E_- state where the entered vortices annihilate with the central antivortex resulting in a $2\Phi_0$ giant vortex, which transforms into a Φ_0-antivortex in the center and three Φ_0-vortices dispatched to the corners of the triangle.

To conclude this discussion, we would like to emphasize that the appearance of antivortices originates from the competition between the infinite order rotational symmetry of the applied field and the finite order point group of the superconducting triangle. Note that our results apply to a finite temperature interval close to T_c where the effect of the nonlinear term in the Ginzburg–Landau equations can be neglected. At lower temperature, the presence of the nonlinear term could give rise to symmetry-breaking effects (see Sec. 2.6).

2.5.3 Rectangles

Up to now we considered samples with quite high symmetry, i.e. disks and loops in Section 2.3 and 2.4, respectively, squares (Section 2.5.1) and equilateral triangles (Section 2.5.2). We shall now analyze the phase boundary of mesoscopic superconductors with lower symmetry, namely a rectangle $a \times b$ $(b < a)$ where the aspect ratio $\zeta = a/b$ deviates from one.

Teniers *et al.* [461] extended the work of Chiboraru *et al.* [90] to the case of a mesoscopic rectangle in the framework of the linearized Ginzburg–Landau theory. They found that, dependent on the aspect ratio ζ, very different vortex patterns are formed, see Fig. 2.17(a–c). For $\zeta = 1$, the known configuration with one antivortex in the center and four vortices on the diagonals is observed (Fig. 2.17(a)). However, a slight increase of the aspect ratio to $\zeta = 1.01$ causes four vortices to appear on the diagonals, one in the center and two anti-vortices next to the central vortex (Fig. 2.17(b)). When ζ increases further on, superconductivity nucleates by forming specific vortex patterns in a row along the longest axis — in this particular case of $\zeta = 1.02$, three vortices are positioned on the longest axis of the rectangle (Fig. 2.17(c)). Therefore, the vortex pattern with an antivortex in the middle, found for a square (Sec. 2.5.1), is very sensitive to deformation, since already a deviation of only two percent between the two axes leads to a strong modification of this pattern. The fields, at which the first vortex enters and at which vortices move into the corners, rise significantly with increasing ζ. When increasing the aspect ratio, the amplitude of the oscillations in the phase boundary is reduced, then the slope is changed and finally the phase boundary evolves into a parabolic dependence on the field, analogous to the $T_c(H)$ curve of a thin line (see Fig. 2.17(d)).

In the following, we shall present the crossover from a square to a rectangle, as studied experimentally by using simultaneously prepared samples[7] with different aspect ratios $\zeta = 1$ $(2 \times 2\,\mu m^2)$, $\zeta = 4/3$ $(1.73 \times 2.31\,\mu m^2)$, $\zeta = 2$ $(1.41 \times 2.83\,\mu m^2)$ and $\zeta = 4$ $(1 \times 4\,\mu m^2)$, but with the same area $S = 4\,\mu m^2$. The experimental phase boundaries of the different rectangles are presented in Fig. 2.18. The $T_c(H)$ curves of the rectangles with $\zeta = 4/3$ and $\zeta = 2$ (Fig. 2.18(b) and (c), respectively) show small oscillations superimposed with a linear T_c dependence on the magnetic field. These curves are remarkably similar to the phase boundary of the square (Fig. 2.18(a)), showing only very small changes in the positions of the cusps: the magnetic

---

[7]Thickness and coherence length of a co-evaporated reference sample were 39 nm and 156 nm, respectively.

field values, at which the vorticity changes from L to $L+1$, are slightly delayed compared to the case of the square. However, no significant modifications in neither slope nor amplitude of the $T_c(\Phi/\Phi_0)$ oscillations can be seen. A very good agreement between the experimental and the theoretical curves [461], for both the position of the cusp and for the amplitude of the oscillations, is obtained for the two rectangles with $\zeta = 2$ and $\zeta = 4/3$. Even a large deformation of the square ($\zeta = 2$) results in negligible changes in the phase boundary. Therefore, the theoretically predicted effect of small changes of ζ on the vortex configuration at the nucleation of superconductivity [461] seems to affect the lowest Landau levels not substantially.

Also in the case of $\zeta = 4$ (Fig. 2.18(d)), a good agreement between the experimental and the theoretical curves is obtained when using a resistance criterion of $R = 0.4\,R_n$. However, for lower criteria, the experimental phase boundary exhibits larger oscillations [334]. Since the increase of ζ leads to a delay of the entry of the first vortex into the sample, the phase boundary is characterized by a parabolic dependence on the field until L changes from 0 to 1. Therefore, for an infinitely long rectangle with finite width, the case of a thin wire or film is eventually recovered. Besides the considered 2D nanocells, 3D nanostructures are also very interesting for studies of the condensate and flux confinement. For example, superconducting spherical nanoshells demonstrate in magnetic fields (due to their curvature) the unusual co-existence of a vortex-free 'Meissner belt' around the spheres' equator with vortex-rich areas around the poles [460].

(a) $\zeta = 1.00$ (b) $\zeta = 1.01$ (c) $\zeta = 1.02$

(d) $H\text{--}T$ phase boundaries

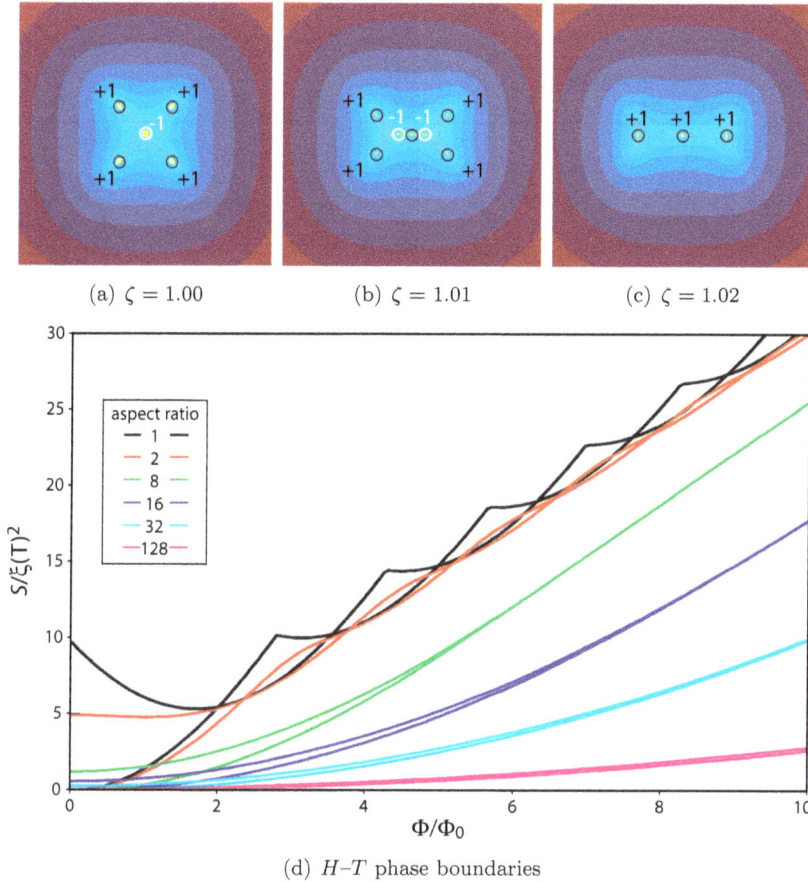

Fig. 2.17   (a–c) The density of the order parameter $|\psi|^2$ in the central region $(a/10 \times b/10)$ of a superconducting rectangle with edge lengths a and b $(b < a)$ at $\Phi = 5.5\,\Phi_0$. Highest and lowest densities are shown in red and yellow, respectively, the latter indicating the position of vortices and anti-vortices. Each different color corresponds to roughly half an order of magnitude. (d) Calculated superconducting $H\text{--}T$ phase boundaries for rectangles with different aspect ratios. For each aspect ratio, the lowest two eigenvalues of the linearized Ginzburg–Landau equation of a mesoscopic rectangle with superconducting boundary conditions are shown (the lowest eigenvalue corresponding to the phase boundary). Lowest and highest curves at zero field of the same color correspond to irrep A and irrep B, respectively (after [461]).

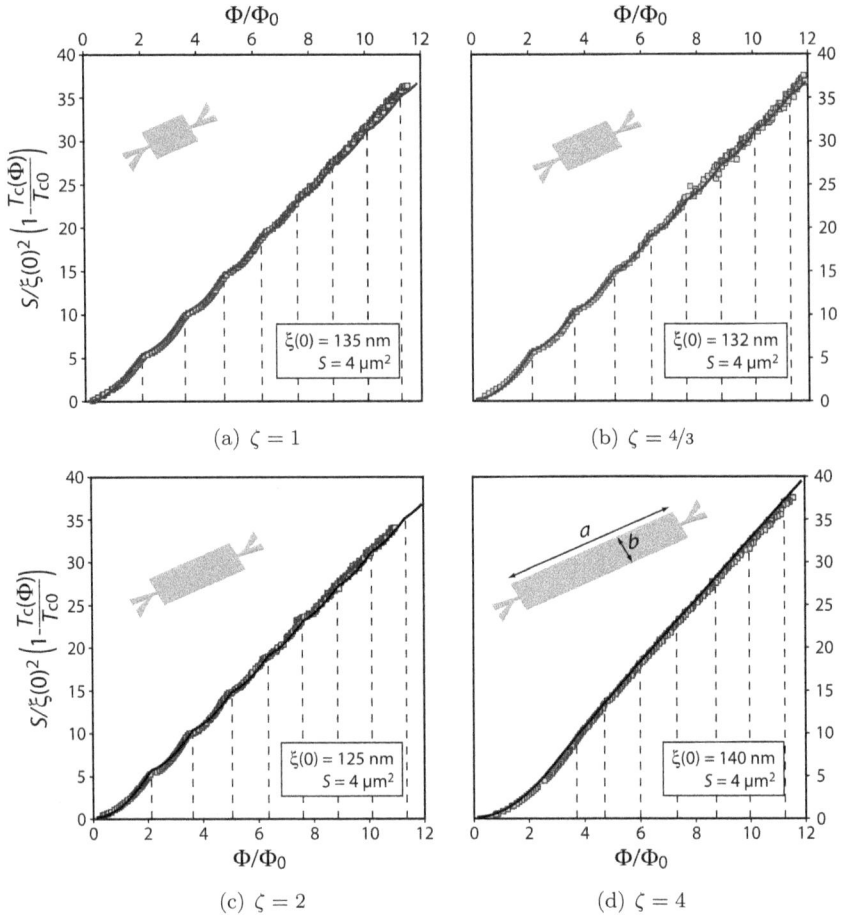

Fig. 2.18   Experimental $T_c(H)$ phase boundaries of rectangles with four different aspect ratios $\zeta = a/b$. Open squares represent the measured values, using a sample size of $S = 4\,\mu m^2$ and coherence lengths $\xi(0)$ is indicated in the panels. The full line is the calculated phase boundary (after [334]).

2.6   The magnetization of singly connected nanostructures

To access the vortex patterns appearing below the $T_c(H)$-line in the framework of the Ginzburg–Landau equations, magnetization measurements can be used. We will start this section on magnetization data for singly connected nanostructures by considering first discs and then triangles and squares. For discs, we shall begin with a discussion of the so-called 'paramagnetic Meissner effect'. A recent review on the paramagnetic Meissner effect and related dynamical phenomena has been published by Li [285].

Another important point is to consider symmetric type-1 superconducting structures well below the $T_c(H)$-line, where thermodynamically stable vortex-antivortex patterns were found in mesoscopic equilateral triangles by solving the full Ginzburg–Landau equations. In type-2 superconductors, similar patterns are unstable deep in the superconducting phase in the T–H plane, where eventually vortex-antivortex pairs annihilate and a conventional Abrikosov vortex lattice is emerging. Interestingly, the non-linear term in the Ginzburg–Landau equation is needed to break the symmetry. As the temperature goes down, we are moving from the symmetry preserving solutions of the linearized Ginzburg–Landau equation, which describes the situation at the superconducting phase boundary $T_c(H)$ and in its vicinity, to the symmetry-breaking solutions of the full Ginzburg–Landau equation with the non-linear term, which should be applied at low temperatures, far from the $T_c(H)$-line. Remarkably this transition from symmetric to non-symmetric solutions for 'vortex molecules' has many similarities with the Jan–Teller transition [238] for normal molecules [96]. At low temperatures, symmetric vortex-antivortex and vortex-giant vortex molecules are all transformed into collections of single quantum Abrikosov vortices. For example, the '+4 − 1' molecule in a mesoscopic superconducting square is turned into conventional cluster of three single quantum vortices, while the '+4 + 2' molecule is converted into an array of six singly quantized vortices, see Fig. 2.19. Depairing critical currents corresponding to different vortex configurations in a square were calculated by Sardella and Oliveira [417]. Experimentally, the recovery at low temperatures of the patterns composed of singly quantized vortices can be accessed by using local vortex imaging techniques and also magnetization measurements on individual superconducting nanocells.

During the recent past, superconducting properties of micron-sized disks have been studied intensively, mostly concentrating on mesoscopic samples with sizes comparable to the coherence length $\xi(T)$. Typically, the phase

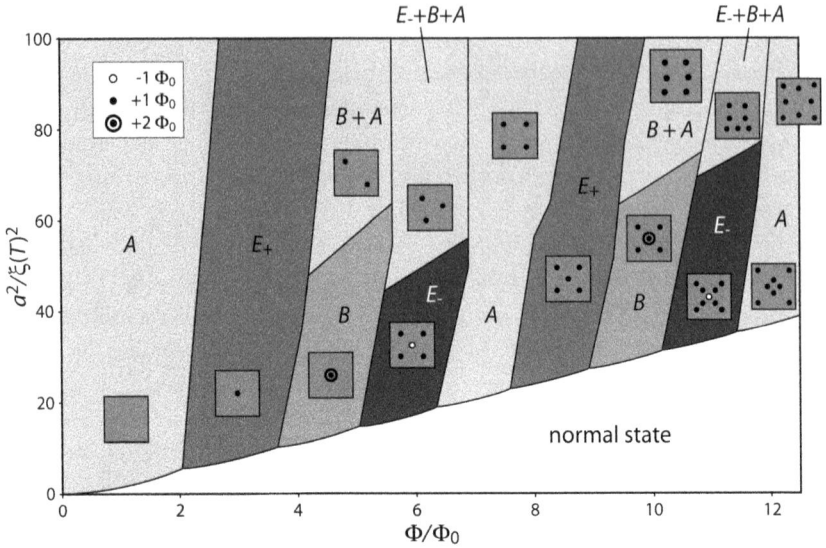

Fig. 2.19   The phase diagram of a superconducting square with superconductor/vacuum boundary condition, obtained by Monte Carlo calculations. For each phase, the vortex structure is shown schematically and the involved irreps are indicated (after [92]).

boundaries of these structures and their properties deep in the superconducting state were found from transport [75] and magnetization measurements [75, 166], respectively.

After its discovery in high temperature superconductors [70, 214, 424], the paramagnetic effect[8] has also been observed in conventional low temperature superconductors [167, 259, 463], and was considered as an intrinsic property of any finite-size superconductor. This effect was first theoretically predicted by Moshchalkov *et al.* in conventional superconductors by using the Ginzburg–Landau equations, treating the flux capture and its compression [352] as the origin of its appearance. Using Hall magnetometers to measure the magnetization of superconducting mesoscopic disks, Geim *et al.* [167] have confirmed that prediction and observed a paramagnetic signal, although a possible origin of the paramagnetic Meissner effect could be, in principle, arising from the measuring technique. Hall magnetometers have been studied theoretically in both the ballistic [383] and in the diffusive [231] regime, demonstrating that for low magnetic fields,

---

[8]The fact that some superconducting samples may attract magnetic field rather than repelling it is referred to as the 'paramagnetic Meissner effect'.

the Hall resistance is linear[9] in H and that it is determined by the average magnetic field in the cross junction — more or less independent on the shape and position of the magnetic field profile. However, it was also shown that the size of the detector has a substantial influence on the magnitude of the measured magnetization, and that it can even reverse the sign of M [443]. The apparent paramagnetic behavior was explained by the fact that for certain magnetic fields, the applied field is exceeded in the center of the sample due to the flux trapped by the Bean–Livingston barrier, meaning that the central region is paramagnetic. Near the edge of the sample, the magnetic field is much lower that the applied field, corresponding to a diamagnetic region. Since the total magnetization of the sample is the integrated magnetization of all regions, the sample turned out to be diamagnetic. However, when calculating the magnetization determined by the detector, the magnetic field in the region outside the disk had to be included, which is strongly paramagnetic. The magnetic field outside the sample is larger than the applied field because of the strong flux expulsion from the disk and the important demagnetization effects in finite thickness films.

The hysteresis transitions [373, 374] and the metastability of the vortex configuration due to the influence of the sample surface [443] were analyzed in the framework of the Ginzburg–Landau theory, which also allows a purely electrodynamic explanation of the paramagnetic effect in finite-size superconductors [514]. The paramagnetic Meissner effect is caused by the presence of vortices inside the sample, since on the one hand, a superconducting current flows around the vortex to screen its internal field from the bulk of the sample, and on the other hand, a second current flows at the boundary to prevent the external field from entering the sample. These two screening currents flow in opposite directions and contribute with opposite signs to the total magnetic moment (or magnetization) of the sample. Therefore, depending on H, the total magnetization may be either negative (diamagnetism) or positive (paramagnetism).

2.6.1 *Disc*

We shall now consider a superconducting mesoscopic disk with radius R and thickness $\tau \ll R$, that is placed in an external magnetic field parallel to the axis of the disc. The magnetization of the sample is measured with a Hall sensor, positioned at a distance $l \sim \tau$ below the disk. From the Biot–Savart

---

[9]Rounded corners did not affect the linear behavior significantly.

law it then follows that at $l \ll R$, the magnetization, measured by a Hall sensor, is approximately independent on the value of l, meaning that we can assume $l = 0$. The area of the sensor we consider here is approximately two times larger than that of the disk, which is why (according to the Biot–Savart law) the sensor measures the total magnetization, i.e. the total difference between the applied field and the local field in the plane of a disk. Therefore, we can map our problem onto the problem of a *two-dimensional* superconducting disk of thickness $\tau < \xi(T) \ll R$, with the effective values of the London penetration depth and the effective Ginzburg–Landau parameter

$$\lambda_{\text{eff}}(T) \sim \frac{\lambda(T)^2}{\tau} \quad \text{and} \quad \kappa_{\text{eff}} = \frac{\lambda_{\text{eff}}(T)}{\xi(T)}, \qquad (2.24)$$

respectively. The effective Ginzburg–Landau parameter takes into account the demagnetization effects [375], and its specific value is to be considered as a fitting parameter.

In order to find the local magnetic field $h(r)$ inside the sample, it can be assumed that the modulus of the order parameter is constant. In this case $h(r)$ can be determined from the second Ginzburg–Landau equation [63]:

$$h(r) = \alpha \sum_{i=1}^{L} K_0(|r - r_i|) + a_0 l_0(r) - 2 \sum_{n=1}^{\infty} [a_n \cos(n\theta) + b_n \sin(n\theta)] I_n(r) , \qquad (2.25)$$

using polar coordinates. In Eq. (2.25), K_n and I_n are the modified Bessel functions, r_i is the coordinate of the ith vortex and L is the number of vortices. The coefficients a_0, a_n and b_n can be calculated from the boundary condition $h(R) = H$. Throughout the following, as well as already in Eq. (2.25), the dimensionless distance r and magnetic flux density h are given in units of $\lambda_{\text{eff}}(T)$ and $H_c/\sqrt{2}$, respectively, with the thermodynamic critical field H_c. Therefore, $H_{c2} = \kappa_{\text{eff}}$ and the magnetic flux quanta $\Phi_0 = 2\pi/\kappa_{\text{eff}}$. The first term at the right-hand side of Eq. (2.25) describes the field of vortices. In the London model we have $\alpha = \kappa_{\text{eff}}^{-1}$ [118], however, the London approximation is only applicable at $\kappa_{\text{eff}} \gg 1$, while in the present case $\kappa_{\text{eff}} \sim 1$. Our approach is to use the expression for the vortex field $h_0(r)$, which was obtained by Clem *et al.* [106] with a variational method:

$$h_0(r) = \frac{K_0\left(\sqrt{r^2 + \xi_v^2}\right)}{\kappa \xi_v K_1(\xi_v)} , \qquad (2.26)$$

where the parameter $\xi_v \sim \xi(T)$ characterizes the vortex core size, and can be found from

$$\kappa_{\text{eff}} \xi_v = \sqrt{2} \left[1 - \frac{K_0^2(\xi_v)}{K_0^2(\xi_v)} \right] . \qquad (2.27)$$

Note that Eqs. (2.26) and (2.27) yield a very accurate description of the vortex field [386]. If the distance between neighboring vortices in a disk exceed by far the vortex core size, we can expand the expression for the vortex field (Eq. (2.26)) in terms of $(\xi_v/r)^2$ and use a usual London formula, as in Eq. (2.25) with the prefactor $\alpha_{\mathrm{eff}}^{-1} = \kappa_{\mathrm{eff}} \xi_v K_1(\xi_v)$. According to the estimates, the results of such approximation are still quite accurate as long as the number of vortices in the disk does not exceed ~ 20. From the boundary condition $h(R) = H$, it follows that

$$a_0 = \frac{1}{I_0(R)} \left[H - \frac{1}{\kappa_{\mathrm{eff}} \xi_v K_1(\xi_v)} \sum_{i=1}^{L} l_0(r_i) K_0(R) \right] \tag{2.28a}$$

$$a_n = \sum_{i=1}^{L} \cos(n\theta_i) \, l_n(r_i) \frac{K_n(R)}{I_n(R)} \tag{2.28b}$$

$$b_n = \sum_{i=1}^{L} \sin(n\theta_i) \, l_n(r_i) \frac{K_n(R)}{I_n(R)} \,, \tag{2.28c}$$

where θ_i are polar angles of vortices. Using Eqs. (2.25) and (2.28), the following expression for the Gibbs energy can be derived:

$$G = LE_0 - 2\pi R H^2 \frac{I_1(R)}{I_0(R)} - \frac{4\pi}{\kappa_{\mathrm{eff}} \xi_v K_1(\xi_v)} H \sum_{i=1}^{L} \left[1 - \frac{I_0(r_i)}{I_0(R)} \right]$$

$$+ \frac{2\pi}{\kappa_{\mathrm{eff}}^2 \xi_v K_1(\xi_v)} \sum_{j=1}^{L} \left[\sum_{\substack{i=1 \\ i \neq j}}^{L} K_0(|r_i - r_j|) - \sum_{i=1}^{L} \frac{K_0(R)}{I_0(R)} I_0(r_i) I_0(r_j) \right.$$

$$\left. - 2 \sum_{n=1}^{\infty} (a_n \cos(n\theta_j) + b_n \sin(n\theta_j)) I_n(r_j) \right] \,, \tag{2.29}$$

with the energy E_0 of one vortex per unit length, which can also be calculated using Clem's model [106].

The positions and the number of vortices at each value of H can be found by minimization of the Gibbs energy. In a disk, vortices will form a shell structure in an equilibrium state as shown by theory [293] and verified experimentally [199]. We assume that vortices in each shell are situated in the vertices of an ideal polygon, but the sizes of polygons and the angles between them are not fixed and are considered as variational parameters. This approximation is rather accurate as long as the number of vortices in the disk is not too large. Once that the positions of the vortices are

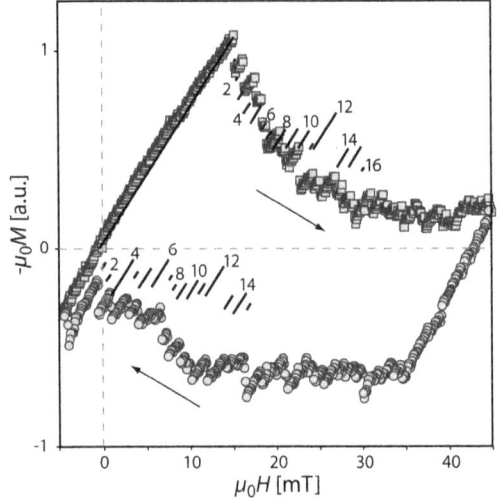

Fig. 2.20 Experimental magneti-
zation curves of a disk at $T/T_c = 0.6$, obtained by Hall magnetome-
tery for an increasing (squares)
and an decreasing (circles) magnetic field. The used Hall-cross
had an active area of $1.4 \times 1.4\,\mu m^2$
(depletion at the edges taken into
account). The theoretical curve
is represented by solid lines (after [43]).

calculated, the magnetization can be obtained by integration of Eq. 2.26
over the sample area. The presence of the sample boundary results in a
surface Bean–Livingston barrier, which prevents vortex entrance and exit
from the sample, leading to the appearance of metastable states in the disk.
Therefore, the experimentally observed magnetization is quite different for
the cases of increasing and decreasing fields (hysteretic behavior), and can
even be positive at the decreasing field (paramagnetic response). For the
penetration of each vortex we use the criterion of a vanishing surface barrier
at a distance ξ_v from the surface. Following Ref. 63, we also assume, for
simplicity, that the positions of 'pre-existing' vortices are not influenced by
the vortex nucleation at the surface. For the vortex exit an approximation
is used, according to which the positions of other vortices remain the same
while a vortex leaves the sample.

The experimental and theoretical results for a superconducting lead disk
of 40 nm thickness, $2.5\,\mu m^2$ surface and a coherence length $\xi(0) = 40\,nm$
are shown for increasing and decreasing fields in Fig. 2.20, where each
jump of the magnetization in the theoretical simulation corresponds to the
penetration of one vortex. We chose $\kappa_{eff} = 2.89$ such that experimental and
theoretical values of the field at the first vortex penetration coincide. In
order to obtain an agreement between the theoretical and the experimental
periodicities of the oscillations in M, we had to assume an effective radius of
the disk of approximately 18% less than the actual one. A possible reason
for the necessity of this correction might be the oxidation of sample edges,

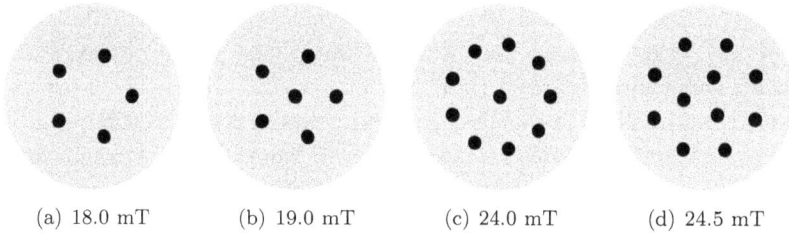

(a) 18.0 mT   (b) 19.0 mT   (c) 24.0 mT   (d) 24.5 mT

Fig. 2.21   Calculated vortex patterns in a superconducting disc for an increasing field with (a) $L = 5$, (b) $L = 6$, (c) $L = 10$ and (d) $L = 11$ (after [43]).

since the used material (lead) oxidizes easily at air, and our sample was protected with a germanium layer from the top, only.

According to the calculations, the state with 12 vortices is the most stable phase since it exists in the widest field range at increasing H. By contrast, in the experimental magnetization curve, the 10th jump is most pronounced. However, since it is not unlikely that two jumps could not be resolved (some of the jumps are rather weak which makes them difficult to separate), we believe that the 10th step in the experimental $M(H)$ curve could actually correspond to the 12th step in the calculated one.

For the calculated magnetization curves of up to the state with 17 vortices, there is a semi-quantitative agreement between theory and experiment, both for increasing and decreasing fields. For decreasing H, a positive magnetization (paramagnetic Meissner effect) is obtained from the calculations. Therefore, a paramagnetic response can be observed not only for mesoscopic samples with sizes of the order of $\xi(T)$, but also for samples of intermediate size between mesoscopic and macroscopic scales. However, as it can be seen in Fig. 2.20, for decreasing H, at least one vortex is inside the sample at negative external fields, while in theory vortex phases are unstable at negative H. Possible explanations for this discrepancy are vortex pinning, deviation of the geometry from a perfect disk or inhomogeneities of the superconducting properties due to oxidization. Apparently, both experimental and theoretical magnetization curves, corresponding to a fixed number of vortices, have the same slope. The physical reason for this is the fact that the radius of the disk is much larger than the London penetration depth, and that the magnetic flux of each vortex is almost precisely equal to Φ_0.

Some vortex patterns corresponding to the case of an increasing field are shown in Fig. 2.21. Apparently, the patterns exhibit a single shell up to $L \leq 5$, while a second shell is formed for $L \geq 6$. Up to the vortex phase of $L = 10$, the inner shell is represented by one single vortex, but for $L > 10$, three vortices are situated in the inner shell. In the 16-vortex state there are three shells in the disk: one vortex in the center, 3 vortices in the second shell and 12 vortices in the third shell. Note that these results are in agreement with the calculation reported in Refs. 81, 264 and analogous to systems of rotating superfluids, such as ^{3}He [207, 384, 464, 491, 492], which also produce stable vortex shells.

From the dependencies $M(H)$ presented in Fig. 2.20, a certain quasi-periodicity of the background of the magnetization curves can be seen in the increasing and decreasing branch for both the theoretical and the experimental data. We attribute these features to the transitions between phases of different shell structures. Since the effective penetration depth $\lambda_{\text{eff}}(T)$ is much smaller than the size of the disk, each vortex in the inner shells gives a contribution close to Φ_0/S to the magnetic induction B, which is why the amount of magnetic flux, carried by the vortices on the outer shell, can be sufficiently smaller than Φ_0. When a new shell is created, more vortices will be located in the center so that their contribution to the induction will increase. On the other hand, the vortices situated in the outer shell will be moved towards the edge, which decreases their contribution to B. Therefore, a change in the shell structure will clearly affect the height of the jumps ΔM. It is interesting to note that ΔM does not strongly depend on the stability of the vortex states since the magnetization curves have the same slope for a fixed number of vortices, meaning that the height of the jumps is solely determined by the distance between two different parallel curves.

2.6.2   *Triangle*

The experimental magnetization curve for a superconducting lead triangle with thickness $\tau = 40\,\text{nm}$, a surface $S = 2.1\,\mu\text{m}^2$ and coherence length $\xi(0) = 40\,\text{nm}$ is presented in Fig. 2.22. Apparently, the critical field is higher in this case compared to the disc (see Fig. 2.20), since superconductivity is enhanced in the corners of the triangle [222]. As for the disk, not only a paramagnetic magnetization can be seen, but also the magnetization curve for a decreasing magnetic field does not return to zero at $H = 0$, meaning that at least one vortex is trapped.

In order to study vortex penetration and expulsion in both disk and triangle, the magnetic field range ΔH between two jumps in the magnetization

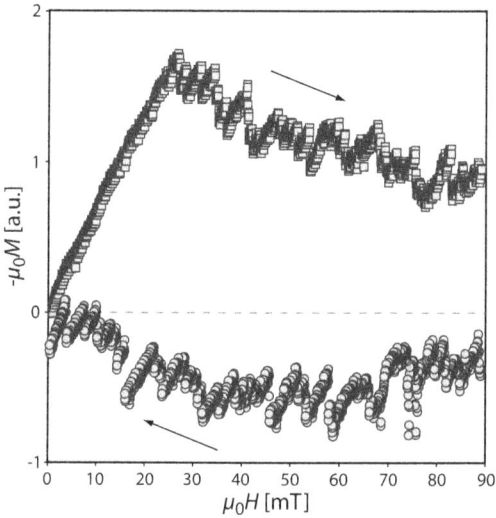

Fig. 2.22 Experimental magnetization curves of a superconducting lead triangle at $T/T_c = 0.6$ for an increasing (squares) and decreasing (circles) magnetic field (after [43]).

curves is presented in Fig. 2.23(a) for the penetration and in Fig. 2.23(b) for the expulsion of vortices. A clear difference for the two structures can be seen as values for ΔH are typically higher for the triangle. A higher ΔH for a certain value of L means that the state of this specific vorticity is stable over a broader magnetic field range. The most probable reason for the stability at some vorticity is the formation of stable vortex configurations in the superconductor (we have to emphasize again that the exact value of L is unknown since it is possible that some jumps were not resolved in the experimental magnetization curve). For the triangle, states of $L = \{6, 9, 14\}$ and $L = \{7, 10, 15\}$ seem to be more stable than others for an increasing and a decreasing field, respectively. Note that these numbers are very close to those vorticities at which vortices can arrange in a triangular lattice, thus keeping the C_3 symmetry imposed by the boundary of the triangle. The vorticities of these symmetric configurations can be given by the numbers $L = 1/2\,n(n+1)$ with integer n, which — among others — results in the values $L = 6$, 10 and 15 for $n = 3$, 4 and 5.

Baelus *et al.* [25] calculated the vortex configurations and their stabilities in mesoscopic triangles. According to their results, at $L = 5$ and 6, vortices form a vortex molecule, consisting of 3 vortices that form a triangle, and a giant vortex with vorticity 2 and 3 in the middle. Further increase of the vorticity leads to a growing central giant vortex. However, the fact that the vortex lattice tries to keep the same geometry as the sample is

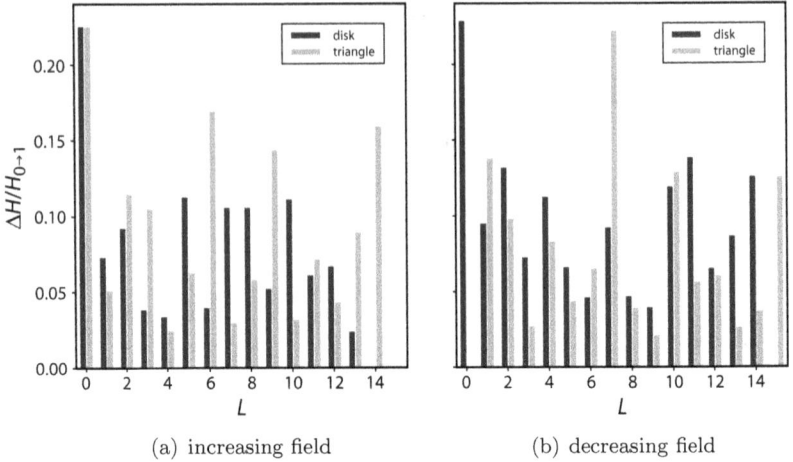

(a) increasing field   (b) decreasing field

Fig. 2.23   The range of magnetic field ΔH between two magnetization jumps, normalized by the magnetic field $H_{0\rightarrow 1}$ needed for the penetration of the first vortex. Dark and light gray bars show the values of ΔH for the disk and the triangle, respectively, derived from the experimental magnetization curves of (a) an increasing and (b) a decreasing external magnetic field.

the only explanation which they found for a stable configuration at $L = 3$. Further stable vortex states at higher vorticity could not be discovered. It is worth mentioning that their calculations are done for the mesoscopic regime where only 13 vortices were present at the critical field H_{c2}, while in the present sample, more than 100 jumps in the magnetization curve were observed. For low vorticity, the formation of giant vortices can thus be, probably, excluded in this case and the observation of stable states in the triangle can be explained by the formation of a triangular lattice of vortices, keeping the symmetry imposed by the confinement. It is likely that this triangular lattice is slightly deformed since the interaction of vortices with boundaries is strongly non-uniform.

For the case of the disc, the stability of different states differed less compared to the triangle, which can be explained by the C_∞ symmetry of the disk. With this geometry, the difference between two neighboring configurations is expected to be much smaller.

2.6.3   *Square*

As a last example, we shall now discuss the magnetization properties of a mesoscopic aluminum square of 62 nm thickness, 14.5 µm² surface and

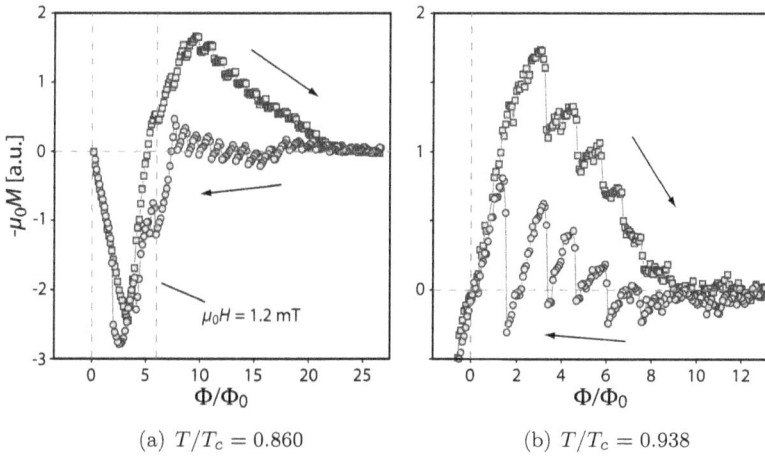

(a) $T/T_c = 0.860$ (b) $T/T_c = 0.938$

Fig. 2.24 Experimental magnetization curves of the Al square for an increasing (squares) and a decreasing (circles) magnetic field at (a) $T/T_c = 0.860$ and (b) $T/T_c = 0.938$. The active area of the used Hall cross was $3.7 \times 3.7 \mu m^2$ (after [334]).

160 nm coherence length, and study the stability of some vortex patterns imposed by the sample geometry. These measurements are important to probe the vortex patterns appearing deeper in the superconducting state due to the importance of the non-linear term in the Ginzburg–Landau theory, see Fig. 2.19.

Figure 2.24(a) shows the magnetization curve for the mesoscopic square at $T/T_c = 0.86$, which exhibits a peculiar behavior at low magnetic fields with a strong paramagnetic signal.[10] More than 15 jumps in both the increasing and the decreasing branch of $M(H)$ are seen, corresponding to a maximal vorticity $L > 15$. Above 1.2 mT, the critical field of the bonding wires is exceeded, and only the magnetization of the square is measured. The slope of the parts with $L = $ const are different in the increasing and in the decreasing branch, which is in good agreement with the calculated magnetization curve of Baelus *et al.* [25], who even found a negative slope (which was not observed in the measurements). The origin of this discrepancy could be different parameters used in their calculations, or small defects at the boundary of the square. For it is a well known fact that

---

[10]Note that such paramagnetic dip has also been observed with empty Hall probes. The origin of the signal was found to be in the aluminum wire bonds, used to connect the current and voltage leads from the Hall probe to the contacts of the sample holder. Below their critical temperature, these wires expel the magnetic field and increase the local field on top of the Hall sensor, giving rise to a paramagnetic response.

the Bean–Livingston barrier [42] is responsible for a hysteretic behavior of the magnetization of superconductors [107, 117, 240, 443]. Geim *et al.* [165] have shown experimentally, by introducing artificial defects in a mesoscopic disk and by measuring the magnetization with a Hall sensor, that defects strongly decrease the Bean-Livingston barrier, leading to a faster penetration and expulsion of vortices. Therefore, small inhomogeneities in the considered sample could eventually lead to a lower value of the penetration field, preventing the observation of the calculated negative slope for a perfect square.

Above $1.2\,\mathrm{K}$, no signal from the aluminum bonding wires was detected, indicating that the critical temperature of the wires is comparable to $T_c^{\mathrm{bulk}} = 1.196\,\mathrm{K}$ of the bulk material. The magnetization of the square at $T > T_c^{\mathrm{bulk}}$ is given in Figure 2.24(b), showing a vorticity of up to $L = 5$. Note that the positions of the transitions from L to $L + 1$ were quite well reproduced. However, the transition of $L = 0 \to 1$ in the increasing branch and from $L = 2 \to 1$ in the decreasing branch occur at approximately the same magnetic field. We should then expect a continuous curve for the state with vorticity 1. This was observed in some measurements but not always, which might be a result of a not perfectly linear response of the Hall voltage as a function of magnetic field. Indeed, above the critical temperature of the square, a weak hysteretic behavior of the Hall voltage was sometimes detected.

The magnetization of the square has also been measured at different temperatures. Monitoring the transitions between states of different vorticities allowed to reconstruct the H–T diagram of Fig. 2.25, showing the expulsion of vortices. Apparently, the lines corresponding to vortex penetration are more or less parallel and equidistant. However, the expulsion of the last vortex from the sample takes place at a much lower field value, meaning that the state of $L = 1$ is very stable. Remarkably, the curve corresponding to the transition of $L = 4 \to 3$ (4 vortices + 1 antivortex) has a quite different shape compared to others, since at low temperatures, vortex expulsion is delayed for this transition. This indicates that — as it could be expected — the configuration with four vortices is stable in the square. Indeed, at $L = 4$, the configuration of vortices will be such that the C_4 symmetry, imposed by the boundary conditions [25, 64, 90], is preserved.

The penetration of vortices shows up almost periodically and is more or less independent on L, while the expulsion occurs later for $L = 4$ and $L = 1$. This indicates that the mechanism of vortex penetration is less dependent on the specific configuration of vortices compared to the mech-

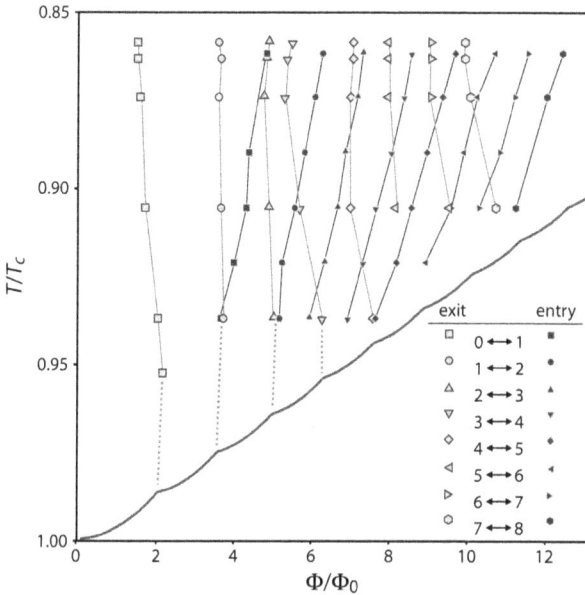

Fig. 2.25   *H–T* phase diagram showing different vortex patterns in the square. Dark and bright lines correspond to vortex penetration and exit, respectively (the dashed lines are a guide to the eye). The transitions are shown from $L = 0 \leftrightarrow 1$ (squares) to $L = 7 \leftrightarrow 8$ (hexagons). The full gray line shows the theoretical $T_c(\Phi/\Phi_0)$ phase boundary of a square, using an area $S = 13.5\,\mu m^2$, a coherence length $\xi(0) = 180\,nm$ and a critical temperature $T_c = 1.28\,K$ (after [334]).

anism of vortex expulsion. Qualitatively, a possible explanation for these two different behaviors is that for the expulsion, vortices that are located in the corners of the square (stable configuration), will first have to move to the center of the corresponding side before crossing the barrier. The necessary motion will require energy since the leaving vortex is repelled by other vortices located at the remaining corners. Crossing the barrier close to the corner without much disturbance of the remaining vortex configuration is another possibility, but in such scenario, the barrier that needs to be crossed is higher.

Contrary to vortex expulsion, during the penetration of vortices, the actual configuration inside the sample will only be affected once the new vortex has already crossed the barrier. Comparing the experimental data (Fig. 2.25) with the theory (Fig. 2.19), an overall qualitative agreement is achieved, but a full understanding of the experimental phase diagram has not been found yet.

2.7   Dynamic effects in mesoscopic structures

Until now we mainly concentrated on the geometry dependent nucleation of superconductivity in individual nanocells, and on the influence of the confinement on the arrangement of vortices. Experimentally, this has been achieved by studying the features of the field dependence of the superconducting/normal-state phase boundaries, using typically low current densities j. In the present section we go beyond this approach and address also dynamic effects that appear at higher values of j. We will start with a discussion of the vortex rectification effects under influence of an applied *ac*-current, and move on to a study of the transition to the normal state due to increasing *dc*-currents.

2.7.1   *Rectification effects in a triangle*

Voltage rectification effects manifest themselves as a non-zero *dc*-voltage that appears even if a zero average *ac*-excitation is applied to the system. In superconductors, this effect can be an indication of a vortex ratchet effect when vortices move in preferential direction under the influence of an *ac*-drive. During the last years, a considerable attention has been attracted to the mechanisms that are responsible for ratchet effects in a broad variety of physical systems, such as colloids [287], granular materials [141], fluids [309], atoms in optical traps [194], electrons in semiconductor heterostructures [289] and Josephson systems [139, 299, 432, 468, 509]. In all of these examples a net flow of particles, driven by a zero average alternating excitation, results from the interaction of the media with an *asymmetric* potential.

Vortex ratchet effects in superconductors have also been predicted theoretically and verified experimentally for the motion of flux lines in superconducting samples with a pinning landscape that lacks inversion symmetry [109, 122, 294, 437, 477, 488, 495]. Alternatively, in superconductors without spatial asymmetry, it is also possible to control the vortex motion by applying a time-asymmetric drive [109]. Moreover, by varying the asymmetry of the drive one can increase or decrease the vortex density at the center of the superconductor. Ratchet effects in superconductors can also be used to decrease the noise that is caused by trapped flux in superconducting devices, by removing trapped vortices with an *ac*-drive [280].

Interestingly, it has been recently demonstrated that voltage rectification in a superconductor does not necessarily imply the motion of vortices

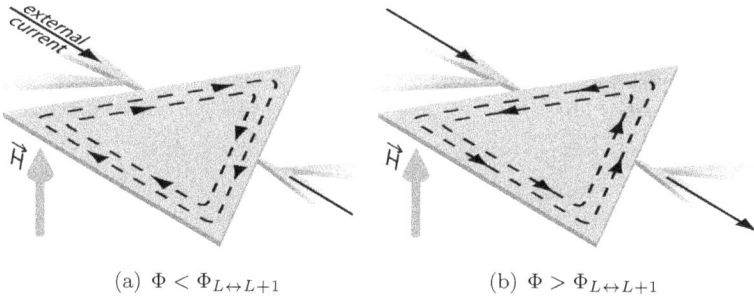

(a) $\Phi < \Phi_{L \leftrightarrow L+1}$ (b) $\Phi > \Phi_{L \leftrightarrow L+1}$

Fig. 2.26 Schematic presentation of the screening currents in a superconducting tri-angle. The screening currents flow in clockwise direction for positive fields below the transition of vorticity $L \to L \pm 1$, and counterclockwise for fields above that transition.

in an asymmetric pinning potential, but rather it might also result from non-symmetric current distributions in the sample [476]. Indeed, Dubonos *et al.* found rectification effects in asymmetric superconducting loops [132]. The observed *dc*-voltage oscillates with the magnetic field in the same way as the persistent current in a loop due to fluxoid quantization [465]. In this case, the rectification is a result of the superposition of the external current and a persistent current [132], causing a difference in the critical current for a positive and a negative externally applied current. Van de Vondel *et al.* [476] showed that in superconducting samples with periodic arrays of triangular antidots, rectification caused by the current compensation effects can coexist with rectification due to ratchet vortex motion.

The *dc*-response can therefore appear in a singly-connected structure under an *ac*-drive. An oscillatory *diode effect* can be seen in supercon-ducting triangles, reflecting the oscillations of the persistent currents in superconducting individual structures [338]. In the following, the influ-ence of the position of the current and voltage probes on this effect is discussed [422], and the results for the case of triangular structures are com-pared with rectification effects observed in a disk. These experimental data are supported by theoretical simulations, done by Kolton and Marconi [422] in the framework of a model system consisting of a loop of N Josephson junctions.

The origin of this alternating diode effect can be found in the persistent currents flowing in the triangle as a consequence of fluxoid quantization effects (see the schematics in Fig. 2.26). Since the current contacts are far from the geometric middle of the sketched triangle, the applied current will flow mainly via the upper part of the structure. When screening currents are induced in clockwise direction, they will compensate a negative applied current (i.e. flowing from right to left). For a positive applied magnetic field, the screening currents in clockwise direction will be the strongest just below the fields of changing vorticity ($L \to L \pm 1$). Upon increase of H, a new vortex enters the triangle, resulting in a persistent current that flows in the opposite direction.

Persistent currents flowing in the upper part of the triangle in the same direction as the applied current I will result in a higher current density than for opposite I. Therefore, when applying an *ac*-current to the system, the structure eventually reaches the normal state during one half-period while staying superconducting during the other half-period. This gives rise to a non-zero averaged voltage over one period and the measured mean *dc*-voltage is a result of a sign-dependent critical current.

The temperature dependence of the diode effect is shown in Figure 2.27 where the sign of the rectified voltage is given as a function of T and H. Apparently, the rectification effect is present for all vorticities. A change of its sign occurs at the transitions of vorticity $L \to L \pm 1$ (given in the graph by dashed lines), as well as between two transitions. At $L \leftrightarrow L \pm 1$, the sign reversal of the measured *dc*-voltage is quite abrupt, indicating that persistent currents change their direction suddenly. However, in between transitions from $L \leftrightarrow L \pm 1$, V_{dc} changes sign rather smoothly since persistent currents evolve gradually from anticlockwise to clockwise flowing currents (for an increasing magnetic field). Typically, the rectification effect is observed below the phase boundary obtained at low current densities $j \ll j_c$, and starts at the onset of the resistive state. It is interesting to note that the alternating sign is measured up to high vorticities. A vortex, entering the triangle, changes drastically the scheme of supercurrent and external current flowing in one direction around the vortex core and in the other direction at the edge [25]. Since our data show alternating *dc*-voltage up to high vorticity ($L = 12$), the applied currents will mostly pass via the upper edge of the triangle where the order parameter is the strongest.

2.7.2   *Reversal of the diode effect*

The above explanation for the diode effect is based on the assumption that the applied current flows mainly along the upper edge of the triangle. In fact, the diode effect in such singly connected structure is comparable to the appearance of the rectified voltage observed by Dubonos *et al.* [132] in an asymmetric ring. In a ring structure, the current flow is well defined and

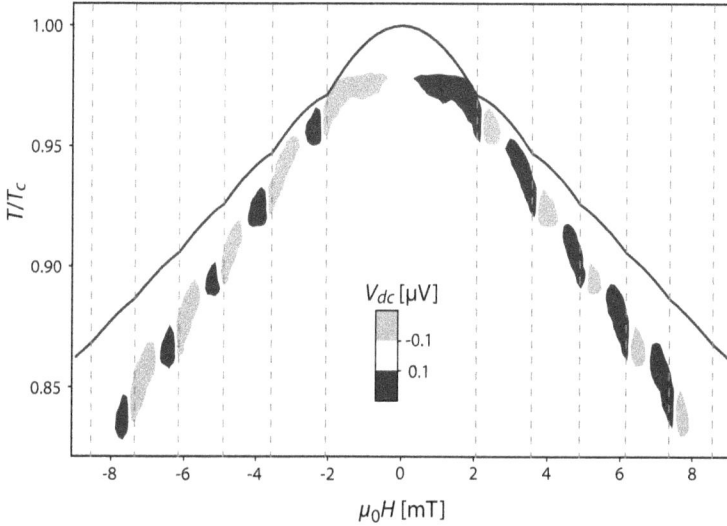

Fig. 2.27 Phase diagram of the rectified voltage V_{dc} as a function of the magnetic field and the temperature. An ac-current of $2\,\mu A$ with a frequency of $1\,kHz$ was used. Lignt and dark gray regions represent $V_{dc} \lessgtr \mp 100\,nV$, respectively. The theoretical phase boundary [90] is indicated by the solid line, using a coherence length of $\xi(0) = 110\,nm$ and a critical temperature of $T_{c0} = 1.34\,K$ [337]. The calculated magnetic fields corresponding to the change of vorticity are shown by dashed lines (after [422]).

the observed dc-voltage can be explained as a results of the superposition of applied and persistent currents, leading to a difference in critical values for positive or negative applied currents. By contrast, in a singly-connected structure such as the triangle, the current flow is not well defined but one can expect a rather homogenous current distribution across the structure. The assumption that the current flows mainly through the upper part of the triangle, following the shortest path where the order parameter is the strongest, is then counter-intuitive. However, in case that this assumption is valid, it implies that the position of the current injection should have an important influence on the rectification effect.

In fact, as we will show in the following, an ac-current that is injected above the geometrical center of a triangle gives an opposite rectification signal compared to current injection below the geometrical center. In addition, a lower signal is obtained if the contacts are attached along a median of a triangle, so that upper and lower part of the triangle are symmetric around this line. Three superconducting equilateral triangles were studied, made of $50\,nm$ thick aluminum, with an area of $S = 2.2\,\mu m^2$

and four wedge-shaped current and voltage contacts. The typical super-conducting coherence length, estimated from reference films, was about $\xi(0) = 120\,\text{nm}$. Three different contacts configurations were studied for current injection [422] along a median, above the geometrical center and at the base of the triangle (see the drawings in Fig. 2.28).

Let us first focus on the interaction between externally applied currents and persistent circular currents induced by the magnetic field. In order to obtain a better resolution of the difference between positive and negative bias currents, an ac-drive of $10\,\mu\text{A}$ peak-to-peak amplitude and $3837\,\text{Hz}$ while measuring the average dc-voltage. If the structures are in the normal state at $T > T_c$ or in absence of screening currents at $H = 0$, the dc-output voltage should be zero, which was observed indeed. By contrast, if screening currents are present as a result of $H \neq 0$, the superposition of applied and circulating persistent currents gives different contributions for addition and substraction of the currents, i.e. a net dc-voltage signal is detected. Figure 2.28 shows the measured dc-voltage V_{dc} as a function of the applied magnetic field and temperature for the triangles with different position of current injection. In order to allow for a reliable comparison of the measured signals, V_{dc} is normalized in each case by the distance between the voltage contacts.

In the case of current leads placed above the geometrical center of the triangle (Fig. 2.28(b)), there is an abrupt change of sign in V_{dc} each time vorticity changes from L to $L + 1$, which is associated with a reversal of the persistent currents. On the other hand, a smooth crossover between two consecutive transitions of vorticity is associated with a progressive re-duction and later inversion of the screening currents. The origin of the rectified voltage is related to an unbalanced distribution of the externally applied current and its compensation (or reinforcement) by persistent cur-rents circulating around the geometrical center. Since the current leads are located above the geometrical center, the applied current is expected to flow mainly through the upper part of the triangle, causing an asymmetry when compensating (or reinforcing) the screening currents more in the up-per part then in the lower part of the structure (see the schematic drawings of applied and circulating persistent currents in Fig. 2.28).

Note that besides the above mentioned voltage sign reversals related to Little–Parks oscillations, an extra sign reversal is observed in Fig. 2.28(b) in the Meissner phase. The fact that this effect is much weaker and not observed for all measured samples indicates that it is sample dependent and, probably, cannot be related to the circulating screening currents.

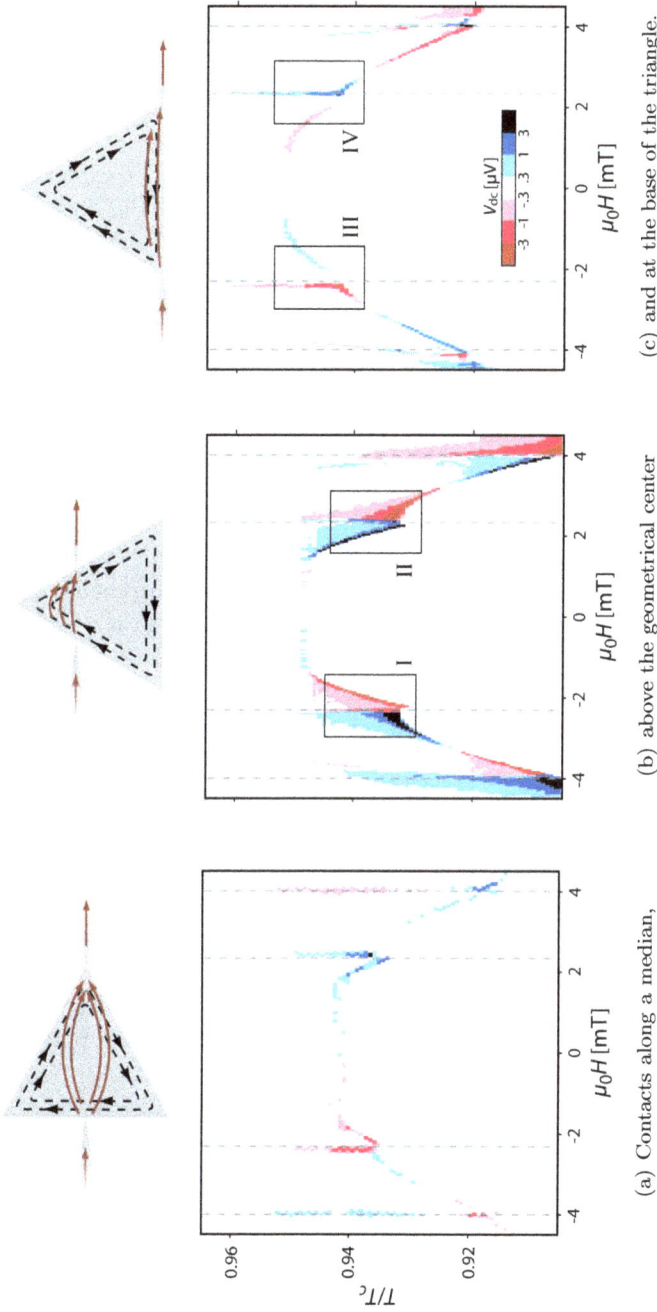

(a) Contacts along a median,

(b) above the geometrical center

(c) and at the base of the triangle.

Fig. 2.28 Rectification signal obtained with an *ac*-excitation of 10 µA and frequency of $\sim 3.4\,\text{kHz}$ as a function of field and temperature for superconducting triangles with contacts (a) along a median, (b) above the geometrical center and (c) at the base of the triangle. Dashed vertical lines indicate the theoretically expected field values for Little-Parks oscillations. The rectified *dc*-voltage is shown with a color scale from positive (blue) to negative (red). The drawings on top of each panel show schematically circulating persistent currents (black) and the applied current (red) for the corresponding contact configuration (after [422]).

According to the described scenario, no rectification effects should be observed if the line along which the current is inserted lies exactly at the center of the circulating persistent currents (Fig. 2.28(a)). Since the upper part of the triangle is in this case a mirror image of the lower part, the current is distributed equally around the center of the structure and no asymmetry in neither compensation nor reinforcement should be expected. This is consistent with the strongly reduced signal shown in Fig. 2.28(a), which is about three times weaker and located in a smaller H–T area compared to Fig. 2.28(b). It is likely that the non-zero signal of Fig. 2.28(a) originates from inevitable minor asymmetries produced by shadow effects during material deposition.

The most compelling evidence that the observed rectification effects in a triangle originate indeed from a direct superposition of external and field induced persistent (circular) currents can be drawn from the data presented in Fig. 2.28(c). In this case, the external current is injected well below the geometrical center of the triangle, and the situation should be reversed in comparison to the results of Fig. 2.28(b). Therefore, at magnetic fields of positive V_{dc} in Fig. 2.28(b), which indicate that a positive current is reinforced in the top of the triangle, an opposite sign is expected in Fig. 2.28(c), meaning that a positive current is compensated at the base of the triangle. This sign reversal between the two cases, shown in Fig. 2.28(b, c), is clearly visible for vorticity $L = 1$, as it can be seen by comparing the sections I and II with III and IV.

At this stage we can conclude that the applied current does not spread equally across the triangle, but rather follows mainly the shortest path. This can also be seen from current-flow simulations for a normal metal triangle as shown in Fig. 2.29, where dark and bright areas indicate regions of low and high current density, respectively. Since the observed effect always appears at the onset of the resistive state, these simulations can give an impression of the current distribution in our studied samples. However, the position-dependent nucleation of superconductivity results in a spatially dependent resistivity, which makes a direct comparison invalid.

2.7.3   *The diode effect in a disk*

So far we have discussed the dependence of the diode effect on the position of the current injection in triangular superconductors. However, the triangular geometry is not necessarily needed to observed this effect. An even more simple geometry for this is a disk [421] with four current and

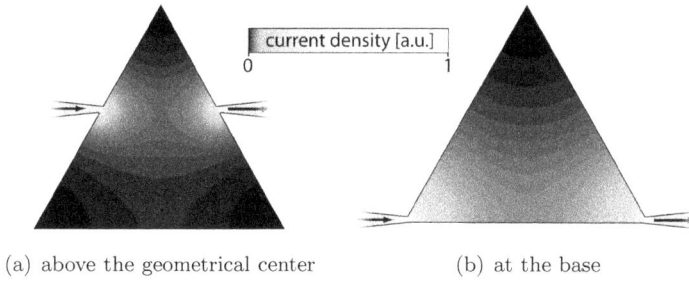

(a) above the geometrical center   (b) at the base

Fig. 2.29 Finite element analysis of the current flow in a normal metal triangle for two different contact configurations. Bright and dark areas indicate regions of high and low current density, respectively. Note the non-linear color scale.

two voltage contacts, as drawn schematically in the upper part of Fig. 2.30. The advantage of such structure is that different cases, namely symmetric current injection (using the contacts I^+ and I^-) and current injection above/below the geometrical center (contacts I^+ and $I^-_{1,2}$, respectively), can be studied in the same sample, thus avoiding any possible differences in geometry. Figure 2.30 shows the recorded dc-voltages corresponding to these three different directions of the applied current. In the symmetrical case (Fig. 2.30(a)) a weak rectification is only observed within a very narrow T–H range at the lowest vorticities. By contrast, when the current is applied through the top op the disk, the rectified signal is observed till higher vorticities. Apparently, the effect is reversed when applying the current through the bottom of the disk.

2.7.4 *Comparison with a theoretical model*

The necessary ingredients for the occurrence of voltage rectification in individual mesoscopic superconductors, as described above, are field-induced persistent currents and off-center injection of external currents. Accordingly, similar rectification effects should also be present in other systems with persistent currents and an asymmetric current path, thus resulting in a difference in critical currents of opposite sign. One example for such system is a closed loop with Josephson junctions. The rectification in a ring with $N > 2$ identical SNS Josephson junctions was recently calculated by Schildermans *et al.* [422], whereas Berger *et al.* analyzed a complementary model with $N = 2$ and unequal Josephson junctions [47]. Without loss of generality, the main effects can be seen in two simple configurations, consisting of a ring with three and five junctions as schematically drawn in Fig. 2.31.

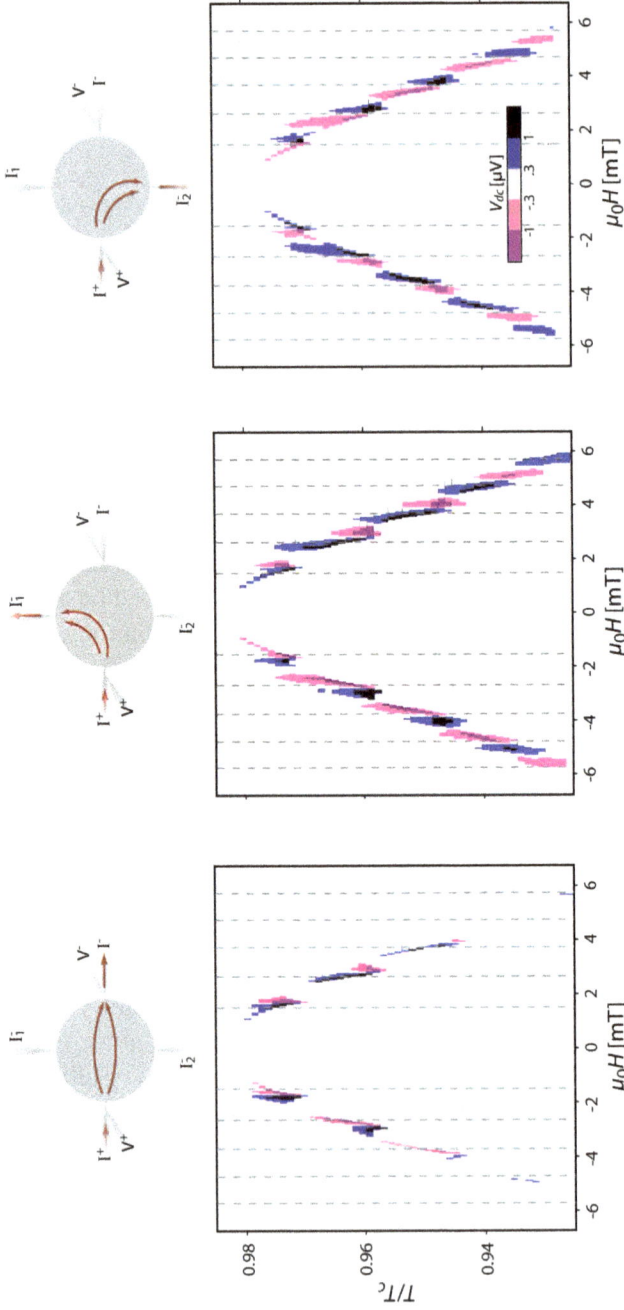

Fig. 2.30 Rectification signal V_{dc} obtained with an *ac*-excitation of $5\,\mu A$ and a frequency of $\sim 1.7\,kHz$ as a function of field and temperature for external current flow from (a) contact I^+ to I^- through the center of the disk, (b) via the upper part from contact I^+ to I_1^-, and (c) via the lower part from contact I^+ to I_2^- (after [421]).

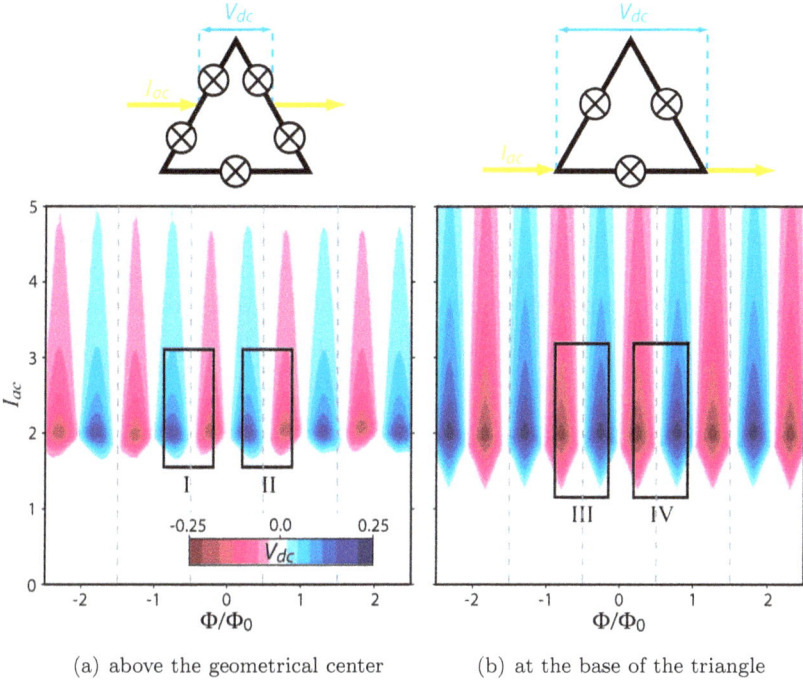

(a) above the geometrical center   (b) at the base of the triangle

Fig. 2.31   Schematic drawing of a superconducting triangle viewed either as a ring of (a) $N = 5$ or (b) $N = 3$ Josephson junctions, depending on the contact positions. Corresponding contour plots show the rectified voltage V_{dc} as a function of magnetic flux Φ/Φ_0 and amplitude of the applied ac-current I_{ac}. Note the opposite voltage signs at $\Phi/\Phi_0 = \pm 1/2$ by comparing the equivalent sections I and II with III and IV (after [422]).

The main assumptions for the calculations are that the superconducting order parameter $\psi(\boldsymbol{r}) = |\psi(\boldsymbol{r})|e^{i\theta(\boldsymbol{r})}$ is such that $|\psi(\boldsymbol{r})| = \psi_0$ with a constant ψ_0 at all islands, $\theta(\boldsymbol{r})$ is spatially constant in each island, and that the weak links between them can be modelled as identical SNS junctions. It is also assumed that the total magnetic field B is spatially and temporally constant. The phase difference at junction $n = 0, \ldots, N$ is $\phi_n = \theta(\boldsymbol{r}_n) - \theta(\boldsymbol{r}_{n-1})$ with $\theta(\boldsymbol{r}_n)$ the phase of the superconducting island centered at

$$\boldsymbol{r}_n = -R\cos\left(\frac{2\pi n}{N}\right) \cdot \boldsymbol{x} + R\sin\left(\frac{2\pi n}{N}\right) \cdot \boldsymbol{y},$$

with the radius R of the ring. The magnetic field contribution to the phase difference a_n is the line integral of the vector potential $\boldsymbol{A}$ between sites n

and $n - 1$:

$$a_n = \frac{2\pi}{\Phi_0} \int_{r_{n-1}}^{r_n} dl\, \boldsymbol{A}(\boldsymbol{l}) \ . \tag{2.30}$$

In the present case, a resistively shunted model is considered, using a resistive channel for the normal electron current in parallel with a Josephson current channel, satisfying Kirchhoffs' laws for the current conservation in each node. A current I is injected between junctions $N - 1$ and 0, and extracted δ junctions further, i.e. between junctions $\delta - 1$ and δ. The resulting set of dimensionless equations for the currents flowing in the ring is:

$$\dot{\phi}_n = I_{\text{up}} - \sin(\phi_n - a_n) \ , \qquad 0 \le n \le \delta - 1 \tag{2.31a}$$

$$\dot{\phi}_n = I_{\text{up}} - I - \sin(\phi_n - a_n) \ , \qquad \delta \le n \le N - 1 \tag{2.31b}$$

$$I_{\text{up}} \equiv \left(1 - \frac{\delta}{N}\right) I + \frac{1}{N} \sum_{n=0}^{N-1} \sin(\phi_n - a_n) \ , \tag{2.31c}$$

which are N first order differential equations for the time evolution of the N phase variables ϕ_n with $n = 0, \ldots, N-1$. It should be noted that each junction interacts with all other junctions via Eq. (2.31c), i.e. the total current in the upper branch of the circuit, which can be interpreted as a mean-field interaction plus a drive. Equations (2.31) can be solved numerically by using the Runge–Kutta method in order to compute the instantaneous voltage drop v between source and drain, which can be expressed as

$$v = \sum_{n=0}^{\delta-1} \dot{\phi}_n = \delta I_{\text{up}} - \sum_{n=0}^{\delta-1} \sin(\phi_n - a_n) \ . \tag{2.32}$$

With this model, the rectified mean dc-voltage $V_{dc} = \langle v \rangle$ as a function of the magnetic field and for an ac-sinusoidal current with different amplitudes I_{ac} can be calculated in the low frequency limit. Currents and voltages are normalized by the single junction critical current I_0 and by $R_n I_0$, respectively, with R_n the resistance of the resistive channel. The results for $N = 3$ and $N = 5$, with the same distance $\delta = 2$ between source and drain, are shown in Fig. 2.31. Apparently, both the $N = 5$ and the $N = 3$ device can rectify, i.e. $V_{dc} \ne 0$ as long as $\Phi/\Phi_0 \ne n/2$ (with integer n) and if I_{ac} is above a critical threshold that is smaller in the case $N = 3$. The maximum value of $|V_{dc}|$ is almost the same in both cases, although for a fixed magnetic field, V_{dc} decays slower as a function of I_{ac} for $N = 3$. It is important to note that qualitatively, the same oscillations are seen as experimentally observed as a function of vorticity.

Typically, experimental results are presented as a function of temperature with a constant applied current, while in the present model, the applied current is changed, keeping the temperature constant. However, the effect is qualitatively similar since both increasing T and I_{ac} have a similar influence on the system, driving it towards the resistive state. In brief, the predicted rectification in this model system is similar to the effects measured in the aluminum triangle. It is worth noticing that from the point of view of the superconducting condensate, the used experimental system can be regarded as a multiply connected structure since the order parameter ψ is maximum at the vertices and minimum at the sides of the triangle. Furthermore, for certain values of field and temperature, $\psi = 0$ at the centers of the sides of the triangle and the system can be thought of as a ring-like structure with SNS-junctions. This scenario is modified by the presence of contact leads which locally enhance the order parameter. Therefore, a triangle with contacts at the sides (Fig. 2.28(b)) can be directly compared with a ring of five junctions, whereas the ring with $N = 3$ imitates the response of a triangle with contacts at the base, as it is shown in Fig. 2.28(c). Indeed, this association can be further justified by noting that in frames I and III (or II and IV) of Fig. 2.31, the different Josephson circuits show opposite responses for the same $\Phi/\Phi_0 = \pm^1/_2$, as it was also found experimentally by comparing corresponding frames in Fig. 2.28.

2.8 Hybrid individual cells

By changing the geometry, topology and size of individual nanostructures, the confinement of the superconducting condensate can be tuned and controlled. This is reflected in the variation of the nucleation field $T_c(H)$ of these structures, which follows the respective variations of the lowest Landau level $T_c(H) \leftrightarrow E_{LLL}(H)$. So far we have been considering confinement effects in superconducting nanostructures in presence of homogeneous magnetic fields. On the other hand, confinement effects can be additionally tuned by creating artificial field inhomogeneities at the nanoscale, for example, by using superconductor/ferromagnet (S/F) hybrids. In these hybrids, the ferromagnetic subsystem can provide the modulation of magnetic field in which the development of the superconducting condensate takes place. An example for this is given in Fig. 2.32, where the compensation of the stray magnetic field of a magnetic dot by an external field leads to very different regions of superconductivity. For the case of S/F hybrids with

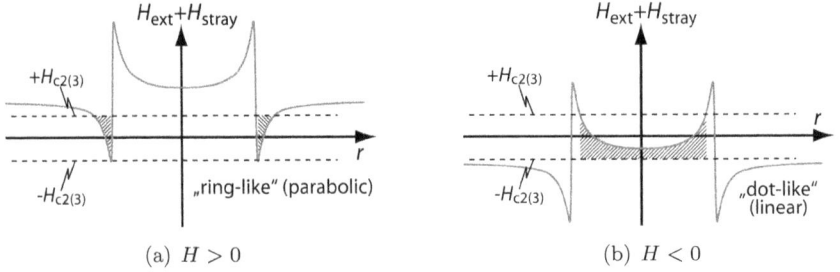

(a) $H > 0$ (b) $H < 0$

Fig. 2.32 Schematics of the compensation of the stray magnetic field H_{stray} of a magnetic dot by an external field H, leading (a) to a ring-like, and (b) to a dot-like region of superconductivity (see the hatched areas).

magnetically hard ferromagnets, the reverse effects of the superconducting condensate on the distribution of magnetic domains and other magnetic properties are quite weak and, therefore, can be neglected. In this situation, the ferromagnetic subsystem generates an additional inhomogeneous field distribution, which can be treated as unchanged in the field range $\pm H_{c3}$ where superconductivity can exist.

In this section we shall present the description of the nucleation of superconductivity and the behavior of the vortex matter in the S/F hybrid nanosystems by analyzing these features in a superconducting square with a cylindrical magnetic dot of perpendicular magnetic anisotropy on top of it.

2.8.1 *Square with magnetic dot*

Similar to the approach followed in Secs. 2.5.1 and 2.5.2 to overcome the difficulties in solving the linearized Ginzburg–Landau equation (Eq. (1.6)), which arise from the explicit presence of $\boldsymbol{A}$, one can apply a vector potential gauge transformation for regular polygons that gives $A_n = 0$ at the samples boundary in the case of a homogeneous external field [90, 91, 93]. The superconducting boundary condition Eq. (1.9b) is then transformed into $\nabla \psi|_n = 0$, and the linearized Ginzburg–Landau equation may be solved by using an analytic basis set, with a corresponding set of basis functions, classified according to the irreducible representations (irreps) of the symmetry group of the problem.

For a square, having C_4 rotational symmetry, these irreps are denoted as A, B, E_+ and E_-, and they correspond to solutions with, respectively, no vortex (vorticity 0), a giant vortex (vorticity +2), a vortex (vorticity

+1), and an antivortex (vorticity -1) in the center of the sample that will be surrounded by a number of vortices (multiple of 4) depending on the applied magnetic field (see Sec. 2.5.1). The calculated vortex patterns are stable close to the phase boundary, but deeper in the superconducting state, the nonlinear term in the Ginzburg–Landau equation (Eq. (1.3a)) becomes important and causes symmetry breaking transitions [96].

The nucleation of superconductivity in a square with a magnetic dot on top (see the inset in Fig. 2.33 for a schematics) can be simulated numerically by applying the procedure described in Sec. 2.5.1. However, in this case, the vector potential results from the contributions of the homogeneous external field and the stray field of the dot, the last can be obtained from magnetostatic calculations. For a dot that is magnetized parallel to the z-direction, the vector potential can be written in cylindrical coordinates as [272]

$$A_\varphi = \frac{Hr}{2} + 4M_{\text{dot}}\sqrt{\frac{R}{r}}\int_0^l \frac{1}{k}\left[1 - \frac{k^2}{2}K(k) - E(k)\right]dz_d , \qquad (2.33)$$

$$\text{where} \quad A_r = A_z = 0$$

$$\text{and} \quad k^2 = \frac{4Rr}{(R+r)^2 + (z - z_d^2)}$$

is a dimensionless variable. The z-dependence of k can be used to account for the presence of a substrate between the dot and the sample to avoid proximity effects. The parameters R, M_{dot} and l in Eq. (2.33) stand for the radius, magnetization and height of the dot, while K and E are elliptic integrals of the first and the second kind, respectively. Figure 2.33 shows the field profile of a magnetic dot for $R = 0.4\,a$, $z = -0.0025\,a$ and four different heights $l = 0.33\,a$, $0.16\,a$, $0.033\,a$ and $0.0033\,a$, where a is the length of the square. Note that for the case of $l = 0.033\,a$, these parameters are comparable with those previously used in experiments [190–192, 326].

The magnetic field dependence of the energy of the lowest Landau level, corresponding to each irrep in presence of the magnetic dot, is given in Fig. 2.34(a) together with the results for the case without the dot (Fig. 2.34(b)). In order to obtain curves that are independent of the sample size, the solutions have been multiplied by the samples surface S, and the magnetic field is presented in units of Φ/Φ_0, where $\Phi = HS$ is the magnetic flux in the sample. These lowest Landau levels $E_{LLL}(H)$ of each irrep define the phase boundaries shown in the lower panels of Fig. 2.34 which, in the case without a dot, are clearly symmetric with respect to the origin [64, 90, 92]. This behavior arises from a well-defined sequence

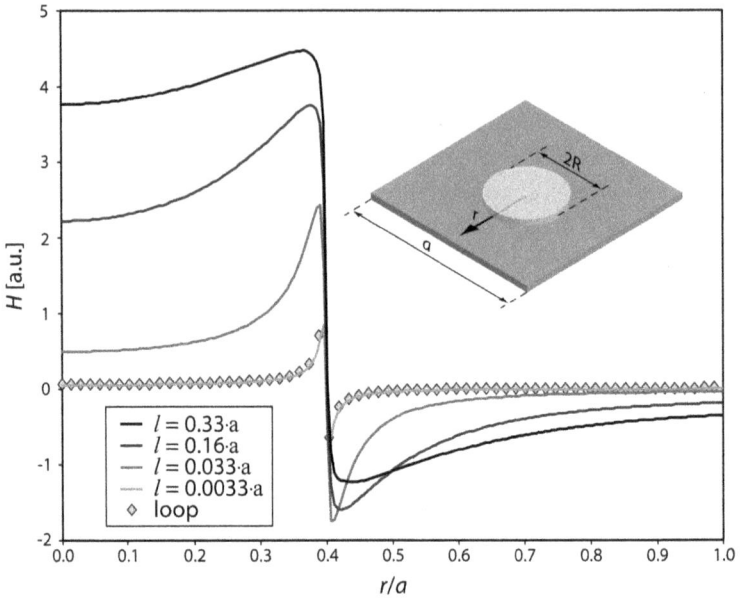

Fig. 2.33   Calculated field profiles of magnetic dots with $R = 0.4\,a$, $z = -0.0025\,a$ and four different heights l (solid lines). Note that in the limit of $l \to 0$, the field profile of a very thin dot approaches the case of a thin loop with a circulating current, see the curve represented by diamonds, which corresponds to the field profile of a loop with radius R and a cross section of $(0.0033\,a)^2$. The inset shows a schematic drawing of the superconducting square with the magnetic dot on top.

of crossings between irreps, $A \to E_+ \to B \to E_- \to A\ldots$, each of them leading to an increase of the total vorticity of the sample, L (that defines the total flux trapped in the square through $L\Phi_0$), by 1 [90, 92]. However, Fig. 2.34(a) clearly illustrates that the dot strongly affects this oscillating behavior of the $T_c(H)$ phase boundary. For instance, the $S/\xi^2(T)$ curves are asymmetric with respect to the polarity of the field, giving rise to a maximum critical temperature at, approximately, $\Phi \simeq -12.5\Phi_0$. This fact clearly indicates the presence of the compensation effect between the stray field of the dot and the applied magnetic field already observed in loops and disks [191, 192, 326], which, as shown by Fig. 2.34(a), in the square also manifests itself in a long-period oscillation regime between $-38.7 \lesssim \Phi/\Phi_0 \lesssim -7.8$. Instead, for positive fields the short-period oscillations characteristic of the no-dot case are progressively recovered. As a result, a crossover from a linear $T_c(H)$ ('dot-like') to a parabolic ('ring-like') $T_c(H)$ background is clearly observed (see Fig. 2.32 and 2.34).

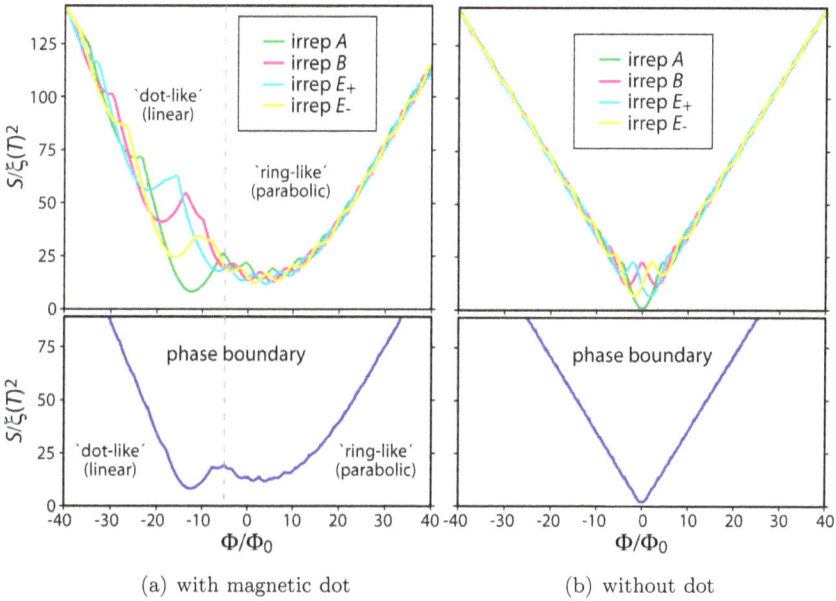

Fig. 2.34 Magnetic field dependence of the energy of the lowest Landau level in a superconducting square, corresponding to each irrep in presence of (a) and without (b) a magnetic dot ($l = 0.33a$, $z = -0.0025a$, $R = 0.4a$ and $M_{\text{dot}} = 18\Phi_0$). The lowest parts of these Landau levels determine the phase boundaries shown in the lower panels. Note the dramatic change in the cusp-like behavior of the $S = \xi(T)^2$ curves, from long-period oscillations between $-38.7 \lesssim \Phi/\Phi_0 \lesssim -7.8$ to short-period oscillations for positive fields (after [84]).

The magnetic dot may also favor energetically one or more irreps with respect to the others, thus leading to changes in the total vorticity of the sample by more than 1 (multiquanta transitions) due to the disappearance of transitions between irreps present in the case without a magnetic dot. The presence of multiquanta transitions in the onset of superconductivity can be observed in Fig. 2.35, where the evolution of the total vorticity of the sample along the phase boundary is presented for different magnetizations of the dot, namely $M_{\text{dot}} = 0$ (no dot), $M_{\text{dot}} = 6\Phi_0$, $M_{\text{dot}} = 12\Phi_0$ and $M_{\text{dot}} = 18\Phi_0$.

In the no-dot case ($M_{\text{dot}} = 0$), $|L|$ shows an almost perfect linear behavior, changing by 1 through field intervals that progressively tend to Φ_0 for increasing Φ [64, 90]. However, this field dependence dramatically changes upon increase of M_{dot}: as a consequence of the compensation between the external field and the stray field of the dot, those values of Φ with $|L| = 0$

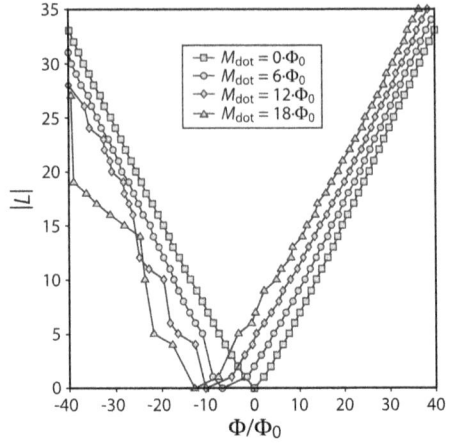

Fig. 2.35 Magnetic field dependence of the total vorticity of a superconducting square with a magnetic dot ($l = 0.33a$, $z = -0.0025a$, $R = 0.4a$ and $M_{dot} = 18\Phi_0$) at the normal-to-superconducting phase boundary, for different values of the magnetization of the dot. For increasing values of M_{dot}, an oscillating behavior of $|L|$ with changes in the vorticity by more than 1 can be seen at negative fields (after [84]).

are shifted to negative fields. In addition, the vorticity oscillates at negative fields with changes in L by more than 1, indicating the simultaneous entrance of several vortices into the sample. For instance, at $M_{dot} = 18\phi_0$ (triangles), one can observe variations in the vorticity by $+4$ from -4 to 0 (along irrep A) and from -14 to -10 (along irrep B), by $+5$ from -10 to -5 (with a change between irreps B an E_-), and by $+8$ from -27 to -19 (along irrep E_+). Note that this last big jump in the vorticity appears approximately at the same field values at which the $S/\xi^2(T)$ curves in Fig. 2.34(a) enter the long-period oscillations regime.

To illustrate that the origin of the effects on the fluxoid quantization in the square (see Fig. 2.35) is the compensation of the applied magnetic field by the stray field of the dot, Fig. 2.36 presents the change in the spatial distribution of $|\psi|^2$ at the vorticity transitions observed at $\Phi = -38.7\Phi_0$ (left panels of Fig. 2.36) and at $\Phi = -7.8\Phi_0$ (middle panels of Fig. 2.36), the magnetic fields that limit the long-period oscillations regime in Fig. 2.34(a). As can be seen, the vorticity transition by $+8$ that occurs at $\Phi = -38.7\Phi_0$ almost coincides (within an interval of $\sim 0.5\Phi_0$) with a profound change in the topology of the order parameter distribution in the square, with the highest values of $|\psi|^2$ (red regions) concentrated in the corners for $\Phi < -38.7\Phi_0$ (analogously to the no-dot case) but with a ringlike structure for $\Phi > -38.7\Phi_0$. This distribution of $|\psi|^2$, clearly related to a compensation of the external field by the stray field of the cylindrical dot (see Fig. 2.32), is preserved when the field increases up to $\Phi = -7.8\Phi_0$. Then $|\psi|^2$ is again higher in the corners, and the well-defined sequence of transitions between irreps is progressively recovered (see Fig. 2.34(a)).

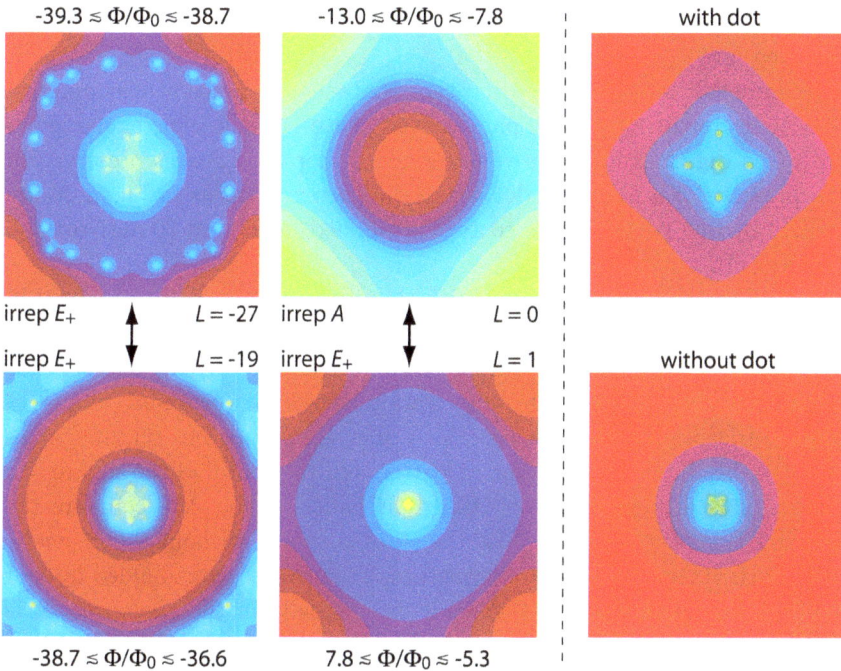

$-39.3 \lesssim \Phi/\Phi_0 \lesssim -38.7$ $-13.0 \lesssim \Phi/\Phi_0 \lesssim -7.8$ with dot

irrep E_+ $L = -27$ irrep A $L = 0$

irrep E_+ $L = -19$ irrep E_+ $L = 1$ without dot

$-38.7 \lesssim \Phi/\Phi_0 \lesssim -36.6$ $7.8 \lesssim \Phi/\Phi_0 \lesssim -5.3$

Fig. 2.36   Change in the $|\psi|^2$ distribution at the vorticity transitions observed for $\Phi = -38.7\Phi_0$ (left) and $\Phi = -7.8\Phi_0$ (middle), the magnetic fields that limit the regime of long-period oscillations in Fig. 2.34(a). These transitions coincide with a change in the topology of the vortex patterns that may favor the simultaneous nucleation of one or more antivortices in the corners of the superconducting square. The two images at the right compare the vortex-antivortex patterns that can be observed at $L = 3$ with and without the magnetic dot. The vortex pattern rotates by 45° and expands dramatically in the presence of the magnetic dot, so that the magnetic dot acts as a sort of a 'magnifying lens' for the compact vortex-antivortex pattern (after [84]).

Most of the multiquanta transitions observed in Fig. 2.35 between $-38.7 \lesssim \Phi/\Phi_0 \lesssim -7.7$ have the common feature of involving the simultaneous penetration of one or more antivortices[11] in each of the corners of the square, where there is no field compensation. These results illustrate the importance of the interplay between the finite rotational symmetry C_4 of the square and the inhomogeneous field of the cylindrical dot with C_∞ symmetry, which may allow topological changes in the distribution of $|\psi|^2$ favoring the nucleation of more than one single vortex along the phase

---

[11] Note that we take as a reference for the vortex-antivortex definition the orientation of the magnetic moment of the dot.

boundary. Similar results have been found for a smaller radius of the dot by using larger values of M_{dot}.

Figure 2.36 also illustrates that the spontaneous formation of symmetry-consistent vortex-antivortex pairs at the phase boundary in regular polygons (see Fig. 2.13 for squares and Fig. 2.15 for triangles) is preserved in the presence of a magnetic dot, since the vortex patterns in Fig. 2.36 are formed by one vortex in the center surrounded by 28 and 20 antivortices, respectively.

In addition, when compared with the distributions of $|\psi|^2$ obtained in the no-dot case in Sec. 2.5.1, the images of Fig. 2.36 clearly demonstrate that the dot can be used to enlarge these vortex-antivortex patterns, thus facilitating their experimental observation with local vortex-imaging techniques. In this case, the magnetic dot works as a sort of lens that magnifies considerably the vortex-antivortex patterns in the center of the square. To further illustrate these results, the right panels of Fig 2.36 compare the distributions of $|\psi|^2$ at $L = 3$ for the cases with and without the magnetic dot. Both vortex patterns are formed by an antivortex surrounded by four vortices that rotate $45°$ and expand dramatically in the presence of the magnetic dot. This effect can be attributed to the singular behavior of the stray field at the edges of the dot, with both positive and negative large peaks (see Fig. 2.33) that seem to attract vortices and antivortices.

The considered hybrid structure clearly demonstrates the most important elements of the interplay between superconductivity and ferromagnetism at sub-micron scales, which can give rise to novel quantum effects in the onset of superconductivity in regular polygons with magnetic dots. These effects include multiquanta transitions between vortex patterns and the formation of larger vortex-antivortex molecules than in the no-dot case. From the point of view of applications, these results open new possibilities to manipulate the flux quantization in superconducting micro- and nanostructures.

2.8.2 *Phase shifter*

Another hybrid S/F nanostructure we shall consider in this chapter is a superconducting ring, which is enclosing a magnetic dot with perpendicular magnetic anisotropy. This dot can generate permanent magnetic flux Φ_{dot}/Φ_0 needed for *phase shifting* in superconducting elements for quantum computing, including the π-shift that corresponds to the flux $\Phi_{\text{dot}} = {}^1\!/_2\Phi_0$ generated by the dot through the ring.

Table 2.4   Characteristics of the considered hybrid S/F nanostructures

Superconducting Component (Aluminium Loop)	Magnetic Component ([Co.4nm Pd1nm]n dot)
$\xi(0) = 138\,\text{nm}$	
$\tau = 46\,\text{nm}$	coercive field: $\sim 150\,\text{mT}$
outer radius: $915\,\text{nm}$	radius: $174\,\text{nm}$ (structure $\mathcal{A}$)
inner radius: $680\,\text{nm}$	radius: $350\,\text{nm}$ (structure $\mathcal{B}$)
$T_c^{\text{max}} = 1.31\,\text{K}$	

Various solid state structures can exhibit quantum behavior that is potentially of interest for quantum computing [363]. A typical example is a superconducting flux qubit based on Josephson junctions, which operates by using the degeneracy of the two equal and opposite persistent currents at half-integer flux $(n + 1/2)\Phi_0$ through the superconducting circuit [97, 160, 333, 370, 371, 478]. The degeneracy is lifted by the charging energy, and the two distinct quantum states $|0\rangle$ and $|1\rangle$ are associated with the opposite circulation of the superconducting condensate in the ring. A resonant external excitation can force the superconducting condensate to oscillate coherently between these two states. Although an external applied field has been used so far in experiments to generate the flux necessary for a π-shift in superconducting qubits, it is thought that the fluctuations of the external flux (flux noise) present a major source of decoherence [281]. Therefore, it has been a challenge to incorporate π-shift in a qubit and enable its operation without an external bias field.

A number of structures using high-T_c superconductors has been proposed, see for example [88, 233, 304, 426, 469]. In high-T_c superconductors, due to the predominant $d_{x^2-y^2}$ symmetry of the order parameter, π-shift can inherently be gained if, for example, the interfaces of the Josephson contacts are made of two superconductors rotated by $\pi/2$ in the $a-b$ plane. However, this particular symmetry poses a fundamental constraint on the overall coherence of the high-T_c superconducting qubits. Since the superconducting gap Δ in high-T_c cuprates is zero along the nodal directions, normal quasiparticles, that are inherently incoherent, are initially present even at very low temperatures. For this reason, the application of high-T_c superconductors is a trade-off in terms of coherence: decoherence due to the flux noise is strongly reduced, but an additional source of decoherence, related to the presence of normal quasiparticles, is introduced.

The hybrid nanosystem, consisting of a superconducting ring with a magnetic dot, offers a new practical realization for shifting the phase in superconducting qubits [193]. The phase shift is achieved by placing a magnetic dot with perpendicular magnetization in the center of a loop that is made of a conventional s-wave superconductor. The flux generated by the dot creates an additional current in the superconducting loop, thus giving rise to a phase shift. This new design has several advantages. First and foremost, the phase shift is a result of a quite basic and general property of superconductors and can be implemented without any limitations. It does not require d-wave or any other specific symmetry of the superconducting order parameter, nor does it put any constraints on the interfaces of Josephson junctions, as with high-T_c superconductors. More importantly, the phase shift is achieved with an s-wave superconductor and, therefore, no additional decoherence appears in the system, as in the case of the d-wave superconductors. Since the magnetic dot is separated from the superconducting loop, it cannot adversely affect the operation of the qubit. The generated flux and, consequently, the persistent current in the loop are stable. Technologically, the fabrication procedures for conventional nanostructured superconductors and ferromagnets have been mastered and can be carried out routinely. By conveniently varying the parameters of the dot it is possible to introduce any phase shift in the loop. Therefore, the magnetic phase shifter can be used as an external current source with high stability. Furthermore, it may well be applied for phase biasing in the experimental study of fractional Josephson vortices [188]. Since the phase shifter can be used with s-wave superconducting circuits, the integrability and scalability of the qubit are not deteriorated.

The superconducting properties of two identical aluminium loops, each of them enclosing a magnetic dot of different size (see the insets of Fig. 2.37 for schematic drawings and Tab. 2.4 for some basic characteristics), were investigated with transport measurements as a function of an external magnetic field, using an ac transport current of $100\,\mathrm{nA}$ and $27.7\,\mathrm{Hz}$. Figure 2.37 shows the critical current of the structures versus the applied magnetic flux, taken in steps of $0.1\Phi/\Phi_0$ within the range $\Phi_0 \leq \Phi \leq \Phi_0$ at $T/T_c = 0.995$. The solid lines are the theoretical curves obtained from the de Gennes–Alexander theory for superconducting micronetworks [149]. The flux has been calculated with respect to the mean radius ($797.5\,\mathrm{nm}$) of the loop, taking the field parallel to the z-direction as positive.

As discussed in Sec. 2.2, a homogeneous mesoscopic superconducting loop with width w and thickness τ much smaller than the coherence length $\xi(T)$ exhibits an oscillatory dependence of the critical current that is given by Eq. (2.8). Close to the zero-field critical temperature, the influence of

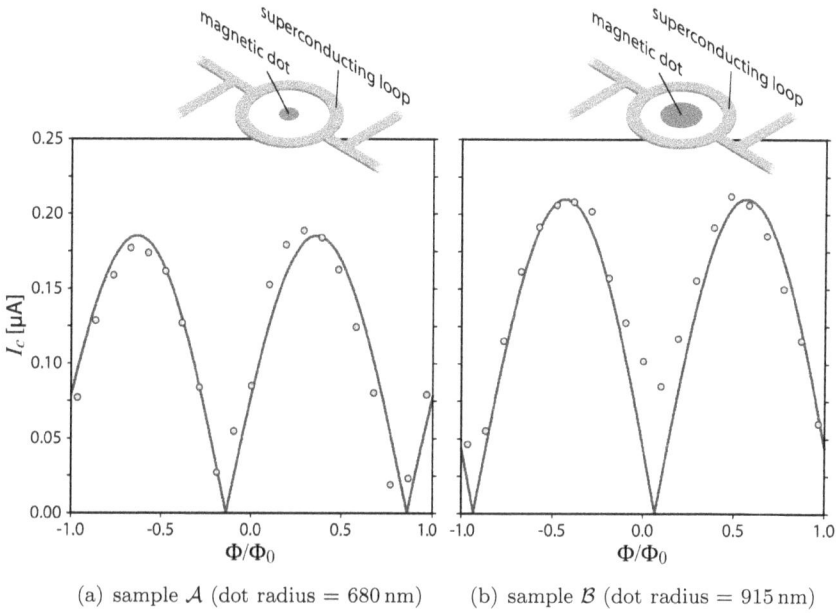

(a) sample $\mathcal{A}$ (dot radius = 680 nm)   (b) sample $\mathcal{B}$ (dot radius = 915 nm)

Fig. 2.37   The critical current I_c of the hybrid S/F nanostructures versus the applied flux $T/T_c^{\max} = 0.995$. The solid line is the theoretical curve according to Eq (2.8). Schematic sketches of the structures are shown as insets (after [193]).

self-inductance can be neglected and the flux through the loop equals the applied flux. Since, in the present case, the coherence length at $T/T_c^{\max} = 0.995$ is approximately 1.9 μm and the width of the loops is 0.235 μm, the considered structures are in the 1D regime and the de Gennes–Alexander theory is applicable. In order to take into account the magnetic dots, Eq. (2.8) has to be modified by adding the flux generated by the dots to the applied flux. Using the saturation magnetization of bulk cobalt, one can estimate with magnetostatic calculations that a flux of $-0.4\Phi_0$ is generated by the dot in structure $\mathcal{A}$ with a dot radius of 680 nm, whereas the corresponding flux is $-1.6\Phi_0$ in structure $\mathcal{B}$ (dot radius = 915 nm). Both values are in a good agreement with the experimental $I_c(\Phi/\Phi_0)$ data shown in Fig. 2.37. Close to $\Phi/\Phi_0 = 0$, sample $\mathcal{A}$ has a finite resistance which is lower than the residual resistance in the normal state, and its critical current has therefore been taken as zero. On the other hand, for the same applied flux, sample $\mathcal{B}$ remains superconducting and has a finite value of the critical current.

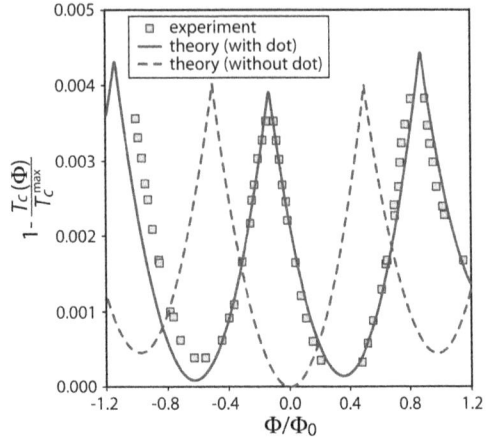

Fig. 2.38 The superconducting phase boundary of sample $\mathcal{A}$, presented as $1 - T_c(\Phi)/T_c^{\max}$ versus the flux Φ/Φ_0. Squares show the experimental data, the solid line is the theoretical fit, and the dashed line depicts the theoretical phase boundary of the loop without the magnetic dot (after [193]).

The loops have a minimum in the critical current in the vicinity of zero applied flux and maxima for finite applied fluxes close to $\pm 1/2\Phi_0$, which is a clear evidence that phase shifts are indeed introduced in the loops. More importantly, given that the superconducting loops are identical and that the minima and maxima are attained for different values of the applied flux, it is evident that the phase shift is governed by the flux of the dot. Neither of the curves displays exactly π shift, but these results clearly demonstrate that a magnetic dot can efficiently be used as a phase shifter, and that the phase shift can be controlled by changing the parameters of the dot. The flux of the dot, necessary for a π-shift, can be tuned by changing the radius of the dot or by varying the number of Co/Pd bilayers forming the dot.

Figure 2.38 presents the superconducting phase boundary of structure $\mathcal{A}$, showing the normalized critical temperature $1 - T_c(\Phi)/T_c^{\max}$ versus the normalized applied flux Φ/Φ_0. Symbols and the solid line represent experimental data and theoretical curve, respectively, whereas the dashed line is the theoretical curve obtained for the reference loop with the same parameters but without magnetic dot. The theoretical $T_c(\Phi/\Phi_0)$ dependence, found from the Ginzburg–Landau theory, shows good agreement with the experimental data.[12] The real dimensions of the loop, as well as the above mentioned values of the flux generated by the dots, have been used in the calculations and the best agreement with the experimental data has been obtained for the coherence length of $\xi(0) = 100$ nm. The discrepancy between this value of $\xi(0)$ and the estimation from the reference sample $\xi(0) = 138$ nm is typically encountered in mesoscopic aluminium structures

[12]See Reference 191 for details of the used method.

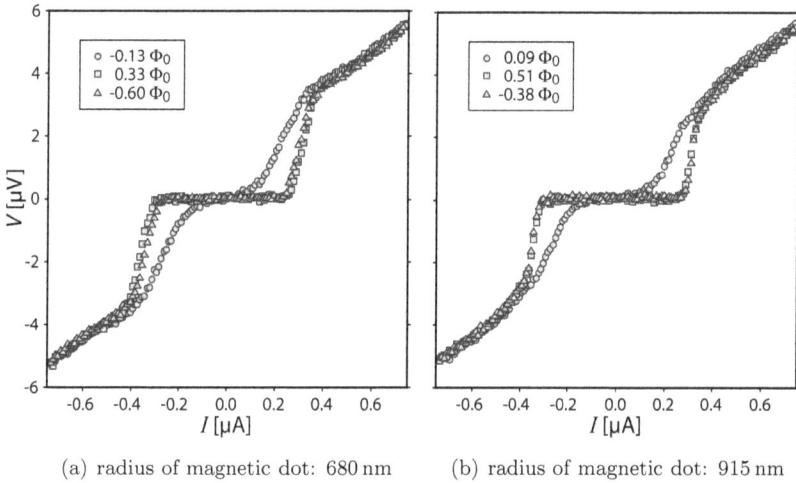

(a) radius of magnetic dot: 680 nm (b) radius of magnetic dot: 915 nm

Fig. 2.39 $V(I)$ curves of (a) sample $\mathcal{A}$ and (b) sample $\mathcal{B}$, taken at the field values corresponding to the minima and the maxima of the phase boundary within the range $-\Phi_0 \leq \Phi \leq \Phi_0$ (after [193]).

and is accounted for by the influence of the contacts, which effectively increase the radius of the superconducting loop, as well as by minor nonuniformity in the width of the loop.

The $T_c(\Phi)$ phase boundary is pronouncedly modified by the stray field of the magnetic dot, since it displays a clear phase shift (Fig. 2.38). It should be noted that, due to its inhomogeneity, the stray field affects the phase boundary in a nontrivial way, resulting in an additional asymmetry in $T_c(\Phi)$ [191]. For this reason, two consecutive minima in the phase boundary of the loop with magnetic dot (corresponding to local maxima of the critical temperature) have different values and appear at different fluxes. The inhomogeneity of the stray field, however, does not prevent the application of the proposed phase shifter. In order to operate, the superconducting qubit has to be driven to the state at which the flux equals $1/2\Phi_0$. Other states are less important for the operation of the qubit, provided that they are well separated from the relevant state so that the interference between them can be ruled out. Irrespective of its inhomogeneity, the stray field of the dot can generate $1/2\Phi_0$, thus providing the necessary conditions for the qubit to operate.[13]

---

[13]Note that the $T_c(\Phi)$ phase boundary may be slightly affected by a displacement of the magnetic dot from the center of the loop.

Figure 2.39 presents $V(I)$ curves taken at the values of the applied flux which correspond to the maxima and minima of the superconducting $T_c(\Phi)$ phase boundary at $T/T_c^{\mathrm{max}} = 0.995$ (values of the fluxes are indicated in the figures). The presence of the phase shifts is unambiguously and directly demonstrated for both samples, as the critical currents for finite applied fluxes are higher than the critical currents close to zero applied flux. Therefore, the considered hybrid S/F nanosystem clearly demonstrates that a phase shift can be induced in a superconducting circuit, as well as that it can be controlled by modifying the parameters of the magnetic dot. Since the magnetic phase shifter can generate an arbitrary phase shift, it can be used as an external high stability current source for superconducting elements, or may serve as a tool to investigate fractional Josephson vortices.

Chapter 3

Clusters of Nanocells

After the previous' chapter description of confinement effects in several individual superconducting nanocells, such as lines, loops, dots and regular polygons, we are ready to move further on to clusters of nanocells 'A' on our way from 'single cell' samples to materials that are nanostructured by the introduction of huge arrays of cells. First, we take 'A' = loop and consider one-dimensional multiloop structures: a 'bola', double and triple loops of aluminium. Second, we consider 'A' = antidot and analyze the properties of a 2×2 antidot cluster. Finally, coupling effects between concentric loops are presented.

3.1 One-dimensional clusters of loops

In this section, we will discuss three different structures, a 'bola', a double loop and a triple loop (see the insets in Fig. 3.1), all of them composed of loops with the same dimensions, leading to the same main magnetic field period $\mu_0 \Delta H = 1.24\,\mathrm{mT}$ of the $T_c(H)$ oscillations. Width and thickness of the strips that form the structures are $w = 0.13\,\mathrm{\mu m}$ and $\tau = 34\,\mathrm{nm}$, respectively. In the experimental data shown in the following, the parabolic background, caused by a finite loop width according to Eq. (2.1) and clearly visible in Fig. 3.1, is already subtracted in order to allow for a direct comparison with the theory. Within the temperature intervals, in which the $T_c(H)$ boundaries were measured, the coherence length $\xi(T)$ is considerably larger than the width w of the strips. This makes it possible to use the one-dimensional models for calculating $E_{LLL}(H)$ and thus $T_c(H)$. The basic idea is to consider $|\psi|$ constant across the strips that form the network while allowing a variation of $|\psi|$ along the strips. In the simplest approach, $|\psi|$ is assumed to be spatially constant (London limit) [7,89], in contrast to

Fig. 3.1 Comparison of the experimental superconducting – normal state phase boundaries $T_c(\Phi)$ of the three different structures shown as insets: a 'bola', a double loop and a triple loop. In the case of the bola, $T_c(\Phi)$ is very similar to the Little–Parks oscillations of a single loop, while the other two samples show additional features. Note that the curves have been shifted along the x-axis for clarity (after [72]).

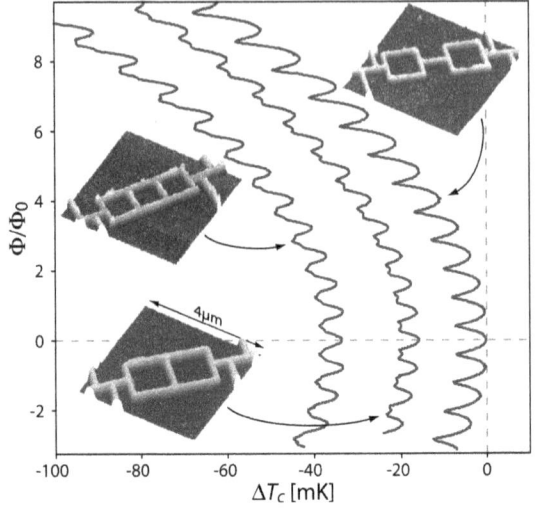

the de Gennes–Alexander approach [7, 119, 152], which allows $|\psi|$ to vary along the strips. The latter approach imposes

$$\sum_n \left(\imath \frac{\partial}{\partial x} + \frac{2\pi}{\Phi_0} A_\|(x) \right) \psi(x) = 0 \qquad (3.1)$$

at the nodes where the current paths join, the sum running over all strips connected to the junction point. Here, x is the coordinate defining the position on the strips and $A_\|$ is the component of the vector potential along x. Equation. (3.1) is often called the generalized first Kirchhoff law for superconducting micronets, ensuring the conservation of current [152]. The second Kirchhoff law for voltages in normal circuits is in the case of superconducting micronets replaced by the fluxoid quantization requirement Eq. (2.4), which should be fulfilled for each closed contour in the superconducting network, i.e. around each loop.

The $T_c(H)$ phase boundaries of the three structures are shown together in Fig. 3.1 for comparison, and separately in more detail in Figs. 3.2–3.4. In the latter, the black lines are the phase boundaries calculated in the London limit, while the solid lines give the results obtained with the de Gennes–Alexander approach. As we discussed in Sec. 2.2 for a mesoscopic loop, attaching contacts modifies the confinement topology, so that the amplitude of the local Little–Parks oscillations is reduced at low magnetic fields. Accordingly, in the present case, the inclusion of leads decreases the amplitude of the $T_c(\Phi)$ oscillations. The gray dashed lines in Figs. 3.2–3.4 show the results of calculations using the de Gennes–Alexander approach

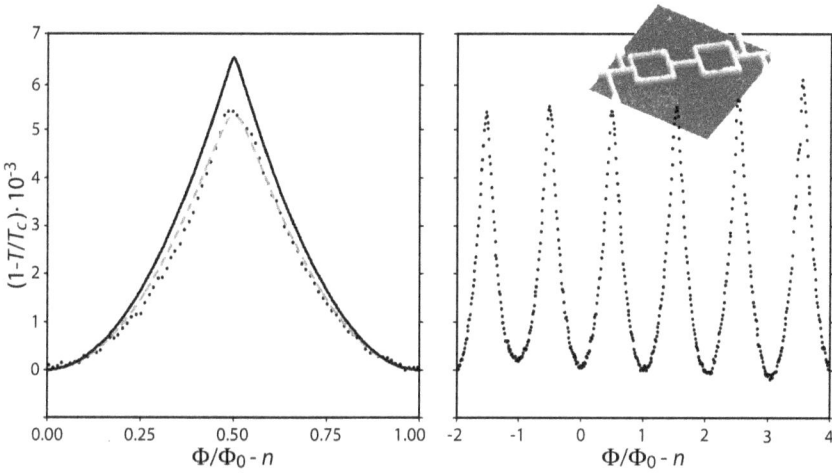

Fig. 3.2 Experimental $T_c(\Phi)$ data for the 'bola' with the parabolic background of Eq. (2.1) subtracted (left and right panels show a single and a few periods, respectively). Experimental data is represented by dots, whereas black and gray lines correspond to the theoretical results obtained in the London limit and with the de Gennes–Alexander approach, respectively. The latter takes the presence of the leads into account (after [345]).

including the presence of the leads. The values for $\xi(0)$, obtained from the fits, agree within a few percent with those found independently from the monotonic background of the $T_c(\Phi)$ curves using Eq. (2.1).

First, in Fig. 3.2, we consider the mesoscopic 'bola' — two loops connected by a wire and not sharing the common sides. Fink *et al.* [152] showed that in the complete magnetic flux interval, the spatially symmetric solution with equal orientation of the supercurrents in both loops has a lower energy than the antisymmetric solution. Similar to a comparison between a mesoscopic loop and a hydrogen atom (Sec. 2.2), we can compare the bola with a hydrogen H_2 molecule, where the symmetric and the antisymmetric solutions correspond to singlet and triplet states, respectively. In fact, the $T_c(\Phi)$ of the bola is the same as for a single loop, provided that the length of the strip connecting the two loops is short, as confirmed by the Little–Parks oscillations observed in the experimental $T_c(\Phi)$ curve in Fig. 3.2.

Now, we will focus on the results obtained by Bruyndoncx *et al.* [72] on the phase boundaries of a double and a triple loop sharing the common sides (see Fig. 3.3 and 3.4, respectively). To facilitate the discussion, we divide the flux period in two intervals: flux regime I for $\Phi/\Phi_0 < g$

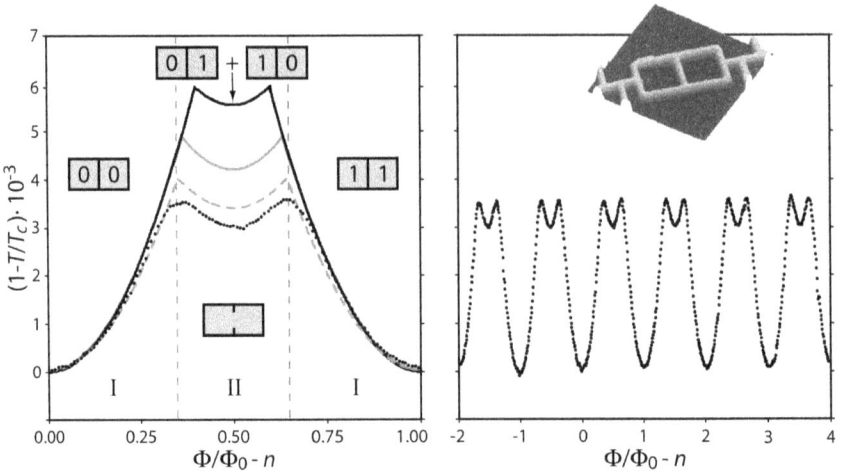

Fig. 3.3   The analogue to Fig. 3.2 for the case of a double loop : experimental $T_c(\Phi)$ data (dots) and results of the different theoretical models. Gray solid and gray dashed lines correspond to the de Gennes–Alexander approach without and with inclusion of the contacts, respectively, whereas the black line is obtained in the London limit (after [345]).

or $\Phi/\Phi_0 > (1 - g)$ and flux regime II for $g < \Phi/\Phi_0 < (1 - g)$.[1] In flux regime I, the phase boundaries predicted by the different models are nearly identical. However, near $\Phi/\Phi_0 = 1/2$ (flux regime II), clear differences are found between the London limit and the de Gennes–Alexander approach. The latter fits better the experimental data with respect to the crossover point g, and the amplitude of the T_c oscillations. With this approach, the spatial modulation of $|\psi|$ and the supercurrents for different values at the $T_c(\Phi)$ boundary has been calculated. In the flux regime I, $|\psi|$ varies only slightly and, therefore, the results of the London limit and the de Gennes–Alexander models nearly coincide. The elementary loops have an equal fluxoid quantum number — and consequently an equal supercurrent orientation — for both the double and the triple loop geometry. This leads in the case of the double loop to a cancelation of the supercurrent in the middle strip, while for the triple loop, the fluxoid quantization condition of Eq. (2.4) results in a different value for the supercurrent in the inner and the outer loops. As a result, the common strips of the triple loop structure carry a finite current.

In the flux regime II, qualitatively different states are obtained from the London limit and the de Gennes–Alexander approach: the states calculated

---

[1] Here, the cross-over point g between regimes I and II is introduced.

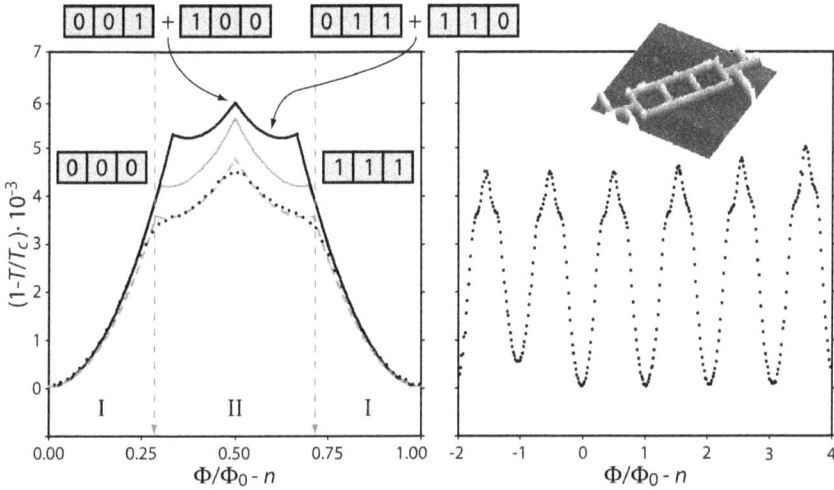

Fig. 3.4   The analogue to Figs. 3.2 and 3.3 for the case of a triple loop: experimental $T_c(\Phi)$ data (dots) and results of the different theoretical models. Gray solid and gray dashed lines correspond to the de Gennes–Alexander approach without and with inclusion of the contacts, respectively, whereas the black line is obtained in the London limit (after [345]).

within the latter have a strongly modulated $|\psi|$ along the strips. This is clearly seen for the double loop, where ψ shows a node ($|\psi| = 0$) in the center of the common strip, the phase φ having a discontinuity of π at this point. This node is a one-dimensional analog of the core of an Abrikosov vortex, where the order parameter also vanishes and the phase shows a discontinuity. In Fig. 3.5 the spatial variation of $|\psi|$ along the strips is shown for $\Phi/\Phi_0 = 0.36$ close to the crossover point g. The dashed curve gives $|\psi|$ in flux regime I, which is quasi-constant. The strongly modulated solution, which goes through zero in the center, is indicated by the solid line. Although there exists a finite phase difference across the junction points of the middle strip, no supercurrent can flow through the strip due to the presence of the node. This node is predicted to persist when moving below the phase boundary into the superconducting state [8, 87].

Already in 1964, Parks [379] anticipated that in a double loop "...*a part of the middle link will revert to the normal phase*", and that "...*this in effect will convert the double loop to a single loop*" (see Fig. 3.6), giving an intuitive explanation for the maximum in $T_c(\Phi)$ at $\Phi/\Phi_0 = 1/2$. Such modulation of $|\psi|$ is obviously excluded in the London limit, where the loop currents have an opposite orientation and add up in the central strip, thus

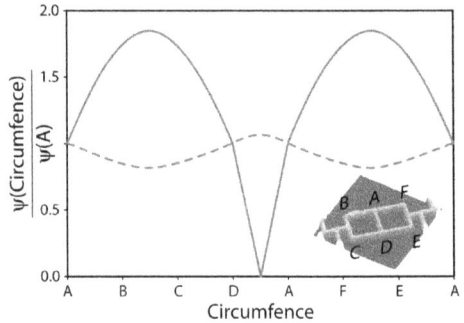

Fig. 3.5 Calculated variation of $|\psi|$ along the circumference of the double loop at the phase boundary for $\Phi/\Phi_0 = 0.36$. The dashed line is the solution with spatially nearly constant $|\psi|$, while the solid line shows the state with a node in the center of the strip that connects the points A and D (after [72]).

giving rise to a rather high kinetic energy. An extra argument in favor of the presence of the node is given by the much better agreement for the crossover point g when the presence of the leads is taken into account in the calculations (see the bright dashed line in Fig. 3.3).

For a triple loop (Fig. 3.4) the modulation of $|\psi|$ is still considerable in flux regime II, but it does not show any nodes. Therefore the orientations of the supercurrent can be found from the fluxoid quantum numbers $\{n_i\}$, obtained from integration of the phase gradients along each individual loop. When passing through the crossover point between flux regime I and II, only the supercurrent in the middle loop is reversed, while an increase of the flux above $\Phi/\Phi_0 = 1/2$ implies the reversal of the supercurrent in all loops.

Surprisingly, the behavior of a microladder with a linear arrangement of N loops appears to be qualitatively different for even and for odd N in the sense that N determines the presence or absence of nodes in the common strips. For an infinitely long microladder, $|\psi|$ was found to be spatially constant below a certain value $\Phi < \Phi_c$ [441], which is analogous to the states in flux regime I. For fluxes $\Phi > \Phi_c$, modulated states of $|\psi|$ with an incommensurate fluxoid pattern were found. At $\Phi/\Phi_0 = 1/2$, nodes appear at the center of every second common (transverse) branch.

A variety of other micronet structures (coupled rings, bolas, a yin-yang, infinite microladders, bridge circuits, like a Wheatstone bridge, wires with dangling branches, etc...), formed by one-dimensional wires, have been analyzed in a series of publications [8, 87, 119, 146–148, 150–152, 208, 277–

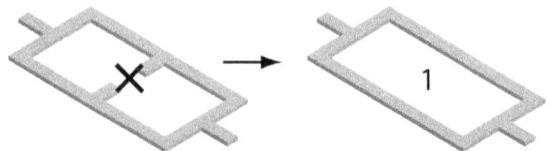

Fig. 3.6 Schematics of a double loop that is converted to a single loop when parts of its middle link are reverted to the normal phase.

279, 441, 442] using the approach pioneered in 1981 by de Gennes [119] and further developed by Alexander [7] and Fink *et al.* [152]. For all these structures, very pronounced effects of the topology on $T_c(\Phi)$ and critical current have been found.

3.2 Two-dimensional clusters of antidots

As a two-dimensional intermediate structure between individual cells 'A' and their huge arrays (see Fig. 1.4), we shall consider a superconducting microsquare with a 2×2 antidot cluster [390, 391]. In this specific case, the symbol 'A' from Fig. 1.3 indicates an antidot. The considered structure consists of a 2×2 μm² superconducting square, made out of 50 nm lead and 17 nm copper² [391] with four square antidots of 0.53×0.53 μm² (see the inset in Fig. 3.7). The Pb/Cu-bilayer behaves as a type-2 superconductor with a $T_c = 6.05$ K, a coherence length $\xi(0) \approx 35$ nm and a dirty limit penetration depth $\lambda(0) \approx 76$ nm.

The $T_c(H)$ phase boundary of a reference sample without antidots [390] revealed characteristic features originating from the confinement of the superconducting condensate by the square geometry (see Sec. 2.3). Therefore, the additional features observed in the $T_c(H)$ phase boundary of the antidot

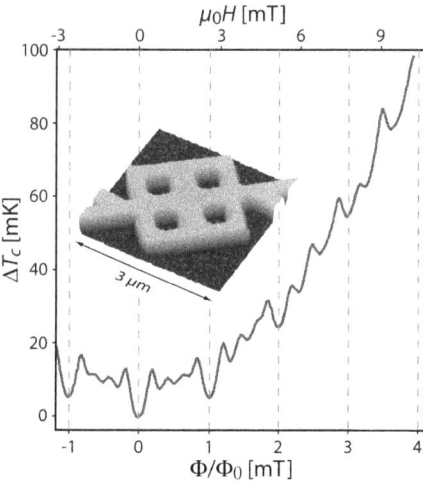

Fig. 3.7 Experimental phase boundary $\Delta T_c(H)$ of the Pb/Cu 2×2 antidot cluster, measured by keeping the resistance of the sample at 10% of its normal state value while varying the applied magnetic field and temperature. The inset shows an atomic force micrograph of the Pb/Cu 2×2 antidot cluster (after [390]).

---

²The thin copper layer was deposited on top of the lead to protect the latter from oxidation and to provide a good contact-layer for wire-bonding to the experimental apparatus.

cluster (Fig. 3.7) can be directly attributed to the presence of the antidots. Strong oscillations are observed with a periodicity of 2.6 mT and in each of these periods, smaller dips appear at approximately 0.75 mT, 1.3 mT and 1.8 mT. As before, the parabolic background, superimposed on the $T_c(H)$ curve, can be described by Eq. (2.1).

Defining a flux quantum per antidot as $\Phi_0 = h/2e = BS$, with $B = \mu_0 H$ and an effective area $S = 0.8\,\mu m^2$ per antidot cell, the minima in the magnetoresistance and the $T_c(H)$ phase boundary at integer multiples of 2.6 mT can be correlated with a magnetic flux quantum per antidot cell: $\Phi = n\Phi_0$. Note that with this definition, the minima observed at 0.75 mT, 1.3 mT and 1.8 mT correspond to the values $\Phi/\Phi_0 = 0.3$, 0.5 and 0.7.

The solutions obtained from the London model define a phase boundary which is periodic in Φ with a periodicity of Φ_0. Within each parabola $\Delta T_c = \gamma(\Phi/\Phi_0)^2$, where the coefficient γ characterizes the effective flux penetration through the unit cell. The value of γ is determined by the combination of λ and the effective size of the current loops. In Fig. 3.8, the first period of this phase boundary, $\Delta T_c(\Phi) = T_{c0} - T_c(\Phi)$ is shown

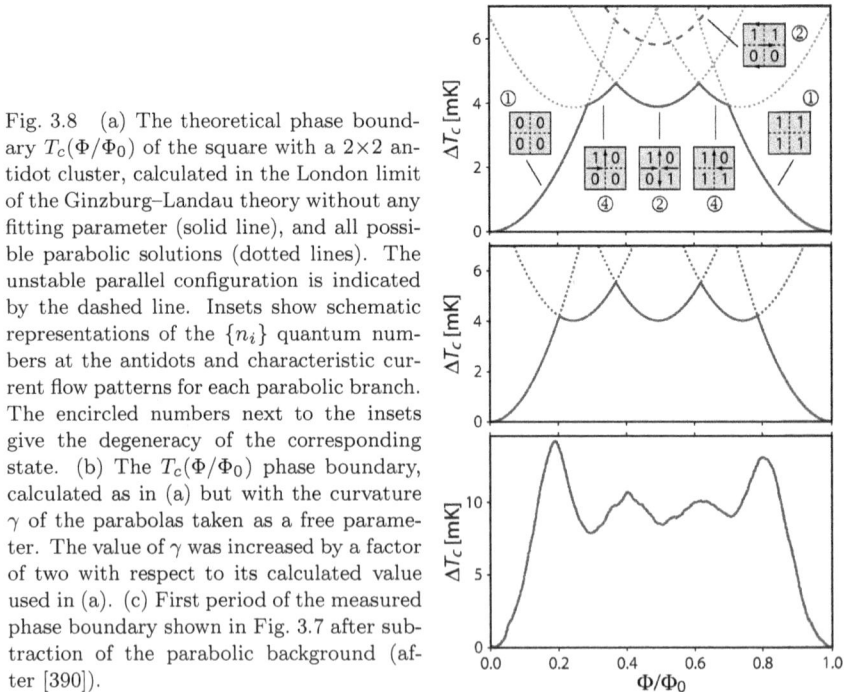

Fig. 3.8 (a) The theoretical phase boundary $T_c(\Phi/\Phi_0)$ of the square with a 2×2 antidot cluster, calculated in the London limit of the Ginzburg–Landau theory without any fitting parameter (solid line), and all possible parabolic solutions (dotted lines). The unstable parallel configuration is indicated by the dashed line. Insets show schematic representations of the $\{n_i\}$ quantum numbers at the antidots and characteristic current flow patterns for each parabolic branch. The encircled numbers next to the insets give the degeneracy of the corresponding state. (b) The $T_c(\Phi/\Phi_0)$ phase boundary, calculated as in (a) but with the curvature γ of the parabolas taken as a free parameter. The value of γ was increased by a factor of two with respect to its calculated value used in (a). (c) First period of the measured phase boundary shown in Fig. 3.7 after subtraction of the parabolic background (after [390]).

as a function of Φ/Φ_0. There are six parabolic solutions that are given by a different set of flux quantum numbers $\{n_i\}$, each one defining a specific vortex configuration. In Fig. 3.8(a), this is indicated by the numbers shown inside the schematic drawings of the antidot cluster. Note that some vortex configurations are degenerate.

From all these possible solutions, for each particular value of Φ/Φ_0, only the branch with a minimum value of $\Delta T_c(\Phi)$ is stable (see the solid line in Fig. 3.8(a)). For the phase boundary, calculated within the one-dimensional model of four equivalent and properly attached squares, no fitting parameters were used since the variation of $T_c(\Phi)$ was calculated from the known values of ξ and the known size of the structure. One period of the phase boundary of the antidot cluster is composed of five branches, and in each branch a different stable vortex configuration is permitted. For the middle branch ($0.37< \Phi/\Phi_0 <0.63$), the stable configuration is the *diagonal* vortex configuration (antidots on the diagonal have equal n_i) instead of the *parallel* state (dashed line in Fig. 3.8(a)).

The net density distribution of the supercurrent, circulated in the antidot cluster for different values of Φ/Φ_0, has been determined using the same approach. Circular currents flow around each antidot. For the states $n_i = 0$ and $n_i = 1$, the flow is in opposite direction, since currents that correspond to $n_i = 0$ must screen the flux to fulfill the fluxoid quantization condition of Eq. (2.4), whereas for $n_i = 1$ they have to generate additional flux. At low values of Φ/Φ_0, currents are canceled in the internal strips and screening currents only flow around the cluster. When the field range corresponding to the second branch of the phase boundary is entered, a vortex ($n_i = 1$) is pinned around one antidot of the cluster (see Fig. 3.8(a)). At the third branch, a second vortex enters the structure and is localized in the diagonal. In the fourth branch of the phase boundary, a third vortex is pinned in the antidot cluster. And finally, the current distribution for the fifth branch is similar to that of the first branch although currents flow in opposite direction [391].

Figure 3.8(c) shows the first period of the measured phase boundary $T_c(\Phi)$ after subtraction of the parabolic background. If we compare it with the theoretical prediction given in Figure 3.8(a), the overall shape can be reproduced although, the experimental plot has two major peaks at $\Phi/\Phi_0 = 0.2$ and 0.8, whereas the theoretical curve only predicts much less pronounced cusps around these positions.

The agreement between the measured and the calculated $T_c(\Phi)$ is improved if we assume that the coefficient γ can be considered as a fitting

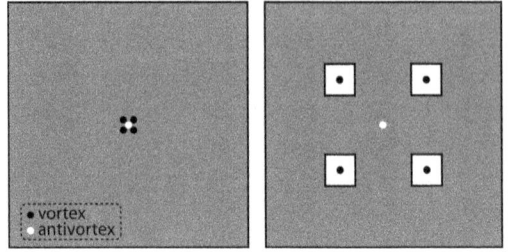

Fig. 3.9 Vortex-antivortex pattern of 4 vortices and 1 antivortex in a superconducting square (left), which is expanded (right) due to the presence of a 2×2 antidot cluster (after [173]).

parameter. This seems to be feasible if we take into account the simplicity and limitation of the used one-dimensional model. Due to the relatively large width of the strips forming the 2×2 cluster, the sizes of the current loops can change since they are rather 'soft' than precisely defined. As a result, the coefficient γ can not be treated as a known constant. If we use it as a free parameter as it was done in Fig. 3.8(b), the curvature of all parabolas forming $T_c(H)$ can be changed and the calculated $T_c(H)$ curve gets closer to the experimental one, although the amplitude of the maxima at $\Phi/\Phi_0 = 0.2$ and 0.8 is still considerably lower compared to the experiment (Fig. 3.8(c)). The discrepancy in the amplitude of the maxima at $\Phi/\Phi_0 = 0.2$ and 0.8 could also be related to the pinning of vortices by the antidot cluster, when potential barriers between different vortex configurations may appear. At the same time, the achieved agreement between the positions of the measured and calculated minima of the $T_c(H)$ curves confirms that the observed effects are due to fluxoid quantization and formation of certain stable vortex configurations at the antidots [390, 391].

An extrapolation of the results obtained from small to larger two-dimensional antidot clusters, such as 3×3, 4×4, etc., gives an idea about possible vortex configurations, which can be expected in superconductors with huge regular arrays of antidots (antidot lattices).

The 2×2 antidot clusters, studied experimentally [390, 391], were later on analyzed theoretically in the framework of the Ginzburg–Landau model [173]. The symmetry-induced vortex-antivortex pattern of 4 vortices + 1 antivortex ('4v+1av') in the square can be strongly expanded by placing four antidots around the center of the square (see Fig. 3.9). The expansion is realized by attracting the four vortices away from the center along the diagonals, until these vortices are pinned by the four antidots. With a 2×2 antidot cluster present, it is possible to substantially enlarge the '4v+1av' pattern, which is otherwise very compact in a reference square without antidots.

3.3 Magnetically coupled loops

In the previous two sections, we discussed clusters of individual nanocells 'A' (loops an antidots) that are *electrically connected*. When *electrically isolated*, a magnetic coupling between two superconductors is still possible: depending on the external field, the local field can be expelled or compressed by one of the two superconductors, and thus alter the total flux through the second superconductor. It is due to this magnetic coupling that two electrically isolated superconductors can interact with each other.

In this section, we shall study the coupling between two superconducting concentric loops by measuring the phase boundary of the outer loop, detecting changes in its $T_c(H)$ line that arise from the electrically isolated inner loop.

Recently, Zhu *et al.* [519] studied the flux state in two magnetically coupled mesoscopic *normal* loops, using a phenomenological model. They addressed the possibility of a ground state with spontaneous magnetic flux trapped in the system, as well as the analogy between flux states in normal metal rings and their superconducting counterparts. The magnetic coupling of an *array* of normal rings has been investigated theoretically by Wang and Ma [496], who calculated the persistent currents for the array with mutual inductance between the nearest neighbored rings. They could show that, while the self-inductance suppresses persistent currents, the effect of the mutual inductance is to enhance it. For a pair of rings, they found that the contribution of the persistent current due to the mutual inductance was about 10%.

Using ultrasensitive susceptibility techniques and scanning Hall probe microscopy, Davidović *et al.* [115,116] studied arrays of electrically isolated superconducting mesoscopic rings. When these rings are biased in an external flux of $1/2\Phi_0$, they can be in either of two energetically degenerated fluxoid states. In one of these two state, supercurrents flow in a clockwise direction with a resulting downwards magnetic moment, while the currents in the other state flow counterclockwise, resulting in an opposite magnetic moment. Produced by the supercurrents in these rings, the magnetic moments are thus analogous to Ising spins and neighboring rings interact antiferromagnetically via their dipolar magnetic fields. It was also shown that the dynamics of the rings is dominated by an energy barrier between the two states, which rises rapidly as the temperature is lowered below T_c.

Using the nonlinear Ginzburg–Landau theory, Baelus *et al.* [27] calculated the magnetic coupling between two concentric mesoscopic supercon-

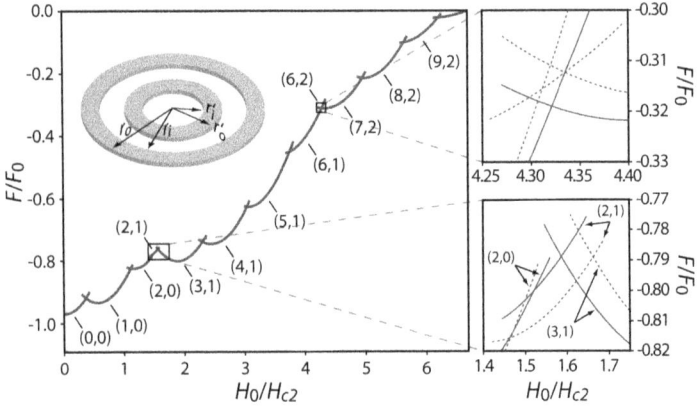

Fig. 3.10   The ground-state free energy of two superconducting ($\kappa = 0.28$) concentric rings as a function of applied magnetic field. The thickness of the structures is $\tau = 1.0\,\xi$, while internal and external radii of the outer and inner ring are $r_{i/o} = 1.5\,\xi/2.0\,\xi$ and $r'_{i/o} = 0.6\,\xi/1.1\,\xi$, respectively. The panels at the right show two of the crossings in more detail, where the dashed curves correspond to the sum of the free energies of the two rings. The notation (L_o, L_i) is used to characterize the different states. (Adapted with permission from Ref. 27. Copyrighted by the American Physical Society.)

ductors. By putting a smaller superconducting ring or disk in the center of a larger ring, they found extra transitions of the ground state where the total vorticity stays the same, but the vorticity of the inner superconductor changes by one (see Fig. 3.10 for the calculated free energy in the case of two concentric rings). Due to the magnetic coupling, the current in the external ring exhibits extra jumps at the transition fields where the vorticity of the inner superconductor changes. For certain temperatures, also a re-entrant behavior and a switching from 'on' to 'off' of the superconducting behavior of the rings were found as a function of the magnetic field.

The Ginzburg–Landau free energy of two concentric superconducting loops can be written as:

$$F_s = F_n + V_i \left(\alpha \left| \psi_i \right|^2 + \beta \left| \psi_i \right|^4 + \frac{m^\star v_i^2}{2} \left| \psi_i \right|^2 \right) + \ldots$$

$$+ V_o \left(\alpha \left| \psi_o \right|^2 + \beta \left| \psi_o \right|^4 + \frac{m^\star v_o^2}{2} \left| \psi_o \right|^2 \right) + \ldots \qquad (3.2)$$

$$+ L_{s,i} I_i^2 + L_{s,o} I_o^2 + M I_i I_o ,$$

with L_s and M the self- and mutual inductance and V the volume of the loop. The indexes i and o refer to the inner and outer loop, respectively.

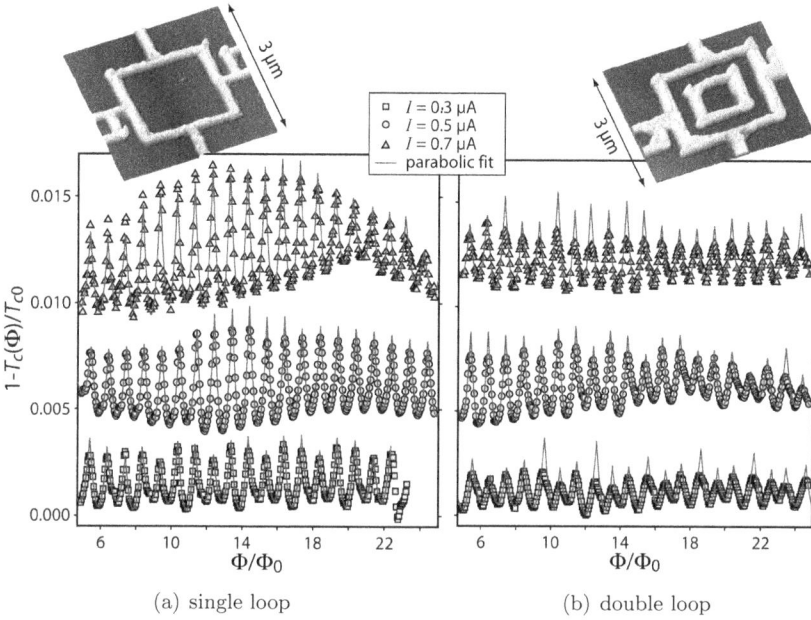

(a) single loop (b) double loop

Fig. 3.11 Measured phase boundary for (a) a single loop and (b) a double concentric loop as a function of the normalized flux $\Phi = S_o \mu_0 H$ threading the outer loop, after substraction of a parabolic background, caused by the finite width of the stripes. Within each period of the oscillations, i.e. between transitions to different vorticities L_o, data points are fitted with a parabola. Insets show atomic force micrographs of the studied structures (after [334]).

The superfluid velocities of the two loops are determined from the fluxoid quantization constraint

$$v_i = \frac{\hbar}{m^\star r_i}\left(L_i - \frac{\Phi + I_o M}{\Phi_0}\right) \tag{3.3a}$$

$$v_o = \frac{\hbar}{m^\star r_o}\left(L_o - \frac{\Phi + I_i M}{\Phi_0}\right) . \tag{3.3b}$$

In order to solve Eq.(3.2), the free energy must be minimized with respect to variations in ψ_i, ψ_o, I_i and I_o. The free energy F_s contains in this case the term $MI_iI_o \propto M|\psi_i|^2|\psi_o|^2$ responsible for the mixing of the two individual order parameters ψ_i and ψ_o.

The measured phase boundaries of two square loops without and with a second smaller loop enclosed are shown in Fig. 3.11(a) and (b), respectively, where the different curves in each panel correspond to different values of the applied transport currents I (see Tab. 3.1 for some characteristics of

Fig. 3.12 Calculated phase boundary $T_c(H)$ of the inner (solid line) and outer loop (dashed, dashed-dotted and dotted lines) for two periods of the inner loop, without magnetic interaction, using the dimensions of the measured structures and a coherence length of $\xi(0) = 103$. The phase boundary of the outer loop is shown for three different transport currents $I = 0\,\mu A$ (dashed line), $I = 0.3\,\mu A$ (dashed-dotted line), and $I = 0.9\,\mu A$ (dotted line). The shift of the $T_c(H)$ curves with increasing I are estimated from Refs. [407,450] with $I_c = I_{c0}(T_{c0} - T_c)^{3/2}$, where $I_{c0} \approx 550\,\mu A$ (after [334]).

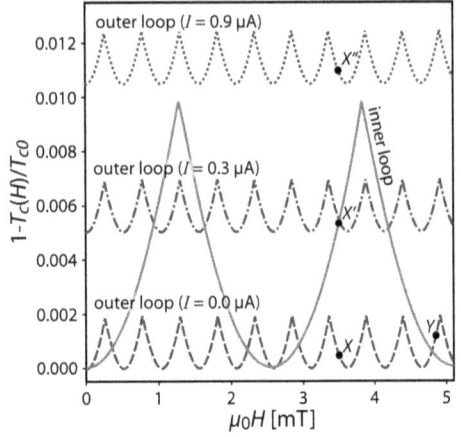

the studied structures). A clear difference in the $T_c(H)$ curves can be seen, since the amplitude of the oscillations is larger in the case of the single loop, and the cusps in $T_c(H)$ obtained from the double loop are rounded[3].

In order to analyze the different states of the inner and outer loops, Fig. 3.12 compares the theoretical phase boundaries of two concentric square loops which are not magnetically coupled but have different side lengths $a_i = 0.9\,\mu m$ and $a_o = 2.0\,\mu m$, thus showing an accordingly larger amplitude of the Little–Parks oscillations in the case of the smaller loop. The period of the $T_c(H)$ oscillations for the inner and outer loop is given by $\mu_0 \Delta H_i = \Phi_0/S_i$ and $\mu_0 \Delta H_o = \Phi_0/S_o$, respectively, with the enclosed areas $S_i = a_i^2$ and $S_o = a_o^2$. Let us consider points X and Y in Fig. 3.12, which are situated at two certain applied magnetic field values on the T–H phase boundary of the outer loop. In the case of point X, the inner loop is

Table 3.1   Characteristics of the studied square loops

Property	Double Square Loop	Single Square Loop
thickness τ [nm]	50	50
stripe width w [nm]	180	180
size of outer loop a_o [µm]	2	2
size of inner loop a_i [µm]	0.9	—
T_c [K]	1.326	1.324
$\xi(0)$ [nm]	105	103

---

[3]Note that according to Fomin *et al.* [154], a possible explanation for the rounding of the cusps is the presence of scattering imperfections in the outer loop.

in the normal state, i.e. it carries no supercurrent, and no flux is coupled to the outer loop. For point Y, on the contrary, the inner loop is in the superconducting state and, accordingly, a supercurrent will flow in both loops in order to satisfy the respective fluxoid quantization constrains of Eq.(2.4). Under these conditions, due to the mutual inductance between the two current loops given by Eqs.(3.2) and (3.3), an influence of the fluxoid quantization in the inner loop on the measured $T_c(H)$ phase boundary of the outer loop is expected.

To extend the flux interval in which the inner loop remains superconducting, a higher transport current can be applied to the outer loop (see the phase boundary $T_c(\Phi)$ of the outer loop, schematically presented in Fig. 3.12 by dashed dotted and dotted lines for $I = 0.3\,\mu A$ and $I = 0.9\,\mu A$, respectively). Following in Fig. 3.12 the line $X \rightarrow X''$ by increasing I, the phase boundaries of the two loops coincide at the point X', where $I = 0.3\,\mu A$. For even higher transport currents such as $I = 0.9\,\mu A$, the inner loop will be superconducting for each point of the phase boundary of the outer loop (see the dotted line in Fig. 3.12). Therefore, an increase of I will not only broaden the interval in which the inner loop is superconducting, but also increase the supercurrent in the inner loop.

Zhang *et al.* [511] have calculated the self-flux of a typical mesoscopic ring. Using their approach, one can show that in the present case at $T/T_c > 0.99$, the self-flux for the inner and outer loop is smaller than 0.4% of the applied flux, and that the additional flux MI_i in the outer loop is less than 4% of the applied flux. However, at lower temperatures of $T/T_c < 0.95$, MI_i can even exceed 20% of the applied flux. The present experimental results were obtained at $T/T_c > 0.99$ and, accordingly, the influence of the inner loop on the measured phase boundary is quite small (see Fig. 3.11).

The modification of the $T_c(\Phi)$ oscillations of the outer loop, due to the coupling between the outer and the inner loops, is still seen in the Fourier spectrum of the $T_c(\Phi)$ line, which is shown in Fig. 3.13. The analysis based on this coupling explains the extra peaks in the Fourier spectrum by the magnetic coupling of the two loops [334]. The systematic shift of the $T_c(\Phi)$ phase boundary of the outer loop with the applied current I induces a well defined evolution of the Fourier spectrum which was indeed found in our experiments. This evolution of the extra peaks in the Fourier spectrum with the applied current gives an experimental evidence for the presence of the magnetic interaction between the two superconducting loops [27].

As suggested in Refs. 27 and 336, direct magnetization measurements deeper in the superconducting state could be helpful to reveal an enhanced

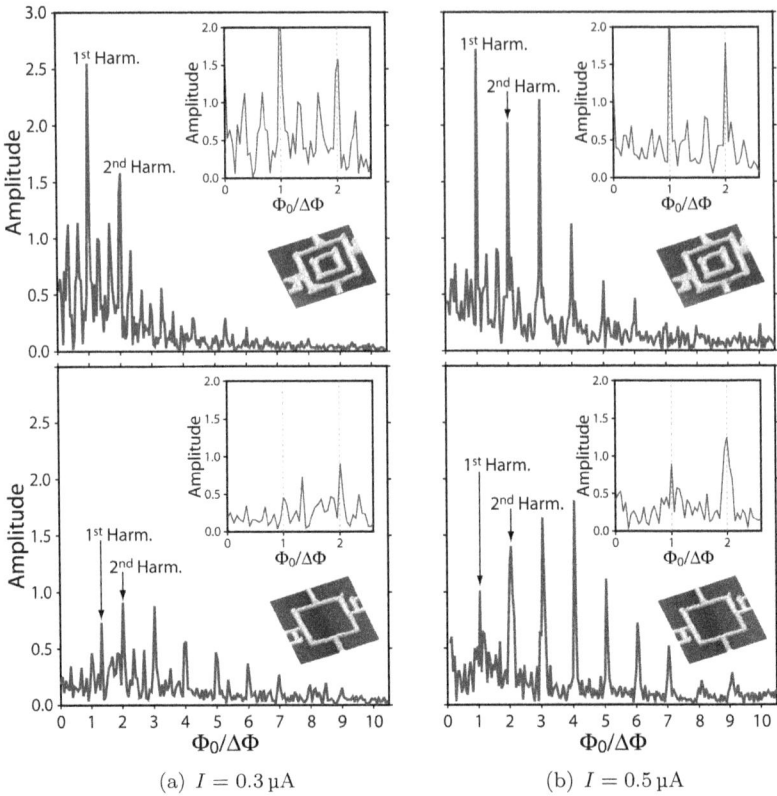

Fig. 3.13   Fourier transform of the phase boundaries shown in Fig. 3.11, after subtraction of the fitted parabolas within each period of oscillation, for a transport current of (a) 0.3 μA and (b) 0.5 μA. Upper and lower panels correspond to the double and single loop, respectively. Insets show the low frequency regions in more detail (after [334]).

magnetic coupling of both loops at lower temperatures. An inner loop made from a different superconductor with a higher critical temperature would certainly increase the magnetic coupling between the two loops. An enhanced critical field is expected in this case for the loop with lower T_c, at the expense of sharing the fluxoid quantization 'burden' with a loop where superconductivity is stronger.

Chapter 4

Laterally Nanostructured Superconductors

The periodic repetition of nanoscopic cells 'A' over a macroscopic area makes it possible to implement the idea of an artificial lateral modulation in nanostructured superconductors. Several different types of elementary cells 'A' have been used for this, such as antidots, where a film is completely perforated by microholes [29,30,213,320,342,343,347,412], blind holes that do not perforate a film completely but rather modulate its thickness at the sites of the blind holes [56,58], or magnetic [303,471], normal metallic [471] and insulating dots [471] covered by — or grown on top — of a superconducting film. These huge regular arrays of the artificially introduced nanoscopic cells were used for systematic studies of the confinement and quantization phenomena in the presence of a two-dimensional artificial periodic pinning potential. We begin in this chapter from the effect of lateral nanostructuring on the $T_c(H)$ phase boundary and then move on to the pinning phenomena, focusing on commensurability effects between the flux line lattice and a periodic pinning array.

4.1 The $T_c(H)$ phase boundary of superconducting films with an antidot lattice

Knowing the flux confinement by an individual nanoscopic cell 'A' — in this case a unit cell containing one antidot — the fundamental question arises to what extent this approach can be applied when composing huge arrays of the order of several mm^2 of such cells 'A', like for example an antidot lattice. This problem is of crucial importance for nanostructured thin films, since repeating laterally a certain nanoscopic cell over a macroscopic area creates an artificial bulk superconductor with improved properties that can be used for numerous applications.

For bulk singly connected samples, the surface nucleation of superconductivity plays only a minor role because the thickness of the surface superconductivity sheath, given by the temperature dependent coherence length
$\xi(T)$, is negligibly small in comparison with the sample size, as it is illustrated in Fig. 1.6(a). The 'bulk' of the sample makes the transition to the
superconducting state when the temperature is decreased below $H_{c2}(T)$
(for a type-2 superconductor). However, in superconducting films with an
antidot array, additional superconductor–vacuum interfaces are introduced
that increase not only the surface to volume ratio, but also the connectivity of the sample. According to a conventional scenario, a 'ring' of surface
superconductivity forms at H_{c3}^* around each antidot[1] and, if the antidots
are sufficiently closely spaced, almost the entire sample becomes superconducting at a field well above H_{c2}, see Fig. 1.6(b). Therefore, in this case,
the nucleation field H_{c3}^* is expected to play the role of the bulk critical
field. Accordingly, it is possible to enhance the critical field above the bulk
value $H_{c2}(T)$ up to H_{c3}^* in laterally nanostructured films in a perpendicular
magnetic field. At first sight it seems impossible to achieve an enhancement
factor $\eta = H_{c3}^*/H_{c2}$ higher than 1.69, but as we show below, even higher
values for η of up to 3.6 can be reached in films with dense antidot lattices.

We shall analyze the $T_c(H)$ behavior in thin lead films with a square
antidot lattice, made by electron-beam lithography on a photoresist layer
and standard lift-off processing [29]. The corresponding AFM pictures of
these samples show that the antidots have the form of squares with rounded
corners (see the inserts of Fig. 4.1). Both films consist of a bi-layer of 50 nm
lead and 20 nm protective germanium, and were laterally nanopatterned
with a square antidot lattice. The studied samples had the same antidot
size $a = 0.4\,\mu m$, but differed in the periodicity d of their antidot lattice,
which was chosen as $d = 1.0\,\mu m$ and $d = 2.0\,\mu m$ (see the inserts of Fig. 4.1(a)
and (b), respectively). Therefore, the influence of the antidot spacing on
the surface superconductivity could be analyzed by comparing the $H_{c3}^*(T)$
curves of the films with a dense and sparse antidot lattice.

The residual resistivity at 8 K of a reference lead film gives the mean
free path $\ell = 26\,nm$. The upper critical field $H_{c2}(T)$ of this film shows the
expected linear variation with temperature according to Eq. (1.11), which
fits to a value $\xi(0) = 38\,nm$. This is quite close to the estimated dirty
limit coherence length $\xi(0) = 0.865(\xi_0\ell)^{1/2} \approx 40\,nm$, calculated from the

---

[1] In order to distinguish between $H_{c3} = 1.69\,H_{c2}$ for a plane interface and the nucleation
field of the specific sample geometries discussed in this chapter, the notation H_{c3}^* will be
used for the latter.

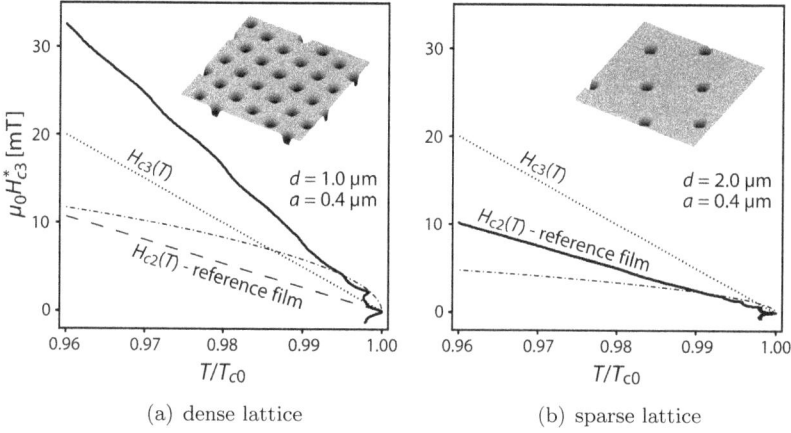

(a) dense lattice   (b) sparse lattice

Fig. 4.1   The measured critical fields $H^*_{c3}(T)$ in a perpendicular magnetic field for lead films with two different square antidot lattice of period d and antidot size a. Dotted lines represent the expected third critical field $H_{c3}(T)$ for a plane superconductor–vacuum boundary with the magnetic field along the boundary ($\eta = H^*_{c3}(T)/H_{c2}(T) = 1.69$). The upper critical field of a reference film without antidots is shown as dashed lines. Dash-dotted curves indicate the square root-like behavior, expected for a wire network with line thickness $w = d - a$. Inserts: AFM images of the lead films with a square antidot lattice with an antidot size $a = 0.4\,\mu m$, and a lattice period d of (a) $1\,\mu m$ and (b) $2\,\mu m$ (after [340]).

known BCS coherence length $\xi_0 = 83$ nm of lead [118]. The superconducting transition temperatures T_{c0} for all three samples range between $7.164\,K$ and $7.186\,K$, meaning that nanostructuring has changed their T_{c0} only slightly compared to $T_{c0} = 7.2\,K$ of bulk lead. The penetration depth $\lambda(0) \approx 46$ nm is calculated for the films from $\lambda(0) = 0.64\lambda_L(0)\sqrt{\xi_0/\ell}$, with $\lambda_L(0) = 37$ nm [118]. The Ginzburg–Landau parameter $\kappa = \lambda/\xi$ equals 1.2, which implies that the studied lead films are type-2 superconductors.

The critical fields H^*_{c3} are shown by the solid lines in Fig. 4.1. The H^*_{c3} values were determined by measuring the resistive transition as a function of temperature at a fixed magnetic field. The used criterion was set at $10\,\%$ of the normal state resistance at $8\,K$.

At low external fields, the critical field curves show a square-root $H^*_{c3} \propto (1 - T/T_{c0})^{1/2}$ background, corresponding to a parabolic suppression of $T_c(H)$ according to Eq. (1.11) that is superimposed with collective oscillations of a periodicity given by one flux quantum per unit cell of the antidot lattice, $\Phi_0/d^2 = 2.07\,mT$ for $d = 1\,\mu m$ and $0.517\,mT$ for $d = 2\,\mu m$. These low field cusps in the $H^*_{c3}(T)$ curves are reminiscent of

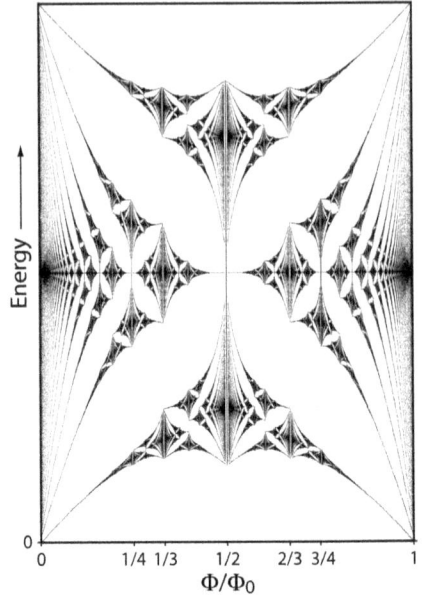

Fig. 4.2 Hofstadter butterfly [220]: The energy spectrum of an electron, restricted to move in a two-dimensional periodic potential under the influence of a perpendicular magnetic field. Φ/Φ_0 is the magnetic flux through the unit cell of the periodic potential, measured in units of the magnetic flux quantum. The lowest energy level corresponds to the $T_c(H)$ phase boundary for a square superconducting network (after [220]).

the $T_c(H)$ phase boundary for superconducting networks, as studied previously [377, 462, 501]. At higher fields, there is a crossover from the square root to a linear H_{c3}^* background. Superimposed with this linear background, single-object-like oscillations [56], reminding the individual antidot oscillatory $T_c(H)$, are observed. Their period is in agreement with approximately one flux quantum per antidot area, Φ_0/a^2. The straight dashed and dotted lines in Fig. 4.1 correspond to $H_{c2}(T)$ and $H_{c3}(T)$, respectively. The dash-dotted line shows the square-root background of H_{c3}^*, which appears due to the finite strip width w (Eq. (2.1)), and is only relevant in the low field regime.

For these superconducting networks, the lowest Landau level of the problem is the lowest level of the Hofstadter butterfly [220] which was nicely demonstrated by the experiments of Pannetier *et al.* [377]. The lowest level of the Hofstadter butterfly, and therefore also the $T_c(H)$ phase boundary of a square superconducting network, has a remarkable structure (see Fig. 4.2). Cusps in $T_c(H)$ show up at specific values of the magnetic field, as a result of commensurability between the vortex lattice and the underlying periodic network. At integer applied flux $\Phi/\Phi_0 = L$ (where Φ is defined per area of a unit cell of the network) the most pronounced cusps appear. At rational flux $\Phi/\Phi_0 = p/q$ smaller cusp-like minima are observed, corresponding to

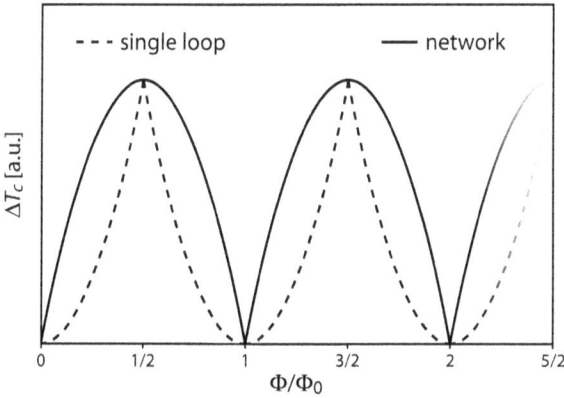

Fig. 4.3 Schematic representation of the $T_c(\Phi)$ phase boundary for a single loop, the Little–Parks oscillatory $T_c(\Phi)$ curve (dashed line), and for a network composed of a large amount of such loops (solid line).

stable vortex configurations, like for example the checker-board pattern at $p/q=1/2$.

The existence of cusp-like minima at integer fluxes can also be interpreted as follows. When placing a superconducting network in a magnetic field that generates a non-integer number of flux quanta per plaquette, two different states $L\Phi_0$ and $(L+1)\Phi_0$ will inevitably be present. Then, in the London limit, for $L < \Phi/\Phi_0 < L+1$, the kinetic energy E, defining the shift of T_c in a magnetic field, becomes

$$E \propto \left(\frac{\Phi}{\Phi_0} - L\right)\left(L+1-\frac{\Phi}{\Phi_0}\right)^2 + \left(L+1-\frac{\Phi}{\Phi_0}\right)\left(L-\frac{\Phi}{\Phi_0}\right)^2$$

$$\propto \left(\frac{\Phi}{\Phi_0} - L\right)\left(L+1-\frac{\Phi}{\Phi_0}\right).$$

(4.1)

Here the factors $\Phi/\Phi_0 - L$ and $L+1-\Phi/\Phi_0$ are the corresponding numbers of $(L+1)\Phi_0$ and $L\Phi_0$ states, respectively. While for a single loop parabolic minima are present in $T_c(H)$ at integer flux $\Phi/\Phi_0 = L$ (see the dashed line in Fig. 4.3), Eq. (4.1) describes the set of intersecting parabolae forming cusps at integer flux for a network. These cusps compose the backbone of the Hofstadter butterfly which adds to the dependence, given by Eq. (4.1), new smaller cusps at rational fields.

It is interesting to consider the evolution of the $T_c(\Phi)$ phase boundary when we increase the size of a two-dimensional network starting from a single nanoscopic cell (a loop) and ending with a huge array composed of such cells (an infinite network), see Fig. 4.4. Here again, it is seen that the parabolic minima at integer flux $\Phi/\Phi_0 = L$ change into cusp-like minima [411]. At the same time, the maximum at $\Phi/\Phi_0 = 1/2$ transforms into a smaller cusp-like minimum when the size of the array is increased.

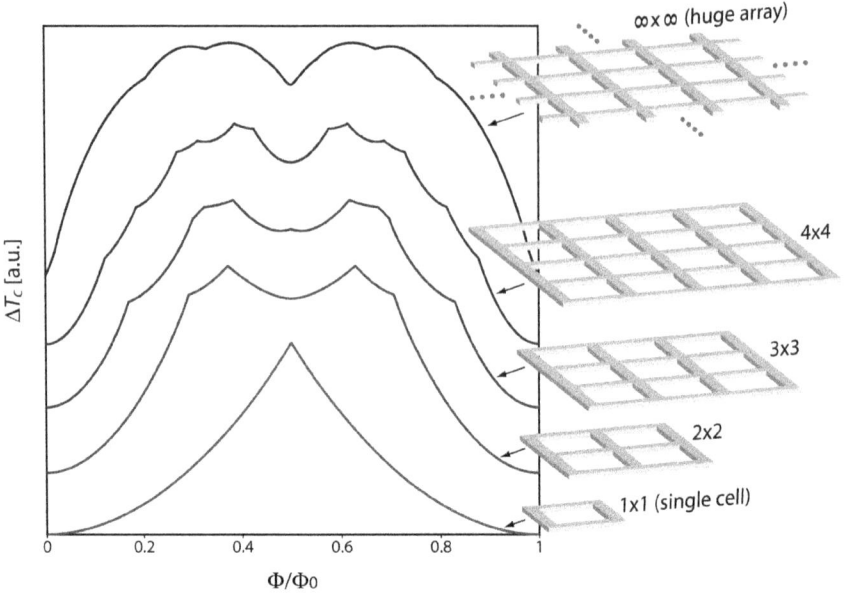

Fig. 4.4 The evolution of the $T_c(H)$ phase boundary (London limit) when increasing the number of nanoscopic cells in a two-dimensional network from one single loop to an infinite array (after [411]).

The parabolic background superimposed with the $T_c(H)$ oscillations, see Fig. 4.1, gives us the possibility to check that, in nanostructured samples with an antidot lattice, the coherence length remains the same as in the reference non-perforated sample. Indeed, from the square-root envelope of the H_{c3}^* curve close to T_c, see the dash-dotted lines in Fig. 4.1, we determine $\xi(0)$ from Eq. (2.1), which is applicable when $w < \xi(T)$ ($w \equiv d - a$ is the width of the strips between the antidots). From this analysis we obtain the same coherence length $\xi(0) = 38 \pm 5$ nm) for both samples. Moreover, the latter coincides with $\xi(0) = 38$ nm determined from the linear slope of the $H_{c2}(T)$ dependence of the reference film. This proves that the parameter $\xi(0)$ was not influenced by nanopatterning. Therefore, the very large values of the enhancement factor (up to $\eta = 3.6$) cannot be attributed just to the reduction of $\xi(0)$ due to the possible presence of defects created by nanostructuring.

At higher fields, a crossover to a different regime takes place where the H_{c3}^* phase boundary resembles the calculated phase boundary for a single antidot. In this regime, the H_{c3}^* curve shows weak oscillations, quasi-

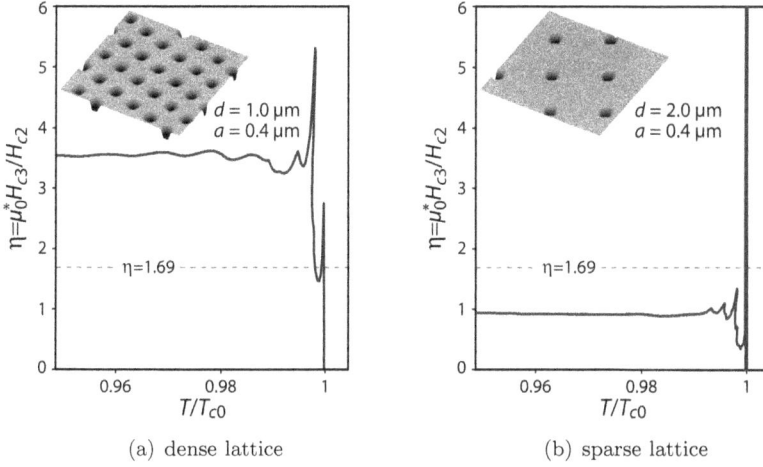

(a) dense lattice (b) sparse lattice

Fig. 4.5   The enhancement factor $\eta = H_{c3}^*(T)/H_{c2}(T)$ as a function of temperature for (a) the dense and (b) the sparse antidot lattice, obtained from the data in Fig. 4.1 after division by $H_{c2}(T)$ of the reference non-perforated Lead film (dashed line in Fig. 4.1). The dashed lines show the value $\eta = 1.69$ (after [340]).

periodic in field and superimposed on a linear background. However, the slope of this linear background appears to be much steeper than expected. In the sample with a dense antidot array, the contribution of the linear background to the H_{c3}^* curve lies significantly higher than the $H_{c2}(T)$ line (Eq. (1.11)) of the reference film (see the dashed lines in Fig. 4.1). By contrast, for the sparse antidot array, the H_{c3}^* curve is very close to $H_{c2}(T)$ of the reference sample. This can be seen more clearly in Fig. 4.5, where the enhancement $\eta = H_{c3}^*(T)/H_{c2}(T)$ is plotted for the two samples with $\{a, b\} = \{1\,\mu m, 0.4\,\mu m\}$ (dense array) and $\{a, b\} = \{2\,\mu m, 0.4\,\mu m\}$ (sparse array). The dashed lines indicate the value $\eta = 1.69$ which, according to the theory, should be the maximum possible enhancement of the nucleation field for a single antidot in a superconducting plain film.

Surprisingly, the enhancement factor η found for the dense antidot array reaches values from 2.8 to 3.6, i.e. up to more than 200% of the expected maximum enhancement. In the sparse antidot array, the usual enhancement $\eta < 1.69$ was retrieved.

In the high field regime, the estimated width of the superconducting ring around two neighboring antidots, i.e. $\xi(T)$, is definitely smaller than the width w of the strips between the antidots. This implies that a simple overlap of these rings cannot be the cause of the transition to the super-

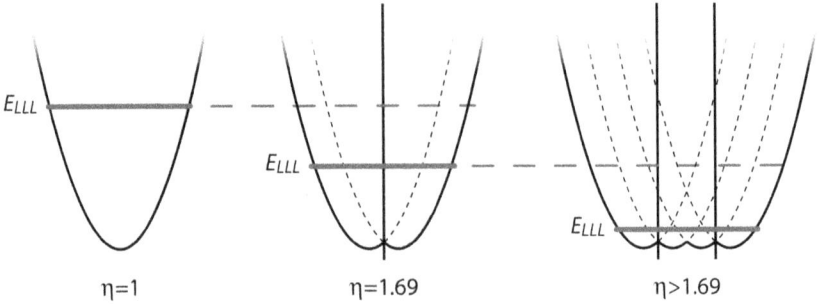

Fig. 4.6   A possible explanation for the enhancement of $\eta = H_{c3}^*(T)/H_{c2}(T)$. The solid horizontal lines represent the lowest Landau levels $E_{LLL} = \eta^{-1}\hbar\omega/2$ for a bulk superconductor with no interfaces (left), a semi-infinite superconducting slab (middle), and a superconducting film with a dense antidot lattice (right). The antidots can introduce many superconductor–vacuum interfaces, resulting in $\eta > 1.69$.

conducting state, seen as a single resistance drop. Moreover, the H_{c3}^* line was measured using a sufficiently low criterion (10% of the normal-state resistance). Therefore, the observed enhancement of $\eta = H_{c3}^*(T)/H_{c2}(T)$ in the high field regime can only be explained by invoking a nucleation of superconductivity in the bulk area between the antidots at much higher fields than expected. For a tentative explanation of the large enhancement of η in that part of the superconducting films with dense antidot arrays, the following scenario might be valid. For a bulk superconductor the lowest Landau level $E_{LLL}(H)$ is the well-known lowest level $\hbar\omega/2$ (i.e. η=1) in a single parabolic potential well (see the left panel of Fig. 4.6). Adding one plane superconductor–vacuum interface creates a mirror image potential (middle panel of Fig. 4.6), thus resulting in a lower level 0.59 $\hbar\omega/2$ (i.e. $\eta^{-1} = 1/1.69 = 0.59$) in a double potential. It is reasonable to expect then a further reduction of η^{-1} if the potential is formed by many overlapping parabolic potentials (see the right panel of Fig. 4.6). The latter might explain the large reduction of η^{-1} and consequently a large increase of η in superconducting films with an antidot lattice, since the antidots introduce many superconductor–vacuum interfaces needed to form the potential shown schematically at the right of Fig. 4.6. To confirm these tentative explanations, however, calculation of the lowest Landau level $E_{LLL}(H)$ is needed for dense antidot lattices, including the presence of broad superconducting strips between the antidots. Since the $H_{c3}^*(T)$ curve is measured using a criterion of 10% of the normal-state resistance, it is clear that there is a global development of the superconducting state in the area between the antidots at substantially higher fields than expected. In this case, for calculating the flux, the integration area in $\Phi = \int b(r)dS$, with the local magnetic field b, is reduced to the effective area of the antidots. This might explain the crossover from the low field regime, where the whole nanoscopic cell is considered, to the high field regime, where the relevant area is just close to the antidot area. Such a change in the relevant area corresponds to the change in the $T_c(\Phi)$ period from Φ_0/d^2 to Φ_0/a^2.

Since the strips forming a network are too narrow to allow the formation of vortices in the lines (interstitial vortices), the connectivity of a super-conducting network is not changed in the presence of magnetic field. When the strips are getting broader, and the opening in the periodically repeated nanoscopic cell becomes smaller, much more superconducting material is left between the openings, thus realizing a regular pinning array rather than a network. In a regular array of antidots as pinning centers, for each char-acteristic size of antidots a, there is a saturation number $n_S \approx a/\xi(T)$ [462]. This number gives the maximum number of flux quanta which can be trapped by an antidot with size a. If the applied field generates a flux per cell (i.e. per antidot in antidot lattices) which is smaller than $n_S\Phi_0$, then all the vortices are pinned by the antidots themselves, leading to the formation of $\Phi_0, 2\Phi_0, \ldots, n_S\Phi_0$ pinned vortex lattices [343]. In this case the connectivity remains the same, and it is given by the number of anti-dots in the film. If the generated flux per plaquette exceeds $n_S\Phi_0$, then the antidots are saturated and the connectivity is spontaneously increased due to the formation of interstitial Φ_0-vortices [29, 413].

To summarize the experimental observations, a dramatic difference is observed for the enhancement factor $\eta = H_{c3}^*(T)/H_{c2}(T)$ in films with a sparse and a dense antidot lattice. In the latter η exceeds by far (more than 200%) the value 1.69, which is the expected maximum value for a single antidot. This enhancement of η may be related to the reduction of the lowest Landau level in a potential formed by many overlapping parabolas (see Fig. 4.6).

Finally, speaking about the enhancement factor η for other geometries, we should note that in the wedge with small angle α, $\eta \propto \alpha^{-1}$ [155]. This divergent behavior of η formally means that η can be tuned to be large by decreasing the wedge angle α.

4.2   Pinning in laterally nanostructured superconductors

Considering in the previous sections the effect of lateral nanostructuring on critical fields $T_c(H)$, we have demonstrated that this important superconducting critical parameter can be tailored by designing a proper topology to confine the superconducting condensate and flux. This concept has been verified on individual nanostructures, clusters containing a small number of mesoscopic cells, and finally their huge arrays in laterally nanostructured superconductors. Systematic studies of the $T_c(H)$ phase boundaries for superconducting structures of the same material but with different confinement topologies have convincingly demonstrated that $T_c(H)$ is determined not only by the choice of a particular superconducting material, but is also very strongly influenced by varying the applied boundary conditions. Therefore, in nanostructured superconductors, the upper (H_{c2}) and lower (H_{c1}) critical fields are not at all well-defined critical parameters, since through nanostructuring (taking for example superconductors with an antidot lattice) we can strongly increase $H_{c2}(T)$ and simultaneously decrease $H_{c1}(T)$, while keeping the thermodynamic critical field $H_c(T)$ almost unchanged. As a result, we should apply renormalized values of $H_{c2}(T)$ and $H_{c1}(T)$ for each given superconductor if it is subjected to artificial nanostructuring. These two renormalized critical parameters can intentionally be designed by changing the condensate and flux confinement through nanostructuring of the same chosen superconducting material. The superconducting state is suppressed by an applied field (see the $H - T$ plane in Fig. 1.11) as well as by the field generated by the currents running through a superconductor. When the generated field reaches the H_{c2} value, superconductivity is destroyed. This gives the maximum possible current, the depairing current I_c^{GL}, see Eq. (1.43).

In type-2 superconductors, which are most perspective for practical applications, the problem of increasing I_c up to its theoretical limit I_c^{GL}, is closely related to the optimization of the pinning of flux lines. The Ginzburg–Landau approach was used by Priour and Fertig [388] and by Shapiro *et al.* [409] to calculate current carrying states at the matching fields and to estimate the vortex core deformations caused by the presence of the periodic arrays of pinning sites. In this section, we shall focus on the advantages offered for the solution of this problem by lateral nanostructuring of the films.

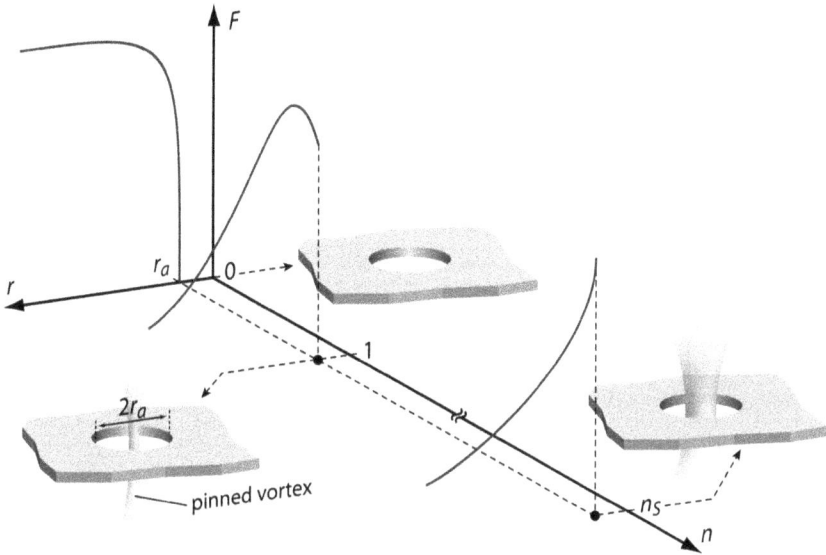

Fig. 4.7   Schematic presentation of the dependence of the free energy F of a vortex on its distance r from the center of a cylindrical antidot with radius r_a and $n_S > 1$, showing the cases with no vortex pinned by the antidot ($n = 0$), with $n = 1$, and with $n = n_S$ flux quanta trapped by the hole.

4.2.1   *Pinning by an antidot or a columnar defect*

It has been shown experimentally and theoretically that a small hole (an antidot) acts as a very efficient pinning center for flux lines. The pinning force of a single cylindrical hole has been calculated by Mkrtchyan and Schmidt in the London approximation (i.e. for high κ) [332]. Buzdin and Feinberg have obtained similar results for a non-superconducting columnar defect, by calculating the electromagnetic pinning interactions using a vortex-antivortex image method [78]. An important conclusion of both studies is that, depending on the radius r_a of the pinning site, more than one flux quantum can be trapped, up to a certain saturation number, n_S, given by:

$$n_S \simeq \frac{r_a}{2\xi(T)} \, . \tag{4.2}$$

The radial distribution of the free energy F of a vortex around the cylindrical antidot with $n_S > 1$ is schematically shown for different values of the number of pinned vortices n in Fig. 4.7. If no vortices are pinned ($n = 0$), the force on a vortex (the gradient of the energy) from an 'empty' antidot is attractive at all distances. As soon as a vortex is trapped ($n = 1$), a potential barrier develops near the edge of the pinning site, which grows as the number n of the trapped flux quanta increases. If n reaches the saturation number n_S, the maximum of the potential barrier reaches the edge of the pinning center and the force on additional vortices becomes repulsive at all distances. We should add that, in case the pinning site is a hole, there exists a Bean–Livingston barrier [42] at the hole edge even if no vortices are trapped in the hole [28]. Moreover, flux trapping is shown to be quite sensitive to the exact boundary condition applied at the boundary of the pinning center [34], so that its efficiency can be tuned by varying parameter b in de Gennes boundary condition, see Eq. (1.9b).

The maximum pinning force per unit length decreases with growing n:

$$f_P^{\max} \simeq \left(\frac{\Phi_0}{4\pi\lambda} \right)^2 \frac{1}{\xi} \left(1 - \frac{n}{n_S} \right) . \qquad (4.3)$$

Note that the possibility of n_S flux quanta being trapped by the pinning site does not necessarily imply that this is energetically the most favorable situation. However, since the pinning force is maximum in the antidot, a vortex will remain pinned by the antidot once it is there. For a triangular lattice of columnar pinning sites, Buzdin has calculated that, after all sites are occupied by Φ_0-vortices, pinning of multiquanta vortices becomes energetically advantageous in case the radius r_a is larger than a critical radius r_c [79]:

$$r_c = \sqrt[3]{\xi(T) a_v^2} , \qquad (4.4)$$

with a_v the distance between the vortices in a perfect triangular lattice. The formation of multiquanta vortices has been confirmed by means of magnetization measurements [343] and has been directly visualized by Bitter decoration of niobium films with circular blind holes [55, 58].

From the above considerations of the saturation number, one expects that relatively small holes or antidots ($n_S = 1$) can trap only one flux quantum. In this case, other flux lines generated by the applied field will be forced to occupy interstitial positions. The presence of interstitial vortices for $n_S = 1$ makes superconductors with an antidot lattice qualitatively different from superconducting networks ($n_S \gg 1$), where the vortex configuration at integer matching fields is always the same as that of the underlying network. In contrast to that, composite vortex lattices (i.e. with vortices at antidots and at interstices) show a remarkable variety of stable patterns quite different from the underlying antidot lattice. Large antidots ($n_S > 1$) can stabilize multiquanta vortex lattices, which do not exist in reference homogeneous superconductors without an antidot lattice. Finally, for antidot lattices with $n_S \gg 1$, the crossover to the regime of superconducting networks takes place.

4.2.2   Regular pinning arrays

When arranging the pinning centers in a regular lattice, matching effects are observed as a consequence of commensurate vortex states in the periodic pinning potential. Since, in a homogeneous superconductor, triangular and square vortex lattices are energetically most favorable (see Fig. 1.14), matching effects are expected to be most pronounced for pinning arrays with a triangular or square symmetry.

The presence of a regular pinning array results in a strong overall enhancement of the $j_c(H)$ and $M(H)$ compared to a reference film without pinning array. Moreover, at temperatures close to T_c, sharp anomalies are observed in the $j_c(H)$ and $M(H)$ curves at specific field values where the vortex lattice matches the lattice of pinning centers. In particular, the first matching field $H_1 = \Phi_0/S$ is defined as the field where the density of vortices equals the density of pinning centers forming the lattice with the unit cell area S. This leads to a one to one correspondence between the vortex lattice and the pinning array. Similarly, matching effects can also occur at integer and rational multiples of H_1, due to a commensurability between the vortex lattice and the pinning array. These matching fields can be denoted as $H_n = n \times H_1$ and $H_{p/q} = \frac{p}{q} \times H_1$, respectively and $\{n, p, q\} \in \mathbb{N}$. For a square antidot lattice with $d = 1\,\mu m$, the matching fields are $\mu_0 H_n = n \times \Phi_0/d^2 = n \times 2.07$ mT. The resulting coherently pinned vortex lattice is very stable, which leads to the observed matching anomalies. For electron-beam nanopatterned pinning arrays, the matching anomalies at H_n are typically in fields up to a few tens of mT. For electrochemically prepared highly ordered porous alumina templates, matching effects are seen in considerably higher fields, up to 2.5 T [209, 500].

4.2.3   Regular pinning arrays with $n_S = 1$

The response of the vortices to the presence of artificial pinning centers is a challenging problem of scientific [102, 113, 172, 213] and technological [111, 466] interest. The 'vortex matter' in superconductors subjected to the action of thermal fluctuations and a random or correlated pinning potential is characterized by a variety of new phases, including the vortex glass, Bose glass, and the entangled flux liquid [59, 60, 67, 361]. These new phases are strongly influenced by the type of artificial pinning centers. Especially random arrays of point defects [101, 171, 480] ('random point disorder') and columnar ('correlated disorder') defects [102, 172] have been intensively studied. The latter are convenient pins to localize the vortices and enhance the critical current density j_c if the vortex density at the applied field $\boldsymbol{H}$ coincides with the density of the irradiation-induced columnar tracks [102].

To understand the behavior of vortex matter in the presence of columnar pins, regular arrays of well-characterized pinning centers have also been studied [213, 296]. One of the most efficient and easiest ways to produce such centers in thin films is to make submicron holes (antidots) [213, 296] using modern lithographic techniques [127, 319, 347].

In the case $n_S = 1$, the well-defined periodic pinning potential is formed by the antidots with a radius r_a much smaller than the period d of the array. The opposite limit ($r_a \approx d$) has been studied before in superconducting networks [377, 462, 479]. In this section, we will focus on perforated films with $r_a \ll d$, which show remarkable enhancement of the critical current at well defined so-called matching fields when the 'attempted' period of the flux line lattice a_v coincides with the period d of a regular pinning array [213, 296, 318, 319, 321, 347].

An antidot of radius r_a can trap at least one Φ_0-vortex [79, 249, 332]. If the antidots are sufficiently large, they can pin multiquanta vortices [79] with the number of the flux quanta n up to the saturation number n_S [332]. For additional flux lines in case of $n > n_S$, the antidot acts as a repulsive center. Because of this saturation effect, the composite flux lattice in fields corresponding to $n > n_S$ is expected to consist of $n\Phi_0$-vortices pinned at antidots (strong pinning) as well as Φ_0-vortices at the interstices (weak pinning), as has been shown theoretically [249].

Magnetization

Figure 4.8(a) shows the magnetization $M(H)$ data at $T/T_c = 0.99$ for a [Pb(15 nm)/Ge(14 nm)]$_3$ multilayer[2] with $T_{c0} = 6.9$ K, $\xi(0) = 12$ nm, and an lattice of antidots with a lattice period $d = 1$ µm and an antidot radius $r_a = 0.15$ µm. A remarkably sharp drop in $M(H)$ at the first matching field

---

[2]In comparison to ordinary films, [Pb/Ge]$_n$ multilayers have the advantage of being type-2 superconductors for any total thickness of the individual Lead layers, provided that each of them is sufficiently thin.

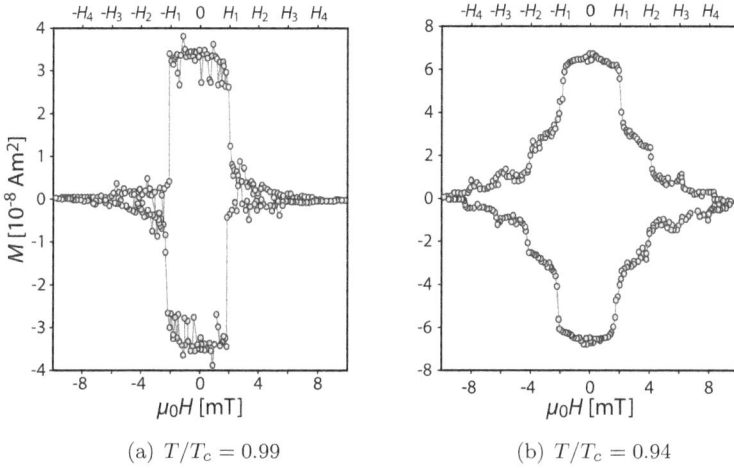

(a) $T/T_c = 0.99$ (b) $T/T_c = 0.94$

Fig. 4.8 Magnetization loops $M(H)$ of a square antidot lattice with a lattice period $d = 1\,\mu m$ and an antidot radius $r_a = 0.15\,\mu m$ at (a) $T/T_c = 0.99$ and (b) at $T/T_c = 0.94$ (after [29]).

H_1 is clearly seen [28,29]. The presence of weaker extra peaks at $H < H_1$ is an indication of the fractional flux phases stabilized by the periodic pinning array [347]. In order to analyze these results, we will use the saturation number n_S and the pinning potential at interstices U_{pi} [249]. At $T/T_c = 0.99$, $n_S = 1$ and only one flux quantum is attracted to the antidot while the second flux quantum is repelled. This situation can be described by the Bose analog of the Mott–Hubbard model for correlated electrons [353]. Since the pinning potential $U_{pi} \propto d/\lambda(T) \to 0$ as $T \to T_c$ [249], the flux lines repelled by the antidots are not localized, and instead they move freely between different very shallow U_{pi} minima at interstitial positions.

The motion of a very small number of excessive interstitial vortices at $H > H_1$ leads to a sharp field-induced first-order phase transition from fully localized at $H < H_1$ ('vortex insulator') to a collective delocalized vortex state at $H > H_1$ ('vortex metal'). The main features of this transition correspond to the Mott metal-insulator transition for the flux line [187,361]. It should be noted that previously, the existence of the temperature-induced first-order transition was derived from resistivity [269], magnetization [18, 329, 381], and heat capacity [423] measurements on high-quality high-T_c single crystals. The $M(H)$ jump at $H = H_1$ is suppressed as T goes down (see Fig. 4.8(b)).

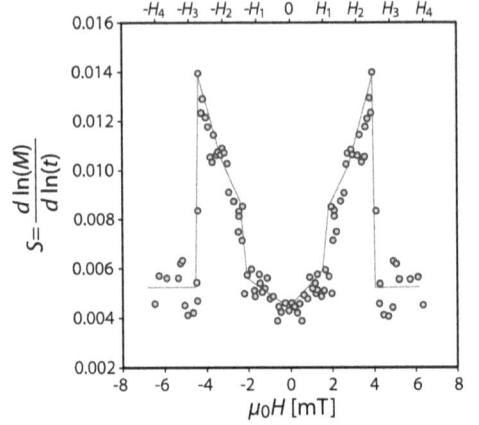

Fig. 4.9 The normalized relaxation rate $S(H)$ at $T/T_c = 0.94$ of a Pb/Ge multilayer with a square lattice of submicron antidots ($d = 1\,\mu\text{m}$, $r_a = 0.15\,\mu\text{m}$). The solid lines are guides to the eye. The onset of the vortex formation at interstices at $H > H_2$ leads to a much higher flux creep rate (after [29]).

The evaluation of parameters H_m^* (matching field) and T_m^* (temperature at which matching is observed) [361] shows that in these samples with $r_a > \sqrt{2}\xi$, H_m^* is very close to the first matching field $H_m^* \approx H_1 \equiv H_\Phi$ and T_m^* is smaller than T_c only by a few mK. In this case the Mott–insulator line H_Φ terminates at temperatures extremely close to T_c, where the relaxation times are very short and therefore the Mott–insulator can be observed. As the temperature goes down, the relaxation times increase and the $M(H)$ anomaly at H_1 is suppressed (Fig. 4.8(b)) due to the equilibrium time problems. Another possibility is that there is no disorder-localized Bose–glass phase at all, sandwiched between the superfluid and the Mott–insulator, as shown by numerical simulations [265].

At a lower temperature (Fig. 4.8(b)), we still have $n_S = 1$, but the pinning potential U_{pi} increases substantially in comparison with $T/T_c = 0.99$ (Fig. 4.8(a)) and enables the localization of vortices at interstices. For fields $H > H_1$, vortices are pushed into interstitial positions which is confirmed by the simultaneous observation of the step-like anomaly in the $M(H)$ curve at $H = H_1$ (Fig. 4.8(b)) and an abrupt increase of the flux creep rate S also at $H = H_1$ (Fig. 4.9). A much higher mobility of the intersticial vortices promotes an easy vortex channeling along the lines between the antidots, which leads to dendrite-like flux penetration clearly seen in magneto-optical measurements [175, 316, 378].

In the field range $H_1 < H < H_2$ the increasing S value corresponds to flux lines loosely bound at interstices. The interstitial positions are completely occupied at $H = H_2$ and a reentrant flux creep rate anomaly shows up indicating the onset of a strongly reduced mobility of the vortices. This

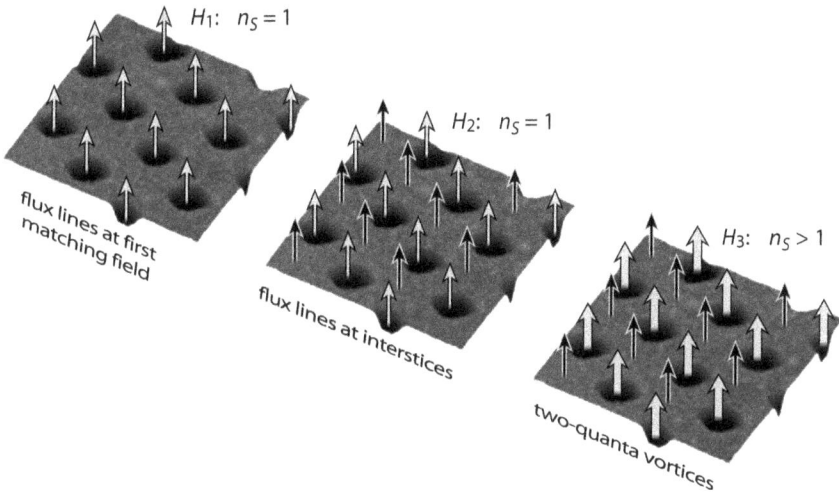

Fig. 4.10 Schematic representation of the evolution of the flux line lattice (gray arrows with black borders at antidots and reverse coloring at interstitial positions) at $T/T_c = 0.94$ as a function of the magnetic field in a Pb/Ge multilayer with a square antidot lattice. At H_3, the $2\Phi_0$-vortices coexist with interstitial Φ_0-vortices.

reduction of mobility signals the onset of the formation of doubly quantized vortices at antidots, as shown in Fig. 4.10. Because of the presence of interstitial vortices the saturation value n_S is expected to increase. This leads to the formation of two-quanta vortices at the antidots at H_3.

Transport measurements

A rather straightforward way to obtain information about the mobility of the two types of vortices — at interstices and at antidots — is to perform low-field magnetoresistance measurements. Since flux motion leads to dissipation, the presence of the antidot lattice is expected to reduce it due to trapping of the flux lines by the antidots and hence to diminish the voltage drop over the sample. In Fig. 4.11(a) a comparison of the field dependence of the resistance $R(H)$ is made between a film with antidots and the reference film without antidots at three different temperatures near $T_c = 4.725\,\mathrm{K}$ and with a fixed ac-current density of $41\,\mathrm{A/cm^2}$. For the reference film, a linear field dependence is measured with a slope which diverges near T_c as $(T_c - T)^{-v}$ ($v \approx 1$). This behavior is typical [54] for high-κ superconductors in the Bardeen–Stephen regime [38] $R = \beta(T)H/H_{c2}(0)$ at low fields ($H/H_{c2} < 0.1$), where $\beta(T)$ is the temperature dependent prefactor. In the

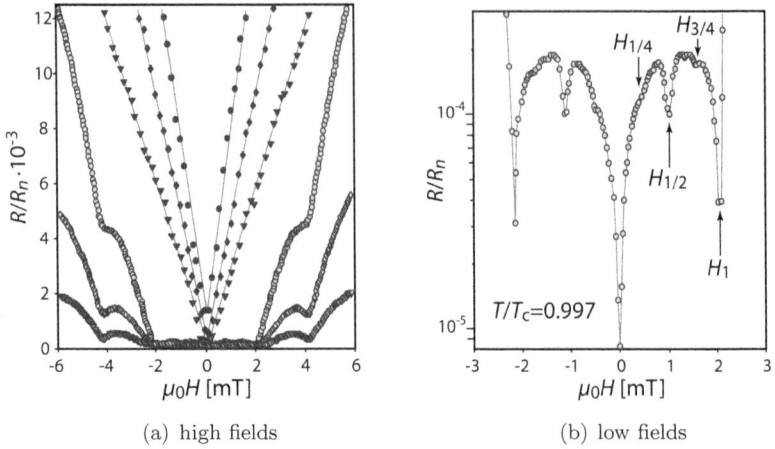

(a) high fields   (b) low fields

Fig. 4.11   (a) Comparison of the normalized resistance R/R_n of a WGe film with a lattice of antidots (open symbols) and a reference film without antidots (closed symbols — the straight lines are a guide to the eye) at different reduced temperatures $T/T_c = 0.997$, 0.996 and 0.995, indicated by circles, squares and triangles, respectively. (b) Magnification of the low resistance part of the $R(H)$ curve (logarithmic scale) of the nanostructured film at $T/T_c = 0.997$. The dips in resistance at rational field values are indicated by arrows (after [413]).

case of the film with an antidot lattice, the resistance is clearly strongly suppressed when the number of vortices is less than that of available antidots ($\mu_0 H < \mu_0 H_1 = 2.07\,\mathrm{mT}$). The strong reduction, with respect to the unperforated film's response, is due to the efficient vortex pinning by the antidot lattice.

A closer look at the $R(H)$ curves for the film with the antidot lattice (see Fig. 4.11(b)) shows that there is even an additional structure present in the low dissipation level. Besides the clear dips at $H = 0$ and $H = H_1$, less pronounced, but still clearly visible R/R_n suppressions, can also be found at

$$H_{p/q} = \frac{p}{q}H_1 \; , \tag{4.5}$$

with $p/q = 1/4$, $1/2$ and $3/4$. The dips in $R(H)$ at rational fields $H < H_1$ are reminiscent of energetically stable rational flux phases [30] (such as, for example, the well known checkerboard configuration [449] at $H = H_{1/2}$, etc.) Above H_1, all the antidots forming a lattice are filled with one flux quantum. Since at $T/T_c = 0.997$, the coherence length $\xi \approx 0.11\,\mu\mathrm{m}$ is quite large, the saturation number [332] $n_S \propto r_a/2\xi \leq 1$ is too small for the formation of two-quanta vortices at the antidots. Moreover, from a crossover in the $j_c(T)$ dependence at fixed field, the reduced temperature $T_{2\Phi_0}/T_c$, at which two-quanta formation becomes energetically favorable, is estimated to be about 0.986 [413]. As a result, at $T/T_c = 0.997$, the extra vortices cannot be accommodated by the antidot lattice and they are forced to occupy interstices between the antidots. In this way the interstitial vortices are caged by surrounding vortices strongly pinned by the antidots. Under the action of the Lorentz force, $F_L \propto j\Phi_0$, perpendicular to the transport current density j, these interstitial caged vortices are easily channeled through the rows of the flux lines at the antidots, leading to a dissipation $V \propto (H - H_1)$. As more interstitial vortices are added, the interaction between them becomes stronger and a deviation from the linear dependence can be observed.

At the second matching field, $H_2 \equiv 2H_1 = 4.14\,\mathrm{mT}$, the interstitial lines are commensurate with the underlying antidot lattice, occupying hereby the positions in the center of the squares with the antidots at the corners, which enhances the energy barrier for the motion of the interstitial flux lines.

Lorentz microscopy on square pinning arrays

The matching effect of vortices in a superconductor having regular pinning arrays, which manifested itself as the appearance of peaks or cusps in the critical current at specific values of the applied magnetic field [29, 296, 319, 344], has been microscopically investigated using Lorentz microscopy [210].

A coherent and penetrating electron beam of a 300 kV field-emission transmission electron microscope [247] allowed a direct observation of statics and dynamics of vortices in a superconducting thin film by Lorentz microscopy [211]. A thin film of niobium ($T_c = 9.2\,\mathrm{K}$) was prepared for transmission observation by chemically etching from a rolled niobium foil of 30 μm thickness, which had been annealed at 2200°C in a vacuum to increase its grain size up to 300 μm. The film had a [110] surface and one or two holes in the middle. In the areas near the hole edges which were ~ 100 nm thick, a square array of pinning centers was produced inside a region 10 μm by 10 μm by irradiation of the niobium film with a focused 30 kV gallium-ion beam. The artificial pinning sites consisted of a pit 30 nm in diameter and a few nm in depth, and of dislocation networks 200 nm in diameter surrounding the pit. The pit with the dislocation networks is called a 'defect' hereafter. Such square lattices of artificial pinning centers regions were repeatedly produced side by side parallel to the film edge at intervals of 4 μm.

First, the static configurations of vortices were observed at various values of magnetic fields to investigate what happens to vortices at specific magnetic fields. When the magnetic field was applied to the sample above T_c and the sample was then cooled down to $T/T_c = 0.49$. Even under this field-cooled condition, vortices moved for a few minutes as a result of the relaxation processes related to the nonuniform film thickness until they reached an equilibrium state. The vortex density changed a little bit from one place to another because of changes in thickness. For the observation, the region, where the film was $\sim 100\,\mathrm{nm}$ thick, was chosen and the vortex density showed no appreciable change. The observation revealed that vortices formed regular configurations at the matching magnetic fields H_n, which are given by $\mu_0 H_n = n \times \Phi_0/d^2 = n \times 2.98\,\mathrm{mT}$ (with $\boldsymbol{H}$ perpendicular to the film plane).

Lorentz micrographs taken at various matching fields are shown in Fig. 4.12. In the case of $H_{1/4}$ (Fig. 4.12(a)), vortices were located at every fourth pinning site with a spacing $4\,d$ in the horizontal direction, where d is the spacing of the defect lattice. Such lines of vortices were stacked in the vertical direction in such a way that adjacent lines were displaced relative to each other in the horizontal direction by distance $2\,d$ to form a centered (4×2) structure. In this case, vortices formed a slightly deformed ($\sqrt{5}/2 \approx 1.12$) triangular lattice. At $H_{1/2}$, vortices occupied every other pinning site in both horizontal and vertical directions, forming a centered (2×2) square lattice (Fig. 4.12(b)). At the first matching field ($n = 1$), vortices occupied all of the pinning sites without any vacancies (Fig. 4.12(c)).

When $H > H_1$, vortices began to squeeze themselves at interstitial positions as 'quasi-bound' [249] or 'caged' [413] vortices. These points were the most stable places for additional vortices to penetrate a square of vortices that were already strongly pinned by defects. In the case of a pit with a larger radius, i.e. $n_S > 1$, two or more flux quanta would be trapped at a single pinning site [343].

Interstitial vortices were found to be randomly distributed until H reached $H_{3/2}$, when vortices entered every other interstice in addition to the square array of vortices all pinned at defects, thus occupying just half of the available interstitial sites (Fig. 4.12(d)). Between H_1 and $H_{3/2}$, the interaction of distant interstitial vortices was too weak to form a regular lattice. Even for $H_{3/2}$, the interstitial vortex that should be situated at the top left portion of the micrograph (see Fig. 4.12(d)) is mislocated at the interstitial position one line lower. At $H = H_2$, all of the interstitial

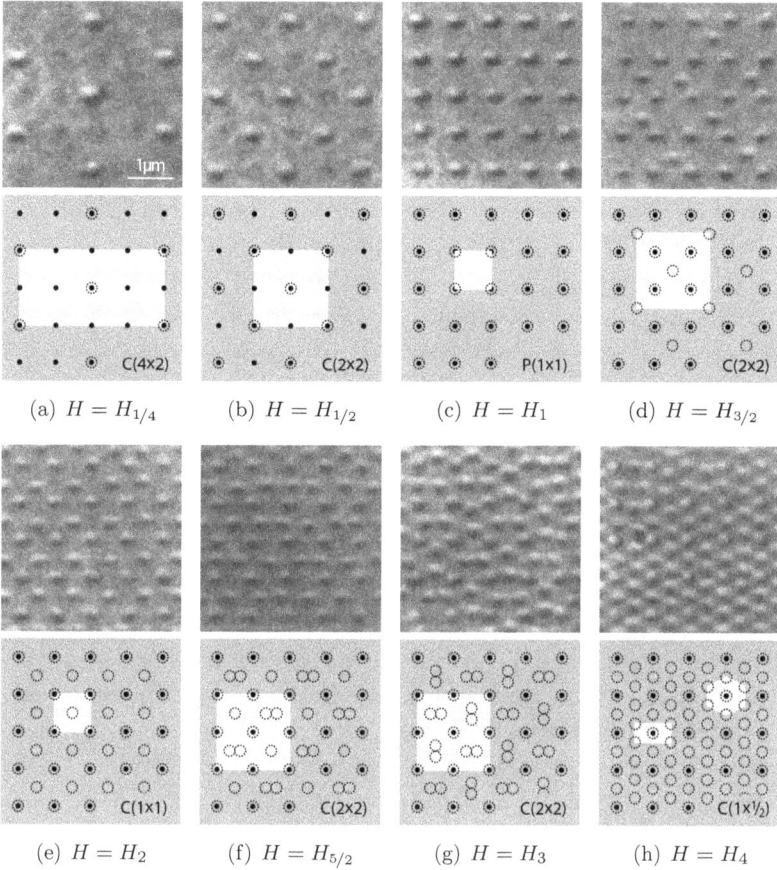

Fig. 4.12 Lorentz micrographs and schematics of the static vortex configuration in a square array of artificial defects at matching magnetic fields H_n for (a) $n = 1/4$, (b) $n = 1/2$, (c) $n = 1$, (d) $n = 3/2$, (e) $n = 2$, (f) $n = 5/2$, (g) $n = 3$, and (h) $n = 4$. Black dots and open circles in the schematic drawings of the Lorentz micrographs show the positions of defects and vortices, respectively. Squares also indicate unit cells of the vortex lattices in each case. Vortices form regular lattices at the first matching magnetic field H_1, as well as at its multiples and its fractions (after [210]).

sites were occupied by vortices, forming a centered (1×1) square lattice (Fig. 4.12(e)).

At $H = H_{5/2}$, additional vortices entered every other interstice of the configuration at H_2, thus forming a centered (2×2) square lattice (Fig. 4.12(f)). That is, one vortex and two vortices alternately occupied interstitial sites in both horizontal and vertical directions. The two interstitial vortices did not overlap but were situated side by side, separated by a distance of $\sim 0.6\,d$. They were aligned parallel to either one of the axes of the square lattice. In the present case, the direction of the two vortices accidentally horizontal.

When H reached H_3, two vortices were located at every interstitial site (Fig. 4.12(g)). The line connecting these two interstitial vortices was not in the same direction but switched alternately from horizontal to vertical. When $H = H_4$, all of the pairs of interstitial vortices were aligned in the vertical direction (Fig. 4.12(h)). Additionally, a vortex was inserted at every middle point between two adjacent sites in the vertical direction (Fig. 4.12(h)). As a result, vortices were arranged in a regular triangular lattice.

Vortices form regular lattices and consequently stable and rigid configurations at H_n. In this case, the vortices do not move easily. Especially for $n = 1$, all of the defects are occupied by vortices and therefore hopping of vortices is forbidden even when the elementary force is exerted on them. In this case, the Mott–insulator phase, introduced by Nelson and Vinokur [361] and by Blatter et al. [59] is realized. In contrast to that, interstitial vortices, appearing at $H > H_1$, cannot be localized by a shallow caging pinning potential at interstices. As a result, interstitial vortices demonstrate a 'metallic' vortex behavior. Both localized vortices at defects and intersticial metallic vortices can be more directly observed by monitoring their dynamics.

For $n_S = 1$, peaks and cusps in the critical current and the magnetization were found at $H = H_{1/4}$ and $H_{1/2}$ but not at $H_{3/2}$ or $H_{5/2}$ by macroscopic measurements [319]. This is reasonable because the pinning potential at defects $(n < 1)$ is deeper than that at interstices $(n > 1)$. In fact, the regular lattice was partially destroyed at $H = H_{3/2}$ or $H_{5/2}$, even in the field of view shown in Figs. 4.12(d) and 4.12(f).

Peaks in the critical current could more directly be explained by the different dynamic behaviors of vortices in two cases in which H is exactly H_1 and H is slightly greater than H_1. In the sample that was field-cooled down to 4.5 K at a field of 3.1 mT, just above H_1 (2.98 mT), the interstitial vortex in the left image of Fig. 4.13 began to hop in the downward direction when H was increased to 3.9 mT. When T was increased to 7 K to shorten the time-scale of the vortex hopping due to thermal fluctuations, the interstitial vortex hopped to the next interstitial site (middle image of Fig. 4.13). This micrograph was taken after the sample had been cooled down to 4.5 K, in order to obtain a high-contrast vortex image, but this cooling procedure

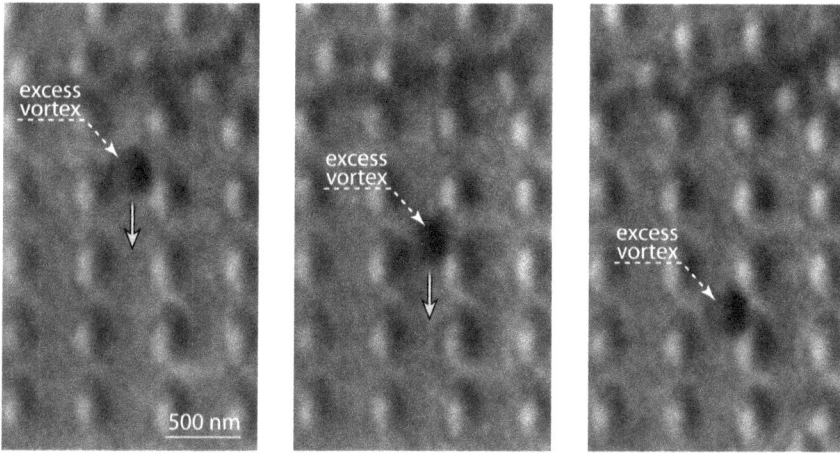

Fig. 4.13   Dynamics of an excess vortex. The excess vortex hopped from one interstitial site to another when H changed from 3.1 to 3.9 mT and T increased from 4.5 K
(left) to 7 K (middle) and then to 7.5 K (right). Arrows indicate the hopping direction
(after [210]).

did not change the configuration of the vortices. When T was further
increased to 7.5 K, the vortex hopped again to the next site (right image of
Fig. 4.13). The interstitial vortex hopping from one site to another reminds
the hopping conductivity of charge carriers in doped semiconductors.

Excess vortices in a regular vortex lattice were observed to hop easily
(see also flux flow results in Ref. 413), whereas a change in magnetic field
two times larger was required to induce hopping of the vortices forming
the lattice. The hopping of 'holes' in a vortex lattice was also observed. A
stronger force was needed to cause a vortex hole to hop than to cause an
excess vortex to do so, because a vortex must be depinned from a stable
defect site. Similar vortex behavior was detected at other matching fields,
such as H_2 and H_3, although it was not as conspicuous as in the case of
$H = H_1$.

The studies by Lorentz microscopy elucidated the microscopic mechanism of the matching effect. When vortices formed a regular lattice, they
could not begin to move unless a force larger than the elementary pinning
force was exerted. At the same time, excess (or deficient) vortices were
observed to move easily when affected by the Lorentz force, thus providing
microscopic explanation for higher critical currents at matching magnetic
fields.

Vortex patterns in presence of a square pinning array

Static and dynamic vortex phases in superconductors with a periodic pinning array have been studied by molecular dynamics simulations by Reichhardt *et al.* [398–401]. They have performed simulations of the vortex
configurations in superconductors with a periodic pinning array of circular
pinning centers with $n_S = 1$. In the equation of motion, the vortex-vortex
interaction for a bulk superconductor (Eq. (1.29)) is used. The vortices
interact with the pinning centers only when they are within a distance λ
from their edge, where the attractive force is proportional to the distance
between the centers of the vortex and the pinning site. Each pinning center
can only pin one flux quantum ($n_S = 1$). The flux-gradient-driven simulations and the simulated annealing (field-cooled) simulations result in the
same vortex configurations. At H_1, a one to one matching between the
vortex lattice and the lattice of pinning centers is established. At H_2, a
square vortex lattice is formed where one flux line is pinned at each pinning
center, and one is 'caged' at each interstitial position. The vortex lattice
at H_3 is highly ordered with pairs of interstitial flux lines alternating in
position, but has neither a square nor a triangular symmetry. The vortex
configurations in a square pinning array have been simulated in Ref. [398]
up to the 28th integer matching field. A variety of ordered, nearly ordered
(e.g. distorted triangular), and disordered vortex lattices are found for integer matching fields up to $H/H_1 = 15$. At $H/H_1 > 15$, no overall order of
the vortex lattice is found. For these high matching fields, ordered domains
are observed which are separated by grain boundaries of defects.

Also at rational multiples of H_1, matching anomalies have been observed, related to a stable vortex configuration in the periodic pinning potential. molecular dynamics simulations [398, 399] for a square lattice of
pinning centers reveal ordered vortex lattices at $H/H_1 = 1/4$ and $1/2$, and
partially ordered vortex lattices at $H/H_1 = 3/2$ and $5/2$. Experimentally,
fractional matching anomalies have been observed in several types of periodic pinning arrays, e.g. antidot lattices in Pb/Ge multilayers [30], or
lattices of sub-micron insulating, metallic or magnetic dots, covered with a
superconducting layer [303, 471].

The matching anomalies at well-defined rational multiples of the first
matching field (Eq. (4.5)) can be explained by the stabilization of a flux
lattice with a larger unit cell in the lattice of pinning centers. We will now
discuss the rational matching configurations in artificial square pinning arrays with period 1.5 μm, consisting of a lattice of sub-micron rectangular

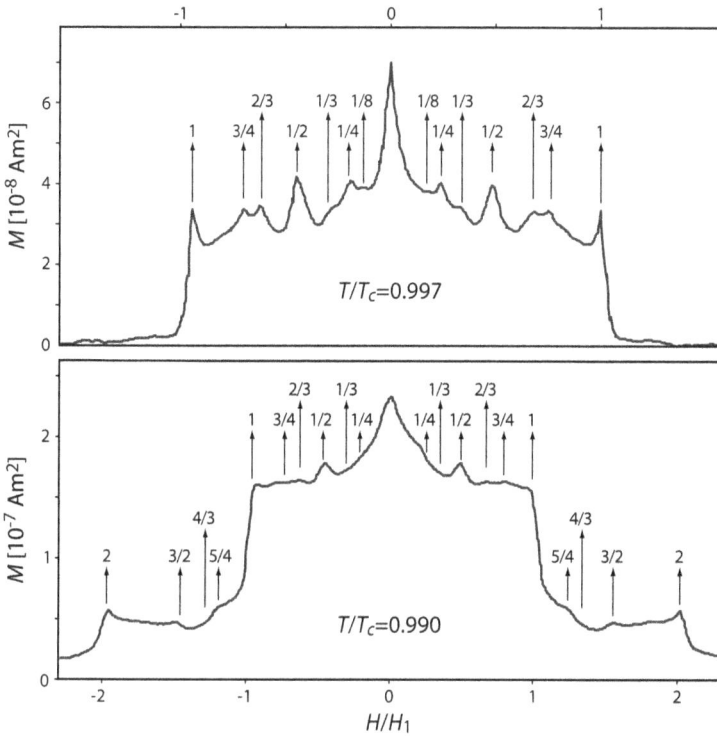

Fig. 4.14 Magnetization curves ($M > 0$) as function of H/H_1 for a 50 nm Lead film on a square lattice of Ge dots with $d = 1.5\,\mu m$ at $T/T_{c0} = 0.997$ (upper panel) and at $T/T_{c0} = 0.990$ (lower panel). Matching anomalies at $|H/H_1| = p/q$ are indicated with arrows (after [471]).

dots (insulating, metallic, or magnetic), covered with a superconducting lead film [471]. In these systems, rational matching anomalies of two different periodicities have been observed, namely the 'binary' fractions ($q = 2^n$, with n integer) of the first matching field: $H/H_1 = 1/8$, $1/4$, $1/2$, $3/4$, $5/4$ and $3/2$, and the 'threefold' ($q = 3$) fractions with $H/H_1 = 1/3$, $2/3$ and $4/3$, which are only observed for $T/T_c > 0.985$.

Similar rational matching anomalies have also been reported by Baert *et al.* [30] in superconducting Pb/Ge multilayers with a square antidot lattice. In these antidot systems, rational matching peaks are observed only within the first period ($|H| < H_1$) at slightly different fields, namely $H/H_1 = 1/16$, $1/8$, $1/5$, $1/4$ and $1/2$. These field values correspond exactly to those where a square lattice of flux lines with a unit cell larger than that of the antidot lattice and if rotated, can be matched onto the square lattice of pinning centers, i.e. when $p = 1$ and $q = n^2 + k^2$, with n and k integer numbers. Only when the radius of the antidots is increased, thus allowing pinning of multiquanta vortices, rational matching peaks are observed in the second period ($H_1 < |H| < H_2$). For very large antidot radii, a crossover to the network behavior is observed. In superconducting networks, anomalies in the critical current are present in all field periods at certain fractional numbers of flux quanta per unit cell: $H/H_1 = 1/4, 1/3, 2/5, 1/2, 3/5, 2/3$ and $3/4$ [376]. Comparison of the series of rational matching fields found in the sample with a lattice of dots with the results obtained for superconductors with a square antidot lattice, and superconducting square wire networks, reveals that these samples exhibit a different behavior. Several of the observed anomalies ($H/H_1 = 1/3, 2/3, 3/4, 5/4, 4/3$ and $3/2$) are not present when small antidots are used as pinning centers. These specific rational matching fields are also not emerging from the simulations of Reichhardt *et al.* [398, 399] for a square lattice of pinning centers. Rational anomalies in the second period have so far mainly been observed for larger antidots, related to the formation of multiquanta vortices ($n_S \geq 2$). The anomalies with $q = 3$ are typical for a superconducting network. Nevertheless, they are clearly visible in the $M(H)$ measurements of a Lead film with a dot lattice, although this sample consists of a continuous superconducting film, far from the network limit [376]. These differences demonstrate that the details and the precise nature of the lattice of pinning centers play an important role in the pinning phenomena and the matching effects.

For the rational matching fields observed in superconducting films with a square dot lattice, the proposed stable vortex configurations are shown in Fig. 4.15. The configuration for $H/H_1 = 1/8$ is a square vortex pattern rotated by $45°$ with respect to the pinning array, which also corresponds to the configuration suggested by Baert *et al.* [30]. For $H/H_1 = 1/4$, the flux lines can almost form a perfect triangular lattice. This configuration has been directly visualized in Lorentz microscopy measurements (see Fig. 4.12(a)). For the anomaly at $H/H_1 = 1/3$, a sequence of alternating diagonal rows (one occupied, two empty, etc. — see Fig. 4.15) is realized, as directly seen in scanning Hall probe microscopy experiments [144, 198]. This vortex configuration is similar to the one at $H/H_1 = 1/3$ in superconducting wire networks [376].

At $H/H_1 = 1/2$, the well-known checkerboard configuration is formed, resulting in a square vortex lattice, rotated over $45°$ with respect to the underlying pinning lattice. For $H/H_1 = 2/3$, an ordered vortex lattice can

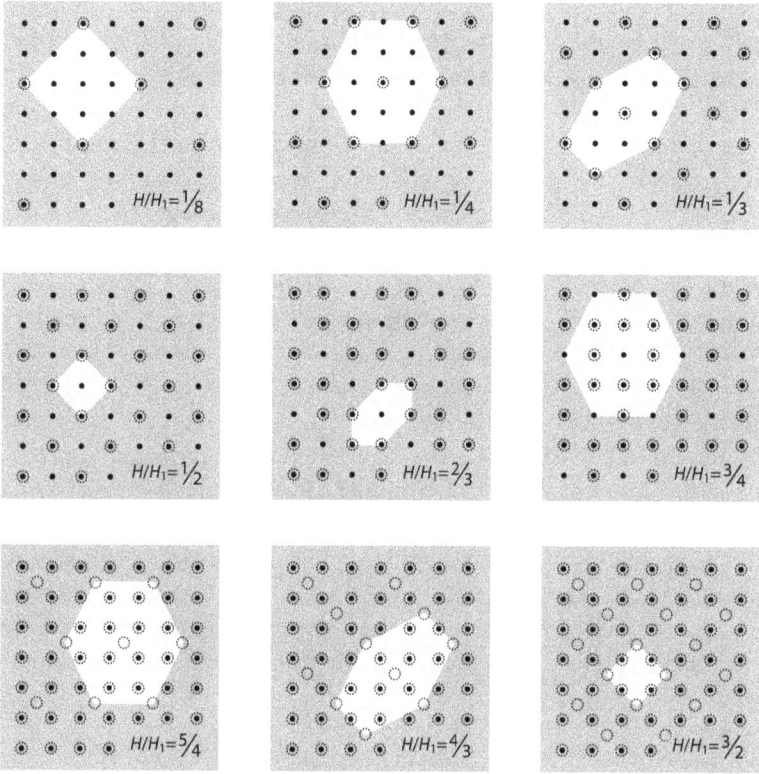

Fig. 4.15   Schematic presentation of possible stable vortex lattices at rational matching fields, indicated by $H/H_1 = p/q$, for a square lattice of pinning centers. The black dots and open circles indicate pinning centers and vortices, respectively. The white lines are guides to the eye, showing the symmetry of the vortex lattice (after [471]).

be formed being the inverse of the one at $H/H_1 = 1/3$, where the occupied pinning sites are replaced by empty ones and vice versa. Similarly, the proposed configuration for $H/H_1 = 3/4$ consists of the inverse of the configuration at $H/H_1 = 1/4$. The vortex patterns at rational matching fields in the second period ($H_1 < |H| < H_2$) can be interpreted by using the following consideration. After at H_1 single flux lines occupy all pinning centers, the minima in the pinning potential at the positions of the dots have turned into a maximum, repelling additional flux lines (Fig. 4.7). This results in a potential landscape characterized by minima at interstitial positions, where vortices can now be caged by the repulsive interaction with the strongly pinned flux lines at the dot positions. The square lattice of interstices now plays the role of the pinning centers in the first field period. In order to obtain the vortex patterns at $H/H_1 = 5/4$, $4/3$ and $3/2$, we can simply consider the superposition of the rational vortex configurations for $H/H_1 = 1/4$, $1/3$ and $1/2$ respectively in the interstitial positions with the completely filled H_1 configuration. This results in the corresponding patterns shown in Fig. 4.15. It should be mentioned that the configuration for $H/H_1 = 3/2$ has been directly observed by Lorentz microscopy, see Fig. 4.12(d).

From the different configurations shown in Fig. 4.15, all stable vortex lattices for $H \leq H_1$ consist of pinned vortices at the positions of the dots. On the other hand, for $H > H_1$, composite flux lattices [29,210,320,342,343,413] are formed, consisting of strongly pinned vortices at dot positions and weaker pinned or 'quasi-bound' vortices at interstices. These interstitial vortices can be depinned much easier, and are responsible for the lower magnetization amplitude (or critical current density) compared to the $|H| \leq H_1$ field range, which was observed in the $M(H)$ curves, especially at relatively high temperatures (see Figs. 4.8 and 4.14). Besides low-T_c superconductors considered in this section, antidot arrays have been also successfully used to control vortex pinning and noise reduction in high-T_c films and rf-SQUIDs made from these films [295,430,505].

Vortex patterns in presence of a triangular pinning array

Also for triangular lattices of pinning centers with $n_S = 1$, stable vortex configurations are found at integer and rational multiples of the first matching field by molecular dynamics simulations [398,399]. Magnetization measurements on a superconducting lead film with a triangular lattice of rectangular submicron magnetic dots revealed a series of matching anomalies which can be directly related to a stable configuration of the vortices in the periodic pinning potential. Assuming that no multiquanta vortices are formed ($n_S = 1$), the stable vortex lattices, giving rise to an anomaly in the magnetization curves, can be identified. The series of field values at which matching effects appear, and for which we will indicate a stable vortex configuration is as follows: $H/H_1 = 1/4$, $1/3$, $3/4$, 1, $5/4$, 2, 3 and 4. The vortex patterns are schematically presented in Fig. 4.16, where the dots and the circles indicate the pinning sites and the vortices, respectively. For some matching fields, a commensurate triangular lattice can be formed, while other stable vortex configurations are ordered but do not form a triangular lattice.

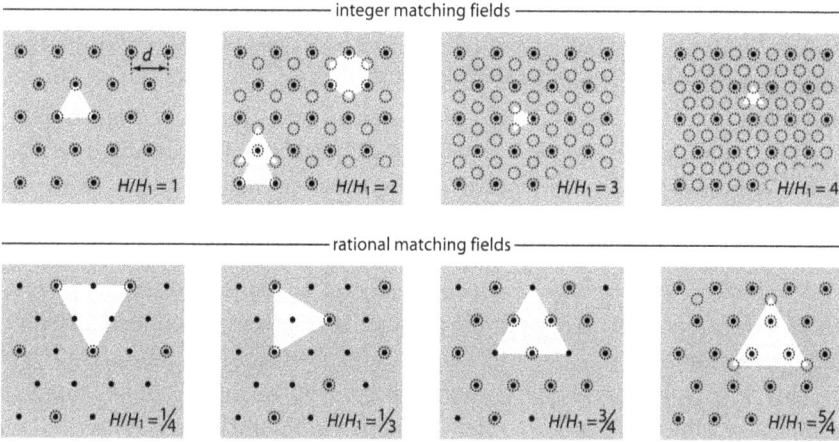

Fig. 4.16 Schematic presentation of stable vortex patterns at integer (upper row) and rational (lower row) matching fields for a triangular lattice of pinning sites with period d. Black dots and open circles represent pinning centers and vortices, respectively. White lines indicate the symmetry of the stabilized vortex lattice (after [398, 399, 471]).

In the absence of pinning centers, the most stable vortex configuration is the triangular Abrikosov vortex lattice. When a triangular lattice of artificial pinning centers is introduced, matching effects will develop at those field values where a triangular vortex lattice with a certain period, a_v, can be formed that is commensurate with the underlying lattice of pinning centers. This condition can be expressed as follows [398, 399]:

$$a_v = d\sqrt{m^2 + n^2 + n \cdot m} \qquad \text{rational matching anomalies} \qquad (4.6a)$$

$$a_v = \frac{d}{\sqrt{m^2 + n^2 + n \cdot m}} \qquad \text{integer matching anomalies} \qquad (4.6b)$$

In these equations, a_v is the period of the triangular vortex lattice, d is the period of the triangular pinning array, and m and n are positive integer numbers. It is assumed that the saturation number of pinned flux quanta per pinning center is $n_S = 1$. After occupation of all pinning sites, additional flux lines are trapped in minima of the pinning potential at interstitial positions. The above conditions are satisfied for the integer multiples $H/H_1 = 1$, 3, 4, 7, 9,... and for the rational fractions $H/H_1 = $ 1/3, 1/4, 1/7, 1/9, etc. The corresponding triangular vortex lattices for the observed matching peaks can be found in Fig. 4.16. In the $M(H)$ measurements of a superconducting Lead film on a triangular lattice of submicron rectangular magnetic dots, no clear matching anomalies are resolved at fields lower than $H/H_1 = 1/4$, and the highest observed matching field is $H/H_1 = 4$. Molecular dynamics simulations of the vortex lattice in a triangular pinning array indeed reveal triangular ordered vortex configurations at the integer matching fields $H/H_1 = 1, 3, 4, 7, 9, 12, 13, 16, 19, 21, 25$ and 28 [398,399].

The anomalies at $H/H_1 = 3/4, 5/4$ and 2 can not be explained by the above argument. However, other ordered non-triangular vortex patterns can be formed at those field values, as is shown in Fig. 4.16. In the suggested vortex lattice for $H/H_1 = 3/4$, a fraction of the pinning centers is occupied in such a way that the inverse of the vortex configuration at $H/H_1 = 1/4$ is formed. (A similar idea was used in the case of a square pinning array for $H/H_1 = 3/4$, and has been experimentally confirmed [210].) In this way, a triangular lattice with period $2d$ of non-occupied pinning centers is defined. In order to find a stable configuration for $H/H_1 = 5/4$, let us consider the pinning potential after all pinning centers are occupied. Minima are then developed at interstices in the center of each triangle of occupied pinning centers. To form a stable vortex lattice at $H/H_1 = 5/4$, the pinning centers are first occupied and the remaining flux lines can form a triangular arrangement at the interstitial positions, similar to the one at $H/H_1 = 1/4$. The resulting configuration is shown in the lower row of Fig. 4.16. For $H/H_1 = 2$, a honeycomb vortex pattern is suggested, consisting of two identical interpenetrating triangular vortex lattices, both with the same period d. In one lattice, the vortices are strongly pinned at the dot positions while the other consists of weaker pinned interstitial vortices. This configuration was also found in the simulations of Reichhardt *et al.* [398,399].

4.2.4   *Multiquanta vortex lattices ($n_S > 1$)*

In this section, regular arrays of relatively large antidots ($n_S > 1$), which can stabilize multiquanta vortex lattices, are considered. Experimental evidence for the existence of such vortex lattices can be found in magnetization data. We analyze the $M(H)$ curves of high-quality nonperforated and perforated $[Pb(10\,nm)/Ge(5\,nm)]_n$ multilayers [411] with coherence length $\xi(0) = 12\,nm$ and penetration depth $\lambda(0) = 260\,nm$, prepared in molecular beam epitaxy apparatus using liquid-nitrogen cooled SiO_2 substrates. Using standard electron-beam lithography, an antidot lattice with a distance $d = 1\,\mu m$ between the antidots of radius $r_a = 0.22\,\mu m$ was fabricated in the film[3] (for details of the sample preparation see Ref. 74). The matching fields

---

[3]The number of antidots per sample unit area is about 10^6 per mm^2.

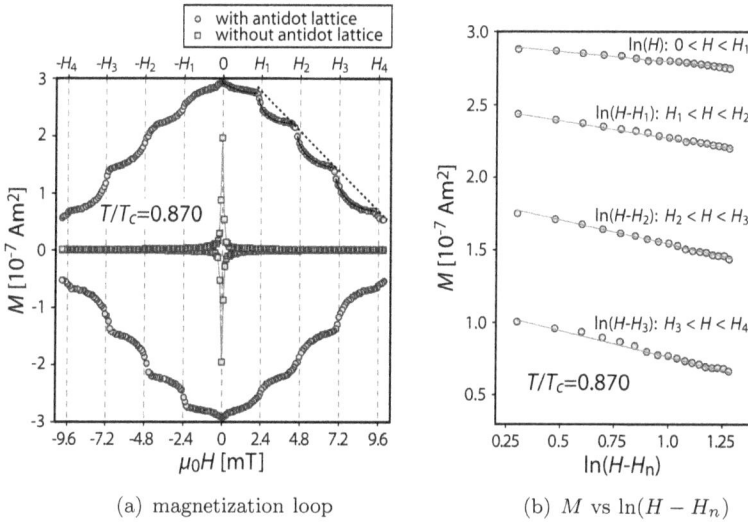

(a) magnetization loop

(b) M vs $\ln(H - H_n)$

Fig. 4.17   (a) Magnetization loop $M(H)$ at $T/T_c = 0.870$ of a $[Pb(10\,nm)/Ge(5\,nm)]_2$ bi-layer with and without antidot lattice. The solid line is a fit with Eq. (4.12). The dashed line is demonstrating the validity of the linear behavior of $M(H_n)$ at the matching fields H_n (Eq. (4.8)). The loops were measured for $M > 0$ and symmetrized for clarity for $M < 0$. (b) The magnetization M versus $\ln(H - H_n)$ at $T/T_c = 0.870$. The different slopes of the solid lines for the different periods are used to determine the effective flux $\tilde{\Phi}_{0n}$ in Eq. (4.12) (after [343]).

H_n of this perforated film with a triangular antidot lattice is $n \times 2.39\,mT$.

A typical magnetization loop $M(H)$ of superconducting films with triangular antidot lattice at temperatures close to T_c is shown in Fig. 4.17(a), together with a corresponding $M(H)$ curve of a reference Pb/Ge nonperforated film. The width ΔM of the $M(H)$ loop of the nanopatterned film is strongly increased due to the efficiency of antidots as artificial pinning centers. At the used temperature close to T_c, the difference between the width of the hysteresis loops ΔM in Fig. 4.17(a) for the as-grown and for the nanostructured film is large enough to attributed the $M(H)$ curve for the latter solely to the contribution arising from flux line pinning by an antidot lattice.

Besides an overall enhancement of ΔM, distinct cusp-like $M(H)$ anomalies are clearly seen exactly at the expected matching fields $\mu_0 H_n = n \times 2.39\,mT$. Strictly speaking, there is always a shift in the position of the matching field H_n extracted from increasing $(H_n^\uparrow)$ and decreasing $(H_n^\downarrow)$ field measurements of $M(H)$. However, sufficiently close to T_c, where the sharp

158 Nanostructured Superconductors

matching $M(H)$ anomalies are observed, the difference $\Delta H_n = H_n^\uparrow - H_n^\downarrow$ is very small. At lower temperatures matching $M(H)$ anomalies are smeared out and simultaneously the difference ΔH_n strongly increases. The radii of the antidots are sufficiently large, so that the nominal calculated saturation number [332] $n_S = r/2\xi(T)$ is approximately 3.3 for the studied film at $T/T_c = 0.870$. In this case, one can assume that cusp-like $M(H)$ anomalies are related to the formation of the multiquanta vortex lattices. One should keep in mind, that n_S, as calculated above, is only a rough estimate of the saturation number since it was derived in the limit of flux lines interacting with a *single* cylindrical microhole [332]. Below, we shall demonstrate, however, that the assumption of the existence of multiquanta vortex lattices is fully supported by a quantitative description of the whole magnetization curve.

Let us consider first the situation when all vortices are trapped by the antidots, meaning that there are no interstitial vortices. We begin with the estimate of the temperature-dependent penetration length $\Lambda = 2\lambda^2/\tau$, where λ is the penetration depth and τ the total thickness of the film. The estimate of Λ shows that $\Lambda(T/T_c = 0.87) \approx 5.8\,\mu\text{m} \gg d = 1\,\mu\text{m}$. This Λ value can be even further increased due to the renormalization of the penetration depth in perforated films, where superconducting volume corrections have to be taken into account [494]. Since $\Lambda \gg d$, the approximation $B = \text{const}$, where B must be averaged at least over several unit cells, can be used to find the magnetization M of the superconductor with the antidot lattice in the field range $H < H_1$ at temperatures close to T_c.

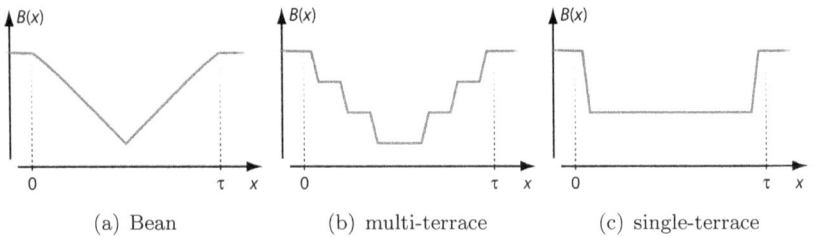

(a) Bean   (b) multi-terrace   (c) single-terrace

Fig. 4.18 Field penetration into a superconducting slab of thickness τ according to (a) the classical Bean model [41], (b) the multi-terrace critical state [110] and (c) the single-terrace critical state [471].

There are at least two arguments in favor of the approximation $B = \text{const}$ [343]. The first one is the particular profile of the calculated field distribution in thin superconducting specimens [32]. For very large Λ the field inside the specimen is constant everywhere, except at the edges and in the center. This field distribution provides a constant field not in the whole sample but at least in a very large part of its area. The second argument in favor of the approximation $B = $ const can be based on the possibility of the existence of a single-terrace critical state in superconducting films with antidot lattices. Indeed, according to Cooley and Grishin [110] the classical Bean model [41] should be substantially modified for superconductors with a regular array of artificial pinning centers which favor a certain integer number of flux quanta pinned by each antidot. In this case, instead of a smooth sand hill-like critical $B(x)$ profile (Fig. 4.18(a)), a multiterrace critical state with several well-defined plateaus, each having its own fixed induction $B = $ const, has been predicted (Fig. 4.18(b)) [110]. The multi-terrace critical state is, in a way, a quantized version of the classical Bean model [343]. The field profile with well-defined terraces is a compromise between a tendency to trap the same number $n\Phi_0$ of flux quanta by each antidot and a formation of a certain average Bean-like $B(x)$ slope. Close to T_c, the latter can be quite small and then an ultimate limit of the multi-terrace critical state — a single-terrace critical state (Fig. 4.18(c)) — could be established. This state, especially exactly at matching fields, implies the flat field penetration profile $B = $ const may be realized in the whole sample, except its surface layer.

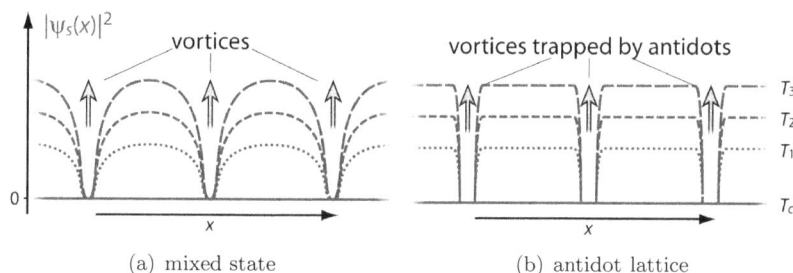

(a) mixed state (b) antidot lattice

Fig. 4.19 The $|\psi|$ modulation at different temperatures in a homogeneous supercon-ductor in (a) the mixed state and (b) in a superconductor with an antidot lattice.

The next important step towards the understanding of magnetization of the multiquanta vortices can be made by taking into account a possible expansion of the London limit $|\psi| = $ const in superconducting films with antidot lattices [343]. Indeed, in homogeneous superconductors in the vicin-ity of the upper critical field $H_{c2}(T)$ the superconducting order parameter $\psi = |\psi|\, e^{i\varphi}$ nucleates with a very strong $|\psi|$ modulation (Fig. 4.19(a)), in agreement with the classical Abrikosov model [2]. Due to this modulation, the London limit can normally be used only in the range $H_{c1} \ll H \ll H_{c2}$, i.e., well below the $H_{c2}(T)$ line. Contrary to that, in a superconductor with an antidot lattice, by making antidots exactly at the expected positions for the penetration of vortices, we are 'helping' the superconducting order parameter to nucleate in a different way (Fig. 4.19(b)). In this case we are forcing $|\psi|$ to be nearly constant in the space between antidots, since, due to boundary conditions at the superconductor-insulator interface [118], the surface $|\psi| = $ const can cross the cylinders, corresponding to the antidot boundaries, only at the angle $90°$. Moreover, when flux lines are guided to penetrate through the antidots, there is no need to create additional zeros of $|\psi(r)|$ (and therefore strong $|\psi|$ modulation) anywhere else. These arguments make it possible to foresee a broader range of validity of the London limit $|\psi| = $ const in films with an antidot lattice.

Following Moshchalkov *et al.* [343], we take into consideration the two important assumptions $B = $ const and $|\psi| = $ const, and use the textbook expression for magnetization [118] with a simple substitution $\xi(T) \to r_a$ for the core radius:

$$M(H) = -\frac{\Phi_0}{4\pi\mu_0\Lambda^2} \cdot \ln\left(\frac{\beta a_v}{\sqrt{e} r_a}\right) , \qquad (4.7)$$

where a_v is the distance between Φ_0-vortices and β is a numerical constant.[4]

First, let us consider a very interesting situation, which occurs exactly at the matching fields $H = H_n$, when we expect that the flux line lattice consists of only one type of vortices, namely, of $n\Phi_0$-vortices forming a regular array coinciding with the antidot lattice. In this case $n\Phi_0$ should be used in Eq. (4.7) instead of Φ_0 and $a_v = d$:

$$M(H_n) = -\frac{n\Phi_0}{4\pi\mu_0\Lambda^2} \cdot \ln\left(\frac{\beta d}{\sqrt{e} r_a}\right) \propto \frac{n\Phi_0}{\Lambda^2} . \qquad (4.8)$$

The difference $M(H_n) - M(H_{n-1}) \propto -\Phi_0/\Lambda^2$ is independent on n and determined only by Λ. All other parameters in Eq. (4.8) are known. Therefore, at $H = H_n$ a linear behavior of $M(H_n)$ as a function of the integer n should be seen. This behavior is in a very good agreement with the observed $M(H_n)$ variation at different temperatures (see the dashed lines in Fig. 4.8(a)). The slope of the dashed lines nicely follows the expected temperature variation of $1/\Lambda^2 \propto (1 - T/T_c)^2$, as it is shown in Fig. 4.20(b). The two-fluid model $1/\Lambda^2 \propto (1 - (T/T_c)^4)^2$ also gives a good fit, but the

---

[4]Note that $\beta = 0.381$ for a triangular vortex lattice in unperforated films [118].

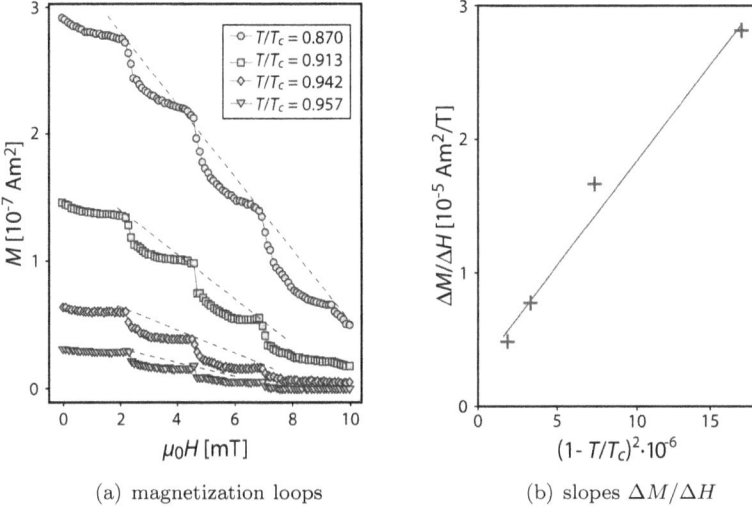

(a) magnetization loops

(b) slopes $\Delta M/\Delta H$

Fig. 4.20 (a) The magnetization loops $M(H)$ for positive M and H at different temperatures for a $[\mathrm{Pb}(10\,\mathrm{nm})/\mathrm{Ge}(5\,\mathrm{nm})]_2$ bi-layer with an antidot lattice. The dashed lines clearly indicate the linear behavior of $M(H_n)$ as a function of the integer n. (b) The slopes $\Delta M/\Delta H$ of the dashed lines as a function of $[1-(T/T_c)]^2$. According to Eq. (4.8) these slopes follow the temperature dependence of $1/\Lambda^2$. The solid line presents the linear fit (after [343]).

Ginzburg–Landau relation

$$\lambda(T) \propto \frac{1}{\sqrt{1-T/T_c}} \tag{4.9}$$

provides a better linearity of the slope $\Delta M/\Delta H$ versus T, see Fig. 4.20(b). Note also that in perforated films, instead of the substitution $\lambda \leftrightarrow \Lambda$, another renormalization of the penetration depth should be considered (see Eq. (4.12) below).

The cusps at $H = H_n$ then can be considered as analogs of the well-known $M(H)$ anomaly at H_{c1} but this time for the onset of penetration of $2\Phi_0, 3\Phi_0, \ldots, (n+1)\Phi_0$ vortices at $H_1, H_2, \ldots, H_n$, respectively. Indeed, the expected lower critical fields $H_{c1}(n\Phi_0)$ are

$$H_{c1}(n\Phi_0) = \frac{n\Phi_0}{4\pi\mu_0\Lambda^2} \ln\frac{\Lambda}{r_a} \; . \tag{4.10}$$

The estimate based on Eq. (4.10) gives very low fields[5], implying that the difference between $H - H_n - H_{c1}[(n+1)\Phi_0]$ and $H - H_n$ can be neglected and, therefore, one should see the logarithmic behavior $M(H) \propto$

---

[5] $\mu_0 H_{c1}(n\Phi_0) \approx 10^{-1}$ mT $\ll \mu_0 H_n$

$-\ln(H - H_n)$ (valid for $H \gg H_{c1}$ in the London limit) in magnetic fields $H_n < H < H_{n+1}$, i.e. between the matching fields. Such behavior is indeed fully supported by the experimental data of Reference [343] (see also Fig. 4.17(b)). This behavior is a result of inserting the relation $a_v \simeq [\Phi_0/(H - H_n)]^{1/2}$ into Eq. (4.7), which is a good approximation in fields $H_n < H < H_{n+1}$ for a $(n+1)\Phi_0$ vortex–vortex distance in a superconductor with an antidot lattice. In this interval the vortex-vortex interaction terms U_{ij} (Eq. (1.26)) in the Gibbs potential are represented by different contributions arising from the presence of two types of vortices: $n\Phi_0$ and $(n + 1)\Phi_0$. Instead of the usual $\Phi_0 \times \Phi_0$ product, these terms will contain other products: $n\Phi_0 \times n\Phi_0$, $n\Phi_0 \times (n+1)\Phi_0$, and $(n+1)\Phi_0 \times (n+1)\Phi_0$ [118]. Due to that, in the interval between the matching fields, H_n and H_{n+1}, the magnetization of superconducting films with an antidot lattice should follow the logarithmic dependence:

$$M(H_n < H < H_{n+1}) \approx -\frac{\tilde{\Phi}_{0n}}{4\pi\mu_0\Lambda^2} \ln\left[\frac{\beta_{\text{eff}}}{r_a\sqrt{e}}\sqrt{\frac{\Phi_0}{\mu_0(H - H_n)}}\right]. \quad (4.11)$$

Here, $\tilde{\Phi}_{0n}$ is an effective flux and β_{eff} is an effective parameter appearing due to the summation of different terms in the vortex-voretx interactions U_{ij} over vortex positions i and j on the antidot lattice. For H close to H_n we may approximate $\tilde{\Phi}_{0n}$ as $\sqrt{n(n + 1)}\Phi_0$.

Plotting M versus $\ln(H - H_n)$, the slopes of the M versus $\ln(H - H_n)$ lines (Fig. 4.17), defining the effective flux $\tilde{\Phi}_{0n}$, can be determined. For a triangular antidot lattice, stabilizing the multiquanta vortices, we have

$$\frac{\tilde{\Phi}_{01}}{\Phi_0} = 1 \ , \qquad \frac{\tilde{\Phi}_{02}}{\Phi_0} = 1.79 \ , \qquad \frac{\tilde{\Phi}_{03}}{\Phi_0} = 2.45 \quad \text{and} \quad \frac{\tilde{\Phi}_{04}}{\Phi_0} = 2.72$$

for $H_{n-1} < H < H_n$ with $n = 1, \ldots, 4$ respectively. The numbers 1.79, 2.45, and 2.72 have to be compared with 1.41, 2.45 and 3.46 estimated from $[\sqrt{n(n + 1)}\Phi_0]/\Phi_0$. Therefore, besides a logarithmic dependence itself, we see also a very reasonable variation of an effective parameter $\tilde{\Phi}_{0n}$ in different field intervals ($H_n < H < H_{n+1}$) with n not exceeding the saturation number n_S. At higher temperatures, the simple argument that $\tilde{\Phi}_{0n}$ should scale as $\sqrt{n(n + 1)}\Phi_0$ fails, since $\lambda(T)$ is spread over many periods and U_{ij}-terms, corresponding to $n\Phi_0 \times (n+1)\Phi_0$ and $(n+1)\Phi_0 \times (n+1)\Phi_0$, should be optimized numerically. These terms can also explain a small increase of magnetization when H is approaching the matching fields from below (see Fig. 4.17(a)).

We also would like to note that the stabilization of the flux lines by the antidot lattice has resulted in a remarkable collective $M(H)$ behavior,

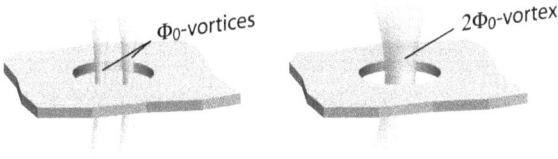

Fig. 4.21 The rearrange-
ment of two single quan-
tum vortices in a bot-
tom layer of a blind hole
(left) in comparison with
a $2\Phi_0$-vortex in an antidot
(right).

Fig. 4.22 Magnetic decoration of the admixture of $2\Phi_0$ and $3\Phi_0$ multiquanta vortices
(left image - adapted from Ref. 58 with kind permission from Springer Science+Business
Media) and of $8\Phi_0$ and $9\Phi_0$ multiquanta vortices (right image — note that the sample
was inclined in this case) in a niobium film with a triangular lattice of blind antidots.
(Adapted with permission from Ref. 57. Copyrighted by the American Physical Society.)

typical for the presence of a $n\Phi_0$-flux line lattice at $H = H_n$ $(n < n_S)$
and a mixture of $n\Phi_0$ and $(n+1)\Phi_0$-vortices between the matching fields
$H_n < H \leq H_{n+1}$. The admixture of the $2\Phi_0$ and $3\Phi_0$ multiquanta vortices
can be clearly seen in the left image of Fig. 4.22, where the results of
the magnetic decoration of the lattice of 'blind' antidots [57] are presented.
Blind antidots help to resolve the n individual Φ_0-vortices at the continuous
bottom layer of a blind microhole, formed by trapping a $n\Phi_0$ multiquanta
vortex by the antidot above the layer. This trick makes it possible to
visualize a $n\Phi_0$ multiquanta vortex as a collection of n single quantum
vortices [57] in a superconducting layer at the bottom of blind antidots, see
Fig. 4.21.

Due to the introduction of the well-defined pinning potential, it is possible to obtain in a multiple-connected superconductor with an antidot lattice
the logarithmic irreversible magnetization behavior which is normally observed as a reversible magnetization of the Abrikosov flux line lattice in

homogeneous superconductors in fields $H \gg H_c$. The irreversibility of the perforated film is possibly caused by its multiple connectivity. The superconducting currents flowing around antidots are similar to a supercurrent in a ring. As it has been already emphasized by Shoenberg quite some time ago (see, for example, Ref. 434 and references therein), a ring, as a multiply connected body, demonstrates a strong irreversible magnetization response. The successful interpretation of all specific features of the magnetization loops for films with an antidot lattice (Fig. 4.17(a)) gives a strong support of the assumption of the existence of the multiquanta vortex lattices. Magnetic decoration experiments [57, 58] have convincingly and directly confirmed the stabilization of the multiquanta vortices by the relatively large antidots (see Fig. 4.22). In these experiments the use of the 'blind holes' has made it possible to decorate the multiquanta vortices at antidots and thus to count the number of the flux quanta trapped by each antidot. Direct experimental observation of multiquanta vortex lattices [57] with no vortices pinned at the interstices confirms the conclusions based on the analysis of the magnetization data [320, 342, 343].

Scanning Hall probe microscopy also reveals the existence of multiquanta vortices [263]. In these experiments, the integration of the measured local fields over the area gives the vorticity. In this way, it was directly verified, for example, that the trapped flux in the $2\Phi_0$-vortex is exactly $2\Phi_0$ per each antidot at $H = 2H_1$. It is interesting to note, that multiquanta (or giant) vortices, which are stabilized in superconductors with relatively large antidots, can also be present in optically trapped Bose–Einstein condensates (BECs) with a geometric arrangement of the laser beams providing pinning in these rotating BECs [174].

Magnetization curves for different antidot diameters

The main experimental observations [29, 30, 319, 320, 342–344] can be summarized as follows:

- artificial arrays of submicron antidots can act as well-defined pinning centers with a controlled size and pattern,
- distinct matching anomalies show up at the expected fields $H_n = n\Phi_0/S$, where S is the unit cell of the antidot lattice,
- increasing the antidot diameter, a clear evidence of a stronger pinning by square and triangular arrays consisting of antidots larger than $\xi(T)$ was found, and

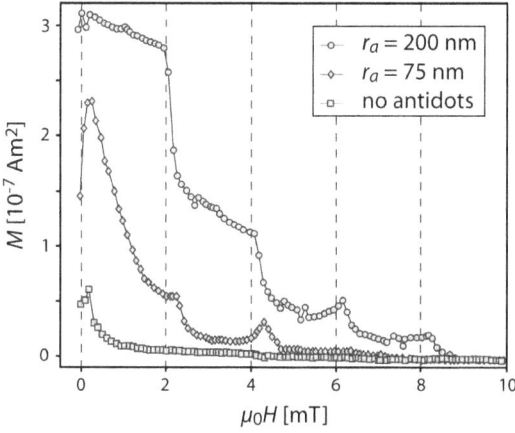

Fig. 4.23 Magnetization curves at $T/T_c = 0.94$ of a [Pb(15 nm)/Ge(14 nm)]$_3$ multilayer with a square lattice of antidots with radius $r_a = 75$ nm and $r_a = 200$ nm. For comparison, the data for a reference multilayers without antidots is also shown. The matching fields $\mu_0 H_n \approx n \times 2.07$ mT are indicated by dashed lines (after [342]).

• similar size dependence of the pinning force is observed as well in other superconducting films.

At the typical used reduced temperature $T/T_c = 0.94$ (see Fig. 4.23) we have $\xi(T) \approx 0.1\,\mu$m. Therefore the radius $r_a \approx 75$ nm of the smaller antidots is quite close to $\xi(T)$ and should give the optimum pinning, if the condition $\xi(T) \approx 2r_a$ is correct in this case and the core pinning potential U_p plays a dominant role. Since for larger antidots pinning has been further enhanced, we are sure that for antidots with $2r_a \gg \xi(T)$ pinning is much stronger than that for $2r_a \approx \xi(T)$. The latter confirms that electromagnetic pinning U_p indeed gives an important contribution: according to Ref. 456 the depth of U_p increases rapidly with r_a and saturates at $2r_a \approx \lambda(T)$. Unfortunately, further quantitative comparison with the theory of Takezawa and Fukushima [456] is not possible, since they have considered the limit of a single pinning center, whereas in this case the circular currents around different antidots are overlapping: in these films $\Lambda = 2\lambda^2(T)/\tau > d$. Here, τ is the thickness of an individual Lead layer or the whole multilayer for decoupled or coupled superconducting layers, respectively. In these films we have $\Lambda > d$, and for qualitative analysis of the data, lattice effects should be taken into consideration in calculations of U_p which has not been done in Ref. 456.

The sharp magnetization anomalies at the matching field were nicely reproduced in calculations of Cooley and Grishin [110], who showed that the appearance of the terrace critical state (Fig. 4.18(b)) results in magnetization jumps with the periodicity corresponding to Φ_0 per antidot lattice unit cell. For small antidot radii $n_S \approx 1$, a substantial reduction of j_c can occur when the antidots are saturated [356] and additional vortices are formed at interstices. In this case very weakly pinned interstitial vortices are much more mobile than the vortices pinned by the antidots [210]. As a result, the motion of interstitial vortices leads to a dissipation and j_c is reduced. The dynamics and plastic flow of vortices in superconductors with coexisting interstitial flux lines and vortices pinned by antidots has been studied by Reichhardt *et al.* [398] by numerical simulations. Due to a higher mobility of interstitial vortices, larger antidots seem to be better for increasing j_c since they can stabilize multiquanta vortices and then there is no need to have loosely bound interstitial vortices, which reduce j_c.

The presented data clearly demonstrate that antidots, with $2r_a$ considerably larger than the coherence length $\xi(T)$, are efficient pinning centers. This conclusion is valid for different superconductors. As a possible explanation for the stronger pinning by larger antidots, we refer to the electromagnetic contribution to pinning, U_p, analyzed by Takezawa and Fukushima [456].

Composite vortex lattices

The saturation number $n_S = r_a/2\xi(T)$ can be varied by changing both the size of the antidots and temperature, since the coherence length $\xi(T)$ increases substantially when $T \to T_c$. Therefore, even for a fixed antidot radius r_a, different types of vortex lattices can be realized by tuning temperature into a proper interval and thus changing n_S. The formation of a composite vortex lattice with $n_S = 2$ ($2\Phi_0$-vortices at antidots and Φ_0-vortices at interstitials) is illustrated by Fig. 4.24, where the coexistence of these two different types of vortices is clearly seen.

We analyze now the magnetization data (Fig. 4.25), taken at $T/T_c = 0.870$ [343], when the width of the hysteresis loop $M(H)$ has made it possible to observe a finite loop width in a much broader field range — up to $20\,\text{mT}$. The $M(H)$ curve for the films with an antidot lattice demonstrates, as we have already seen before, four successive cusp-like anomalies at $H = H_1, \ldots, H_4$ corresponding to the formation of multiquanta vortex lattices carrying $\Phi_0, \ldots, 4\Phi_0$ flux quanta. Therefore, it is reasonable to assume that in this case the saturation number is at least four, which is close to the value expected from the calculated ratio $r_a/2\xi(T) \approx 3.3$ at $T/T_c = 0.870$. However, in higher fields, at $H = H_5$, the next matching anomaly does not show up as indicated by the vertical arrow in Fig. 4.25. This 'mysterious' disappearance of the matching cusp can be interpreted as

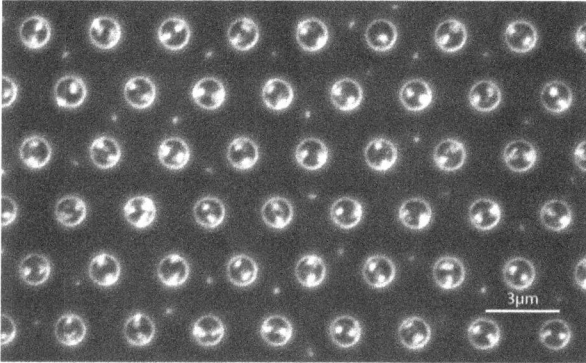

Fig. 4.24 Magnetic decoration pattern illustrating the coexistence of $2\Phi_0$-vortices at blind antidots and Φ_0-vortices at interstitial positions in a Nb film with a triangular lattice of blind antidots. (Adapted from Ref. 58 with kind permission from Springer Science+Business Media.)

an indication of the onset of pinning of vortices by interstices. Indeed, the position 'I_2' in the center of the parallelogram $\overline{A_1 A_2 A_3 A_4}$ (see the sketch of Fig. 4.25) formed by the antidots is a saddle point in the pinning potential produced by the antidots $A_1 - A_3$ (see the diagram of Fig. 4.25). As a result, interstitial vortices will be pinned at the two equivalent positions I_1 and I_3 corresponding to the two minima of the pinning potential. The number of flux lines required to fill in the antidot lattice all positions of the type I_1 and I_3 is twice as large as the number of antidots. All interstitial positions I_1 and I_3 will only be filled by flux lines at $H = H_6$, leading to a cusp at H_6 (Fig. 4.25). This explains the 'mystery' of the missing $M(H)$ cusp at $H = H_5$ (Fig. 4.25) and gives a strong evidence that at $H = H_4$ antidots are indeed saturated, i.e. that $n_S = 4$ at $T/T_c = 0.870$.

On the other hand, the presence of interstitial vortices can also be detected from a change of the effective flux, $\tilde{\Phi}_{0n}$, which is given by the slope of the M versus $\ln(H - H_n)$ curve (see Eq. (4.12)). By comparing the slopes of these plots for the different field intervals $H_n < H < H_{n+1}$, the presence of interstitials is revealed by a changed slope in the interval $H_{ns} < H < H_{ns+1}$ [343]. The mobility of the interstitial vortices and vortices trapped by the antidots was also probed with ac-susceptibility measurements [393], where the transition from intravalley vortex motion (Campell regime) to intervalley motion (critical state regime) was observed. The presence of the antidots substantially broadens the H–T area where the Campell regime with vortices oscillating around the same pinning site,

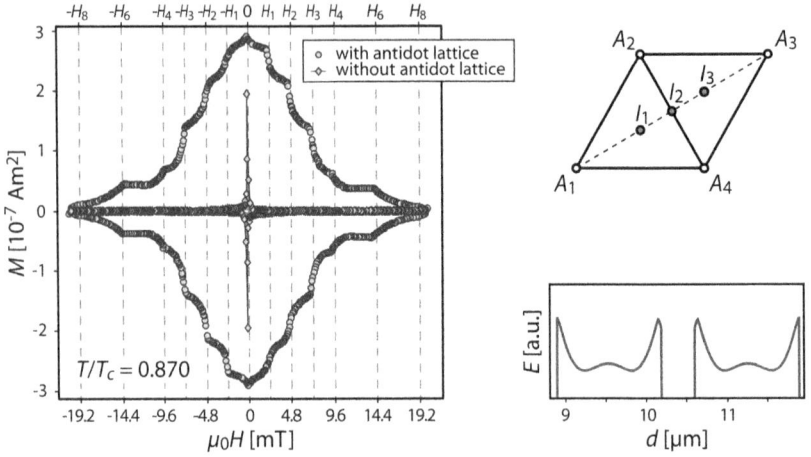

Fig. 4.25   Magnetization loop $M(H)$ at $T/T_c = 0.870$ of a $[\text{Pb}(10\,\text{nm})/\text{Ge}(5\,\text{nm})]_2$ bi-layer with and without antidot lattice measured in a broad field range (left). The sketch (right) shows the parallelogram $\overline{A_1 A_2 A_3 A_4}$ formed by the antidots. Interstitial flux lines will be pinned at positions I_1 and I_3. The diagram at the right shows a cross-section along the line $\overline{A_1 A_3}$ of the energy surface of one flux line in the antidot lattice, calculated with $\kappa = 15$, $d = \lambda$ and $r_a = 0.2\,\mu\text{m}$ for saturated antidots (after [343]).

is applicable. The corresponding linear ac-dynamics of these vortices was modelled in Ref. 120.

The assigned saturation number $n_S = 4$ at $T/T_c = 0.870$ can also be checked by investigating the variation of the sequence of matching anomalies in magnetization curves at different temperatures. For a triangular anti-dot lattice, the saturation number n_S simply defines the cusp-like match-ing anomaly at $H = H_{n_S}$ followed by the missing cusp at $H = H_{n_S+1}$. Therefore, the decrease in n_S caused by the increase of $\xi(T)$ at higher tem-peratures, should also lead to a shift of the missing $M(H)$ cusp at H_{n_s+1}. Figure 4.26(a) ($T/T_c = 0.942$, $n_S = 3$) and Figure 4.26(b) ($T/T_c = 0.971$ K, $n_S = 2$) convincingly demonstrate that this is indeed the case: at $T/T_c = 0.942$ there is no matching peak at H_4 (therefore, $n_S = 3$), and at $T/T_c = 0.942$ a very sharp cusp at H_2 precedes a missing cusp at H_3 ($n_S = 2$). The variation of n_S with temperature is summarized in Fig-ure 4.28 where the temperature dependence of n_S correlates well with the temperature dependence of $1/\xi(T) \propto (1 - T/T_c)^{1/2}$ (dashed line).

The pinning potential at antidots and at interstices is very different and it is also very sensitive to the temperature variation. As a result, as $T \to T_c$, the minima of the pinning potential at interstices become so shallow that

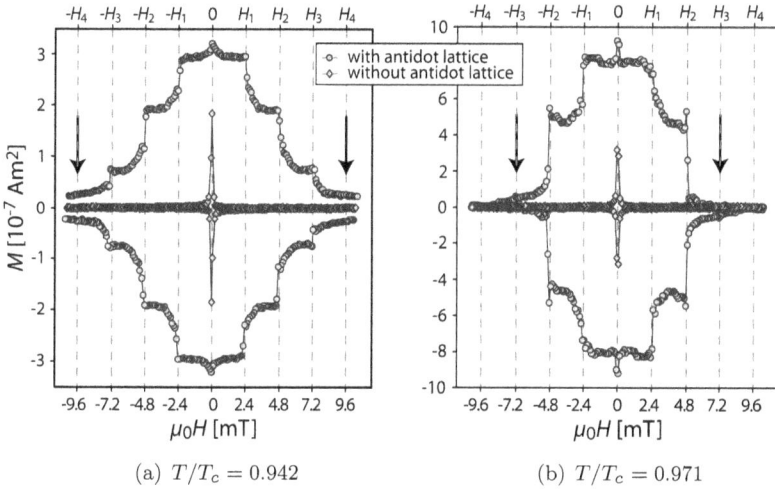

(a) $T/T_c = 0.942$ (b) $T/T_c = 0.971$

Fig. 4.26   Magnetization loops $M(H)$ of a $[Pb(10\,nm)/Ge(5\,nm)]_2$ bi-layer with and without a (triangular) antidot lattice at (a) $T/T_c = 0.942$ and (b) $T/T_c = 0.971$. Arrows indicate the missing $M(H)$ cusps (after [343]).

they cannot pin any longer the interstitial vortices, [30] which will then form an *interstitial flux fluid*. Therefore by changing temperature, one can induce at interstices the crossover from the flux solid to the flux fluid.

Summarizing, we would like to emphasize the unique possibilities to stabilize novel flux phases in superconducting films by making regular arrays of pinning centers (in our case, antidots). These phases are:

- *Multiquanta vortex lattices:* They can be realized if the antidots are sufficiently large and if the field does not exceed the value H_{n_S} defined by the saturation number n_S.
- *Composite vortex lattices:* These flux lattices are observed when the normalized radii of the antidots $r_a/2\xi(T)$ are sufficiently small and H exceeds the limiting field H_{n_S}. Also the pinning potential at interstices should not be very shallow to provide a weaker, but still sufficient pinning to form a softer interstitial flux solid. The composite vortex lattices are characterized by the coexistence of the two-weakly and strongly pinned-interpenetrating lattices at interstices (Φ_0-vortices) and antidots ($n\Phi_0$-vortices), respectively [29, 413].
- *Multiquanta solid vortex lattice with interstitial fluid of Φ_0-vortices:* This flux phase is formed when $T \to T_c$ and the inter-

stitial pinning potential becomes very shallow and thus cannot prevent the melting of the caged interstitial pinning Φ_0-vortices. The melting transition as a function of magnetic field is then observed exactly at $H = H_{n_S}$. For example, in Fig. 4.26(b), $n_S = 2$ and a very abrupt first-order phase transition is observed at $H = H_2$, whereas in Fig. 4.8(a), $n_S = 1$ and the transition is observed at $H = H_1$. In other words, the homogeneous $n\Phi_0$-vortex solid with $n = n_S$ produces in this case only the $B = $ const background while the melting of the interstitial Φ-vortex phase causes the very sharp magnetization drop at $H = H_{n_S}$. The interstitial vortex fluid reversibly responds to the field variation and this results in a zero width of the hysteresis loop in fields $H > H_{n_S}$.

The flux phases listed above can exist at temperatures not too far from T_c, since at lower temperatures the tendency to form a conventional Bean profile (Fig. 4.18(a)) starts to dominate and matching anomalies are suppressed.

In addition to the novel vortex phases listed above and already observed experimentally, the existence of intersticial giant vortices and symmetry-induced interstitial vortex-antivortex patterns (Fig. 4.27) in superconductors with a square antidot lattice has been predicted theoretically in the framework of the nonlinear Ginzburg–Landau model [46].

4.2.5   *Crossover from pinning arrays to networks ($n_S \gg 1$)*

The systematic studies of the efficiency of antidots, as artificial pinning centers, as a function of their radius r_a (Fig. 4.29(a)) have revealed [342] that for the core pinning combined with the electromagnetic pinning the optimum size of the antidots is not $\xi(T)$ at all, but rather $2r_a \gg \xi(T)$ [344]. As a result, the highest critical currents have been obtained for the multiquanta vortex lattices that can be stabilized by these sufficiently large antidots, since their saturation number is $n_S \approx r_a/2\xi(T) \gg 1$. At the same time it is quite evident that by increasing the antidot diameter we are inducing a crossover to another regime (Fig. 4.29(b)) when eventually $2r_a$ becomes nearly the same as the antidot lattice period d. In this case the width of the superconducting strips w between the antidots is so small that at temperatures not too far below T_c the superconducting network regime $w \leq \xi(T)$ can be realized. For this regime the $M(H)$ curves are characterized by the presence of very sharp peak-like anomalies at integer matching fields H_n (Fig. 4.29(b)) and a reproducible structure between H_n, which may correspond to rational matching peaks. Both integer and rational matching peaks have been also observed in various superconducting networks [206, 305, 377] and Josephson networks [134]. Visually the $M(H)$ curves in the *network regime* are quite different from those in the *multi-quanta vortex regime*: the former demonstrate the narrow $M(H)$ *peaks* at $H = H_n$ (Fig. 4.29(b)), while the latter show pronounced *cusps* at integer matching fields (Fig. 4.29(a)).

In the regime of a superconducting network, critical currents in moderate fields are already smaller than for the regime of the multiquanta vortex lattices, see Fig. 4.29(a). At higher fields, however, at least for the radius $r_a = 0.3\,\mu\text{m}$, critical currents are higher for the largest studied antidot diameter, i.e. the optimum antidot size for pinning is field-dependent. The reduction of the width of the superconducting 'stripes' between the anti-

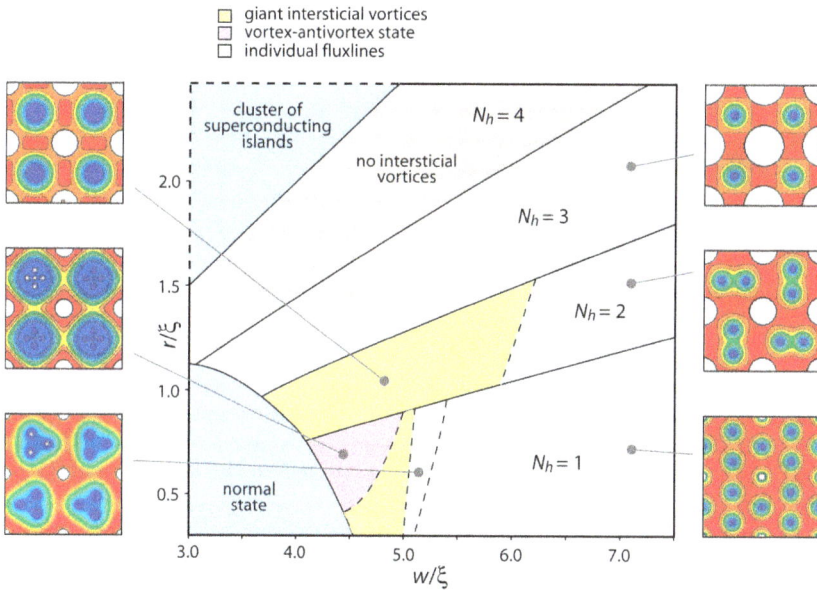

Fig. 4.27 Equilibrium vortex patterns at $H = H_4$ in a superconductor with a square antidot lattice, as a function of the antidot radius r and the period w of the lattice. Solid and dashed lines denote the first order transitions between the states with different antidot occupation number N_h, and second order configurational transformations, respectively. The insets show Cooper-pair density plots, where blue/red or dark-gray/medium-gray denotes low/high density (darkest color denotes vortices; white, antidots) of the corresponding states. Open dots are a guide to the eye, indicating the position of the zeros of the order parameter. (Adapted with permission from Ref. 46. Copyrighted by the American Physical Society.)

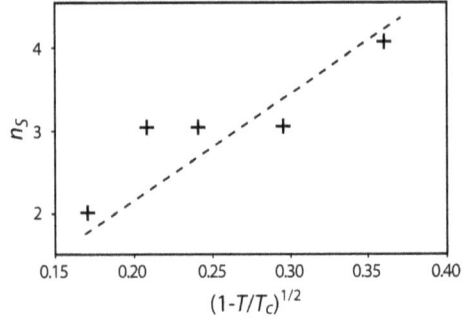

Fig. 4.28 The temperature dependence of the saturation number n_S correlates with the temperature dependence of $1/\xi(T) \propto (1-T/T_c)^{1/2}$, as indicated by the dashed line (after [343]).

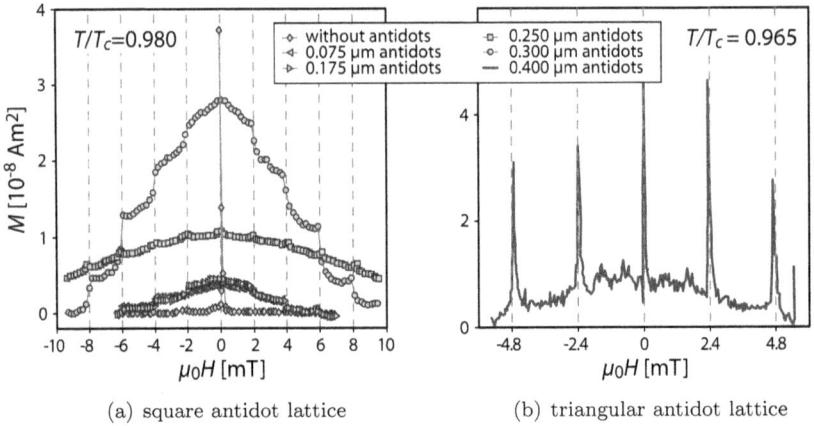

(a) square antidot lattice   (b) triangular antidot lattice

Fig. 4.29   (a) Magnetization curves at $T/T_c = 0.980$ of a single 60 nm thick WGe film with a square lattice of antidots with radius $r_a = 0.075 - 0.3\,\mu m$. For comparison, the data for a reference film without antidots is also shown. The matching fields $\mu_0 H_n \approx n \times 2.07\,mT$ are indicated by dashed lines. (b) Upper magnetization curve at $T/T_c = 0.965$ of a [Pb(15 nm)/Ge(14 nm)]$_3$ multilayer with a triangular antidot lattice of $r_a = 0.4\,\mu m$ (after [342]).

dots, needed to obtain high j_c in high fields, reflects actually a well-known designer rule for making superconducting cables, which usually consist of a bunch of very fine superconducting filaments embedded into a normal metallic matrix. Therefore, the optimum size of the pinning centers turns out to be field dependent: smaller antidots are more efficient in low fields, while in higher fields larger antidots are needed to optimize j_c. The efficiency of enhancing j_c by combining different antidot sizes in a single sample will be illustrated in the next section, where composite antidot lattices are considered.

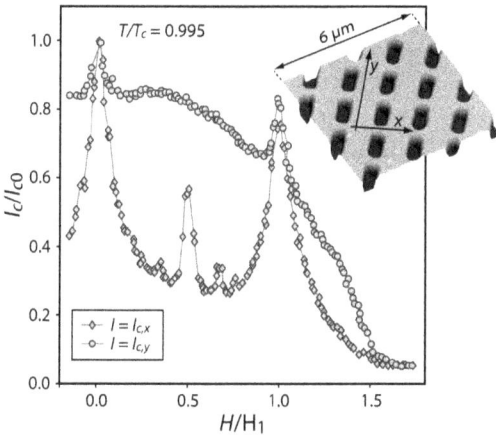

Fig. 4.30 Normalized critical currents $I_{c,x}$ and $I_{c,y}$ as a function of the external field for a 150 nm thick lead film with a square lattice of rectangular ($0.6 \times 1.1\,\mu m^2$) antidots, measured with a current in the x-(diamonds) and y-direction (circles). The inset shows an atomic force micrograph of the considered sample (after [482]).

Summarizing the analysis of the multiquanta vortices, it is worth noting that a superconducting film with a lattice of relatively large antidots seems to demonstrate the single-terrace critical state which appears due to the multiple connectivity of the film and the stabilization of the $n\Phi_0$-flux lattices. The separation of the areas where flux penetrates from those where the superconducting order parameter nucleates provides a kind of a 'peaceful coexistence' of flux lines pinned by antidots with the superconducting condensate in the space between them. Fabricating an antidot lattice to let flux go through, we are thus helping the order parameter between the antidots to sustain much higher applied currents and magnetic fields. The presence of the antidot lattice also broadens the validity of the London limit. Using the two essential assumptions: $B = $ const (single-terrace critical state, especially at matching fields $H = H_n$) and $|\psi| = $ const, close to T_c leads to a convincing quantitative description of the magnetization of the multiquanta vortex lattices, including linear behavior of $M(H)$ at matching fields $M(H_n) \propto -n\Phi_0/\Lambda^2$ and logarithmic behavior elsewhere: $M(H_n < H < H_{n+1}) \propto -\ln(H - H_n)$. By varying the saturation number $n_S = r_a/2\xi(T)$ through the use of different $\xi(T)$, one can demonstrate that the missing matching cusp at $H = H_{n_S+1}$, signaling the onset of the formation of the interstitial vortices at $H > H_n$ is systematically shifted to lower matching fields as $T \to T_c$.

Since vortex pinning can be controlled by varying the antidot size, an artificial in-plane anisotropy of the pinning force and the critical current can be introduced by making a periodic array of rectangular antidots. In these samples, a distinct anisotropy in the pinning properties is clearly observed: a higher critical current, see Fig. 4.30, and a sharper $I-V$ transition from normal to superconducting state [482] are seen when the current is applied along the long side of the rectangular antidots. Interestingly, depending on the exact trajectories of the depinned interstitial vortices [401], the transport in two in-plane perpendicular directions can still be anisotropic. This anisotropy, according to molecular dynamics simulations, is very pronounced for triangular, honeycomb and kagomé pinning arrays, but can be also observed even for the square pinning arrays. This anisotropy is expected to be strongly suppressed if, instead of interstitial vortices, multiquanta vortices are formed, thus strongly reducing the possibilities of in-plane anisotropic dynamical vortex states based on the presence of the interstitial vortices.

4.2.6   *Composite antidot lattices*

In this section, we consider a composite antidot array, consisting of two interpenetrating square lattices with the same period but different antidot size $a_1 > a_2$. The two sublattices are shifted with respect to each other by half a unit cell along x and y directions, so that the smaller antidots are placed in the centers of the unit cells of the lattice of larger antidots (see the insert in Fig. 4.31). This arrangement of antidots corresponds to the vortex lattice configuration at the second matching field in a sample with a single square array of antidots with $n_S = 1$. The purpose of introducing this arrangement of antidots is to enlarge the field range where the critical current density $j_c(H)$ is enhanced by having efficient pinning sites exactly at the locations where the mobile interstitial vortices would appear were the smaller antidots not present there. From the measured resistivity $\rho(T/T_c = 1.04) = 5.33\,\mu\Omega\cdot\text{cm}$, we estimate an elastic mean free path of $\ell = 9\,\text{nm}$, and therefore a superconducting coherence length $\xi(0) = 25\,\text{nm}$ in the dirty limit. These values are smaller than those obtained for the reference antidot sample where $\ell = 27\,\text{nm}$ and $\xi(0) \approx 40\,\text{nm}$. Since $\ell = 27\,\text{nm}$ was obtained in a film without antidots, co-evaporated with the sample containing the composite antidot lattice, this difference seems to be caused by the more complex lift-off procedure due to the presence of the small holes.

Knowing the mean free path ℓ and using the London penetration depth for the bulk material [369], we obtain $\lambda(0) = 71\,\text{nm}$. Due to the perforation, the effective penetration depth increases and, therefore, λ should be

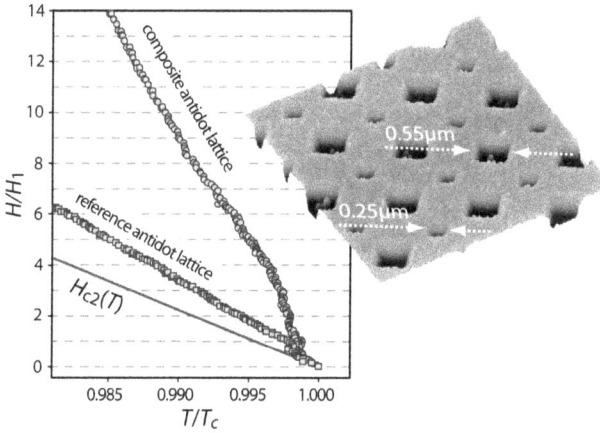

Fig. 4.31 $T_c(H)$ phase boundary (circles) for a 50 nm lead film with a composite antidot array, using a resistance criterion of $10\% R_n$. The two interpenetrating square lattices have the same period $d = 1.5\,\mu$m but different antidot size $a_1 = 0.55\,\mu$m and $a_2 = 0.25\,\mu$m. Squares show the phase boundary of a reference antidot lattice with $d = 1.5\,\mu$m and $a = 0.55\,\mu$m, using the same criterion. The solid line is the calculated linear $T_c(H)$ phase boundary for a plane film with the same coherence length $\xi(0) = 40$ nm as the reference antidot patterned film. The field axis is normalized to the first matching field $H_1 = 9.2\,$G (after [439]).

modified according to [494]

$$\Lambda(0) = \frac{\lambda(0)}{\sqrt{1 - \dfrac{2S_a}{S_t}}}, \qquad (4.12)$$

where S_a and S_t are the area of the holes and the total area per unit cell, respectively. As a result, the Ginzburg–Landau parameter $\kappa = \Lambda(0)/\xi(0)$ amounts to $3.4 > 1/\sqrt{2}$, meaning that the sample is a type-II superconductor.

Figure 4.31 shows the critical temperature $T_c(H)$ as a function of field for both the sample with a composite antidot lattice and the reference antidot film. The solid line depicts the expected upper critical field boundary of a plane film with the same coherence length as the reference antidot sample according to Eq. (1.11). The measured boundary of the reference antidot film is close to the $H_{c3}(T)$ line corresponding to the surface nucleation of superconductivity around the holes, whereas the solid line represents the bulk superconducting transition $H_{c2}(T)$.

Due to the presence of the antidot array, matching features appear in the $T_c(H)$ curve with a periodicity of $H_1 = \Phi_0/d^2 = 0.92\,$mT, corresponding to the lattice parameter $d = 1.5\,\mu m$. Local maxima are visible in the $T_c(H)$ curve of the composite array for all integer matching fields H_n with $n = 1, \ldots, 6$, whereas no evidence of rational matching features is observed. Thus, the addition of the extra antidot in the center of the unit cell of the array with large antidots leaves the matching period unchanged. This is an important observation, since the composite antidot lattice can also be regarded as a square lattice, tilted by 45°, with a unit cell twice as small as that of the original lattice. If this were the periodicity felt by the vortices, the matching period would amount to $1.84\,mT$, which is twice as large as the observed period. In that case, one would expect the local maxima at even matching fields $H_{2,4,\ldots}$ in Fig. 4.31 to be more pronounced than the ones at odd matching fields $H_{1,3,\ldots}$. Since this is not the case, one can conclude that all these peaks correspond to integer matching fields, indicating that the main period felt by the vortices is the period of the lattice with large antidots.

Further information can be gained from the $R(T)$ transition width as a function of H

$$\Delta T_c(H) := T(R \equiv 0.03 R_n) - T(R \equiv 0.01 R_n), \qquad (4.13)$$

as it is shown in Fig. 4.32. In this plot, three different regimes can be clearly distinguished:

- *Collective regime:* For $H < H_4$, the coherence length is larger than the width of the strands, thus leading to a parabolic background in the $T_c(H)$ phase boundary. In this so-called 'collective' regime, the $R(T)$ transition width remains almost constant.
- *Interstitial regime:* For fields higher than H_4, an increase of the transition width can be observed, superposed with matching features at H_5 and H_6. The sudden increase in the transition width can be interpreted as a crossover to the regime where interstitial vortices appear in the sample. The interstitial regime is indicated by the hatched area in Fig. 4.32 for the composite array. This regime ranges up to $3.6\,\xi(T) = d - a$, i.e. up to $\sim H_8$, where a change in the $T_c(H)$ slope can be observed.
- *Single object regime:* For higher fields, the single object regime is entered, where a linear phase boundary slightly distorted by an oscillation with period $H^* = \Phi_0/a_1^2 \sim 6.9\,mT$ is expected [412]. Although the linear phase boundary is indeed observed, single object oscillations are difficult to resolve in the narrow field range investigated.

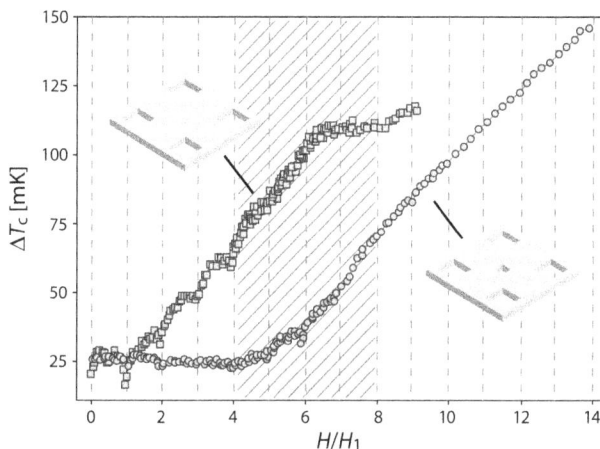

Fig. 4.32   The transition width (Eq. (4.13)) of a 50 nm lead film with a composite antidot array ($d = 1.5\,\mu$m, $a_1 = 0.55\,\mu$m and $a_2 = 0.25\,\mu$m — circles) and a reference antidot patterned film ($d = 1.5\,\mu$m, $a = 0.55\,\mu$m — squares). The hatched area indicates the 'interstitial' regime for the composite array (after [439]).

For comparison, Fig. 4.32 also shows $\Delta T_c(H)$ for the reference antidot sample. From this curve it is easy to see that if the smaller additional antidots are absent, the crossover to the interstitial regime occurs at $H \sim 1.5 H_1$. Therefore, the presence of the smaller antidots has substantially delayed the appearance of interstitial vortices. From the $\Delta T_c(H)$ curve, we thus conclude that the total number of trapped flux quanta per unit cell of the antidot lattice is at least four.

The observed behavior can be explained as follows. Up to H_1, the vortices will be pinned by the large antidots, while between H_1 and H_2, the small antidots are also occupied by vortices. Due to their size, these small antidots trap at most a single quantum vortex. Therefore, they will be completely saturated at H_2, thus creating repulsive potentials at the positions of the small antidots. Figure 4.33 shows a schematic evolution of the potential landscape along a diagonal of the array (see the dashed line in the inset at the left) that would be experienced by a vortex for $H = 0$, $H = H_1$ and $H = H_2$. Since the large antidots pin one flux quantum, at $H = H_1$ a surface barrier has emerged at the antidot edges. For $H = H_2$, the contribution to the potential of the small antidot at the center of the unit cell is strongly repulsive. When additional vortices enter the sample at $H > H_2$, they will be pushed towards the large antidots, leading to an increase of their effective saturation number. In other words, the additional repulsive potential at the small antidots helps to increase the saturation number of the larger antidots. We therefore conclude that of the four flux quanta trapped per unit cell of the composite antidot lattice, one is pinned by the small antidot while three are pushed into the larger holes. This leads to a substantial broadening of the field range where a strong enhancement of $T_c(H)$ is observed. A similar picture was introduced by Doria *et al.* [130] to explain the multiple trapping of vortices at high fields, as a result of the pressure exerted by the external vortices into the pinning site.

In order to study the vortex dynamics within the superconducting state, we now turn to isothermal critical current measurements close to $T_c(H)$, see Fig. 4.34. The absolute value of the critical current density at zero field for the composite antidot array amounts to $I_{c0} = 6.8 \times 10^8 \, \text{A/m}^2$ at $T/T_{c0} = 0.974$. This value is a bit lower than the critical current density obtained for the reference antidot lattice, in part due to the specific geometry of the lateral nano-patterning which influences the current distribution throughout the film, hereby also affecting the critical current I_{c0}. In order to compare the pinning properties of the film with the composite antidot lattice (circles in Fig. 4.34) and the reference antidot lattice (solid lines) measured at the same reduced temperature, we have normalized the critical current by I_{c0}. Notice that since the saturation number n_S is mainly determined by the coherence length [332, 365] $\xi(T)$, which in turn depends solely on the reduced temperature T/T_c, it is enough to compare the results obtained on these samples at the same reduced temperature, without the necessity of normalizing the field.

The $I_c(H)/I_{c0}$ curves for the film with a composite antidot array (Fig. 4.34) have been measured for negative fields (circles) and symmetrized

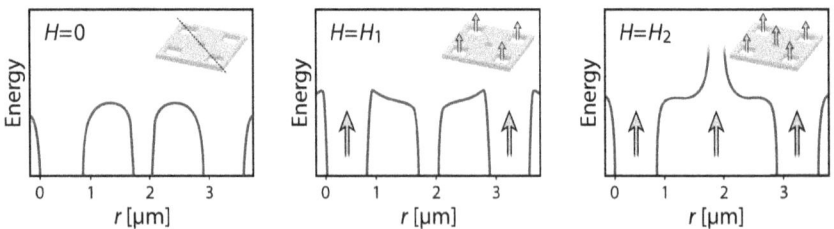

Fig. 4.33 Schematic representation of the potential along a diagonal of the composite antidot lattice (see the dashed line in the inset at the left), experienced by a vortex entering the sample for $H = 0$, $H = H_1$ and $H = H_2$. Pinning by large antidots is enhanced by placing a repulsive potential at small antidots after their saturation (after [439]).

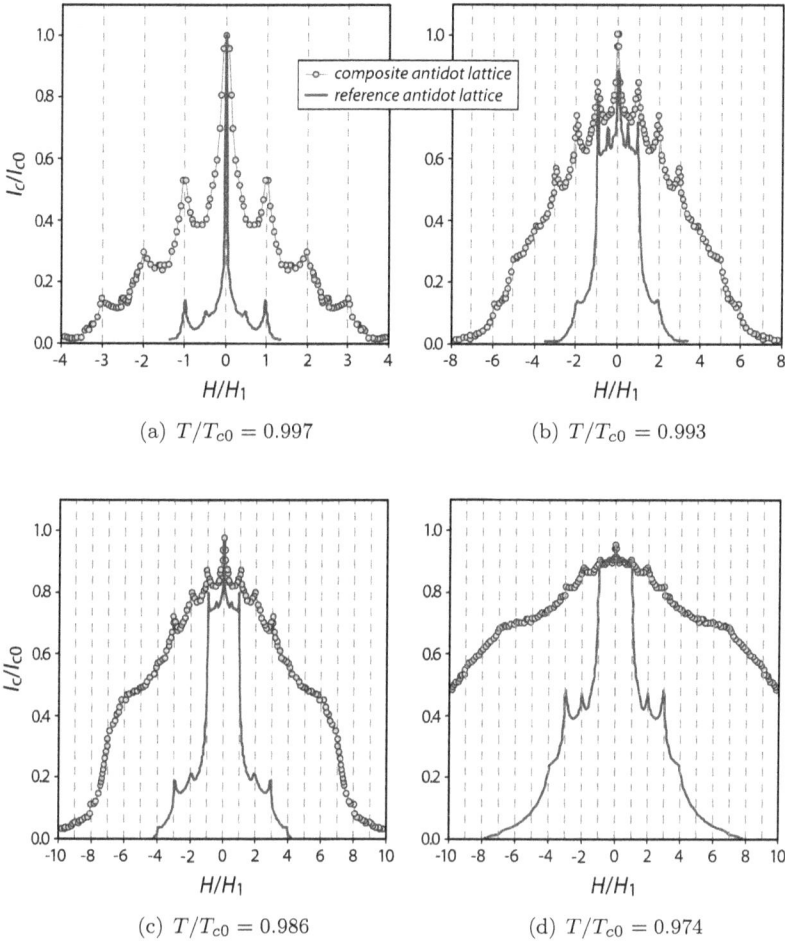

(a) $T/T_{c0} = 0.997$

(b) $T/T_{c0} = 0.993$

(c) $T/T_{c0} = 0.986$

(d) $T/T_{c0} = 0.974$

Fig. 4.34 Normalized critical current I/I_c at (a) $T/T_{c0} = 0.997$, (b) $T/T_{c0} = 0.993$, (c) $T/T_{c0} = 0.986$ and (d) $T/T_{c0} = 0.974$ of a film with a composite antidot array. The curves were measured for $H > 0$ (circles) and symmetrized for $H < 0$. For comparison, the solid lines show the normalized critical current obtained for a film with the reference antidot lattice (after [342]).

for $H > 0$ for clarity. All curves show distinct periodic matching features, with a period H_1 corresponding to the unit cell of the lattice with the large (or the small) antidots ($d = 1.5\,\mu\text{m}$. At $T/T_{c0} = 0.997$ (Fig. 4.34(a)), the $I_c(H)/I_{c0}$ curve for the film with a composite antidot lattice shows sharp maxima at H_1, H_2 and H_3. This behavior is expected at temperatures sufficiently close to T_{c0}, where it is not possible to have interstitial vortices in the superconducting strands between the antidots. As we mentioned above, interstitial vortices appear in the sample only for $T/T_{c0} \leq 0.994$. Accordingly, at a lower temperature of $T/T_{c0} = 0.993$ (Fig. 4.34(b)), and all temperatures below that (Fig. 4.34(c) and (d)), a strong enhancement of $I_c(H)/I_{c0}$ in the film with a composite antidot lattice can be found for fields higher than the first matching field H_1, compared to the reference antidot lattice. The reason for this lies in the ability of the composite antidot lattice to pin more flux quanta inside the antidots compared to the reference antidot array. It should be noted that the field range where the film has a finite critical current, i.e., where the film remains superconducting, is considerably broader for the film with the composite than for the reference antidot array.

The appearance and sharpness of the matching features in the $I_c(H)/I_{c0}$ curves are temperature-dependent. At $T/T_{c0} = 0.993$ (Fig. 4.34(b)), every integer matching peak up to H_6 can be clearly seen. The maxima at H_1, H_2 and H_3 are very pronounced. At H_4 and H_5, one finds cusps rather than local maxima in $I_c(H)/I_{c0}$, whereas the matching feature at H_6 is again peak-like. This indicates that the vortex patterns formed at H_4 and H_5 are less stable than the vortex configuration at H_6.

When the temperature is lowered down to $T/T_{c0} = 0.986$ (Fig. 4.34(c)), we find again sharp matching features in $I_c(H)/I_{c0}$ at H_1, H_2 and H_3, and only very weak cusps at H_4 and H_5. At H_6, the local maximum has developed into a pronounced cusp, after which a substantial change in the $I_c(H)/I_{c0}$ slope occurs. A second smaller slope change can be seen at H_7. At the lowest measured temperature of $T/T_{c0} = 0.974$ (Fig. 4.34(d)), the only matching features left are the sharp local maxima at H_1, H_2 and H_3, together with one pronounced cusp at H_7. It appears that, at this temperature, the seventh matching field H_7 plays the same role as the sixth matching field at $T/T_{c0} = 0.986$. This fact indicates that at $T/T_{c0} = 0.974$, the total number of trapped flux quanta per unit cell of the composite lattice increases, most probable, from four to five.

It is worth noticing that the normalized critical current at the first matching field H_1 reaches approximately the same value for the film with the composite and with the reference antidot lattice, except for the $I_c(H)/I_{c0}$ curve obtained at $T/T_{c0} = 0.997$. This fact makes the film with the reference antidot array a good candidate to compare its pinning properties with those of the composite antidot array.

Molecular dynamics simulations were performed to obtain the vortex patterns at the matching fields H_5, H_6 and H_7. To model the composite

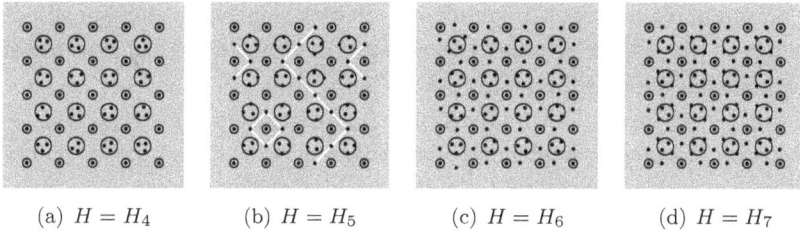

(a) $H = H_4$ (b) $H = H_5$ (c) $H = H_6$ (d) $H = H_7$

Fig. 4.35 Suggested vortex pattern at (a) $H = H_4$, (b) $H = H_5$, (c) $H = H_6$ and (d) $H = H_7$. All patterns have been obtained by molecular-dynamics simulations by an annealing procedure, except the one at $H = H_4$. Open circles and black dots represent pinning sites and single quantum vortices, respectively. Note that molecular dynamics simulations are done only for Φ_0-vortices. In view of that, n vortices trapped by bigger antidots correspond in fact to $n\Phi_0$ multiquanta vortices (after [439]).

vortex lattice, two interpenetrating arrays of Gaussian sites with a different radius and a different pinning force were used. Figure 4.34 shows the vortex configurations we suggest for H_4, H_5, H_6 and H_7. multiquanta vortices are represented in this model by a multiple occupation of a pinning site with (repulsive) single-quantum vortices. Since in the experiment the pinning sites consist of real holes in the film, the vortices trapped in the same pinning site will be interpreted as multiquanta vortices, even though they are depicted as separate single flux quantum entities in the plots. The vortex pattern at H_4, which is drawn schematically and was not calculated, depicts all antidots occupied with the maximum number of vortices. The large antidots trap $3\Phi_0$-vortices, while the smaller antidots trap only one Φ_0-vortex. No interstitial vortices are present in the sample. At H_5, there is one interstitial vortex per unit cell of the array. A tendency can be seen of interstitial vortices forming diagonal lines, which make zigzag traces across the sample, indicated by white lines in Fig. 4.34(b). Thus the obtained pattern is consequently not very stable. At H_6, a highly symmetric vortex pattern is formed. In this case, two interstitial vortices are present per unit cell, which are positioned approximately at the center of the line connecting two neighboring large antidots. Due to its high symmetry, the vortex pattern at H_6 can be very stable. This result is consistent with the fact that the matching features at H_6 in $T_c(H)$ or $I_c(H)/I_{c0}$ are more pronounced than at H_5 (see for example Figs 4.32 and 4.34(c)). At H_7, the calculations were not able to produce a regular vortex pattern with an occupation of three at the large pinning sites and one at the smaller pinning sites. This indicates that the vortex pattern at H_7 is not very stable [394, 516].

In summary, a composite pinning array with smaller antidots, positioned exactly at the centers of cells of an array of large antidots, can stabilize the vortex lattice at several matching fields going from H_1 to at least H_7. Measurements of the critical temperature $T_c(H)$ and current $I_c(H)/I_{c0}$ as a function of magnetic field have demonstrated that the composite antidot lattice can trap a considerably higher amount of flux quanta per unit cell inside the antidots, compared to a reference antidot film without the additional small antidots. This means that the appearance of weakly pinned interstitial vortices in the composite antidot lattice is delayed to higher magnetic fields. The presence of the smaller antidots therefore increases the effective saturation number of the large antidots, which leads to a considerable expansion of the field range in which an enhanced critical current is observed. Since smaller antidots pin better in low fields while larger antidots in higher fields, a broad distribution of different sizes of the pinning centers should be introduced in a superconducting material in order to enhance its critical current in a broad field range.

So far, we have analyzed vortex behavior in presence of periodic pinning arrays (antidots). We have seen that the critical current at matching fields can be dramatically increased, but at the same time, between the matching fields, large vortex arrays/lines/domains can be depinned, thus reducing the critical current. To prevent this vortex depinning at non-commensurate fields and to block channeling of interstitial vortices, a variety of aperiodic, fractal and quasi-periodic arrays have been used [248, 263, 330, 331, 402, 438, 487], including Penrose lattices. In particular, molecular dynamic simulations of vortex matter, interacting with a five-fold two-dimensional Penrose array of pinning centers, predict higher critical currents in comparison with periodic pinning arrays. These predictions were confirmed experimentally [248, 438], and stronger critical currents can indeed be attributed to the suppression of vortex channeling and of the depinning of large vortex domains/rows. Moreover, vortex-vortex interactions on the Penrose lattice cause also the formation of 'vortex corals', where a giant $2\Phi_0$-vortex in the center is surrounded by a ring of Φ_0-vortices [263]. Additionally, admixtures of randomly distributed insulating nanocomposites, like nanodots [196, 205], can also strongly enhance the pinning force.

4.3 Ratchet effects in antidot lattices

From the point of view of classical thermodynamics, it is not possible to induce directed motion of particles by using equilibrium fluctuations only, otherwise it would become a realization of a *perpetuum mobile* of the second kind [445]. Nevertheless, non-equilibrium fluctuations, such as periodic excitations or a 'colored' noise, are allowed to take advantage of the asymmetry of a periodic ratchet potential to promote motion of particles in a preferential direction [403]. New solid state-based ratchet systems are currently being developed for controlling the motion of electrons [288] and fluxons, as well as for particle separation [309]. In particular, ratchet pinning potentials in superconducting devices may be very useful to control the dissipative motion of fluxons, which causes undesired internal noise.

Modern lithographic techniques make it possible to fabricate periodic arrays of vortex pinning sites with size and shape that can be easily tuned, thus opening new perspectives for making different asymmetric pinning potentials. In this context, several ideas to control flux motion by applying an *ac*-excitation have been proposed [280, 368, 495, 517, 518], and several of them have been realized experimentally [488, 504]. One realization has been recently implemented on a niobium film with a square array of nanoscopic triangular magnetic dots [488]. The authors reported rectification of the *ac*-driven vortices due to the asymmetric shape of the dots. Nevertheless, the detailed dynamics of vortices in such structures is not yet completely understood. Simpler model vortex ratchet systems can be made on the basis of asymmetric antidots. Interestingly, besides asymmetry in space pinning sites, drive asymmetry in time can also be used to realize vortex ratchets [109].

4.3.1 *Vortex rectification in films with asymmetric pinning*

Now we shall consider a composite square array of pinning sites, with its unit cell consisting of a small and a big square antidot placed close to each other and separated by a narrow superconducting wall, as a vortex rectifier. As demonstrated by both the *dc*- and *ac*-transport measurements at several fields and temperatures, this configuration is able to break the reflection symmetry of the total effective pinning potential and promote flux quanta rectification [477]. Moreover, these data reveal a remarkable hysteresis in the current-induced pinning–depinning process. This gives an apparent inertia to the system with important consequences in the overall dynamics. We will also discuss how standard overdamped models fail in describing these results, and propose an underdamped ratchet model that provides a very good fit to our experimental data.

The pattern was made by electron-beam lithography on a SiO_2 sub-strate. The superconducting aluminium film of 38 nm thickness was pre-pared in a $5 \times 5 \, mm^2$ cross-shaped geometry in order to allow four-point electrical transport measurements in two perpendicular current directions. The central part of the cross consisted of two strips of 300 µm width, con-taining the nano-engineered array of asymmetric pinning sites with a period of $d = 1.5 \, \mu m$ (see the inset in Fig. 4.36). For this specific geometry, a value of 0.92 mT corresponds to the first matching field. From the measured val-ues of the coherence length and the penetration depth (see Tab. 4.1), we can conclude that the patterned film with the composite array of square antidots is a type-II superconductor ($\kappa = \lambda(0)/\xi(0) = 2.89 > 1/\sqrt{2}$).

Figure 4.36(a) shows a contour plot of the magnetic field and current dependence of the difference in induced voltage for the *dc*-currents flowing in the positive and negative x direction, $V(I_x, H) - V(I_{-x}, H)$, at 97.3 % of the critical temperature. The voltage drop in the x direction is a measure of the vortex velocity in the y direction. Since the I_{-x} and I_x *dc*-currents induce a Lorentz force in the negative and positive y directions, the vortices, driven by this force, will probe the asymmetry of the pinning sites. We see that for all fields in the range of $H = [0, 1.1 \times H_1]$, there is a *dc*-voltage difference confirming the presence of an asymmetry in the pinning potential. This difference is the largest for fields just below the first matching field, which can be explained by the high symmetry of the vortex configuration that cancels out the vortex-vortex interactions. In this case the resulting forces acting on vortices originate exclusively from the asymmetric pinning potential.

To obtain the data presented in Fig. 4.36(b), a 1 kHz sinusoidal current was applied to the sample and the output signal was measured with an oscilloscope used as an analog-digital converter. Integration of this signal

Table 4.1   Characteristics of the aluminium film with asymmetric pinning lattice

Superconducting Property	Method of Investigation
$T_{c0} = 1.469 \, K$	$R(T)$ measurement, using a criterium of 10% R_n
$\ell = 3.9 \, nm$	residual resistivity at 4.2 K
$\lambda(0) = 195 \, nm$	dirty limit expression
$\xi(0) = 67.5 \, nm$	dirty limit expression

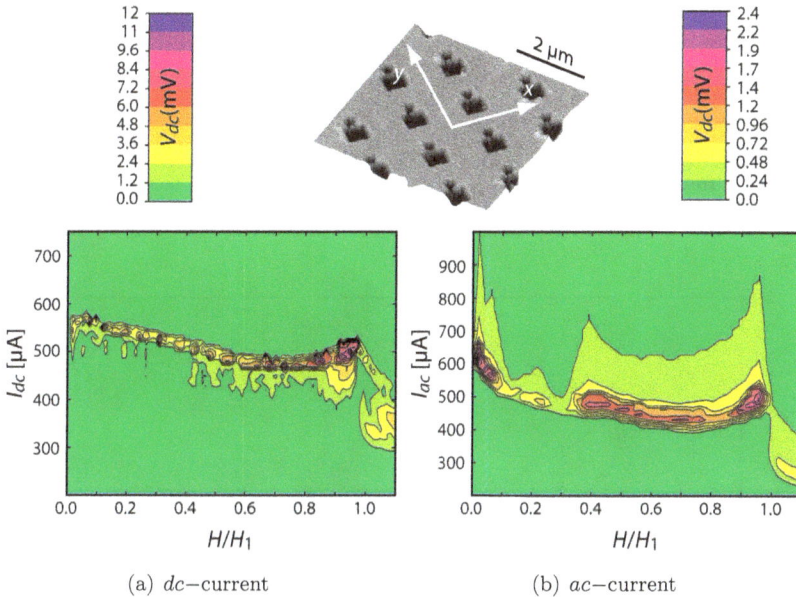

Fig. 4.36 (a) Contour plot of the magnetic field and *dc*-current dependence of the voltage difference for *dc*-currents applied in the x and the $-x$ direction $V(I_x, H) - V(I_{-x}, H)$ at $T/T_c = 0.973$. (b) Contour plot of the net *dc*-voltage $V_{dc}(I_{ac}, H)$ as a function of the magnetic field and amplitude of an *ac*-sinusoidal current at a frequency of 1 kHz and $T/T_c = 0.973$. The inset shows an atomic force micrograph of a 5×5 μm^2 area of the asymmetric pinning sites (after [477]).

gives the average *dc*-response, which is plotted as a function of the normalized magnetic field, H/H_1, and of the amplitude of the applied current I_{ac}. As before, a ratchet effect can clearly be seen at all fields between 0 and $1.1 \times H_1$. This effect is maximum just below H_1 and $1/2 H_1$, where vortex-vortex interactions are mostly canceled. Measurements carried out at other temperatures between $T/T_c = 0.967$ and 0.980 showed similar features. Nevertheless, the rectification effect is weaker at higher temperatures, which can be explained by an increase in $\xi(T)$ and $\lambda(T)$, making the size of the vortices big compared to the antidot size. In this case the vortices cannot sense the asymmetry of the pinning potential anymore.

As it is seen from the vertical scans in Fig. 4.36(b), the net *dc*-voltage increases sharply with the *ac*-current amplitude up to a maximum value before decaying smoothly to zero (see also Fig. 4.37(e)), which is a typical behavior of *ac*-driven objects in a ratchet potential in the adiabatic regime. At low frequencies, the net *dc*-flow $\langle v \rangle$ of particles increases mono-

(a) 477 µA   (b) 502 µA   (c) 544 µA   (d) 704 µA

(e) Net *dc*-voltage

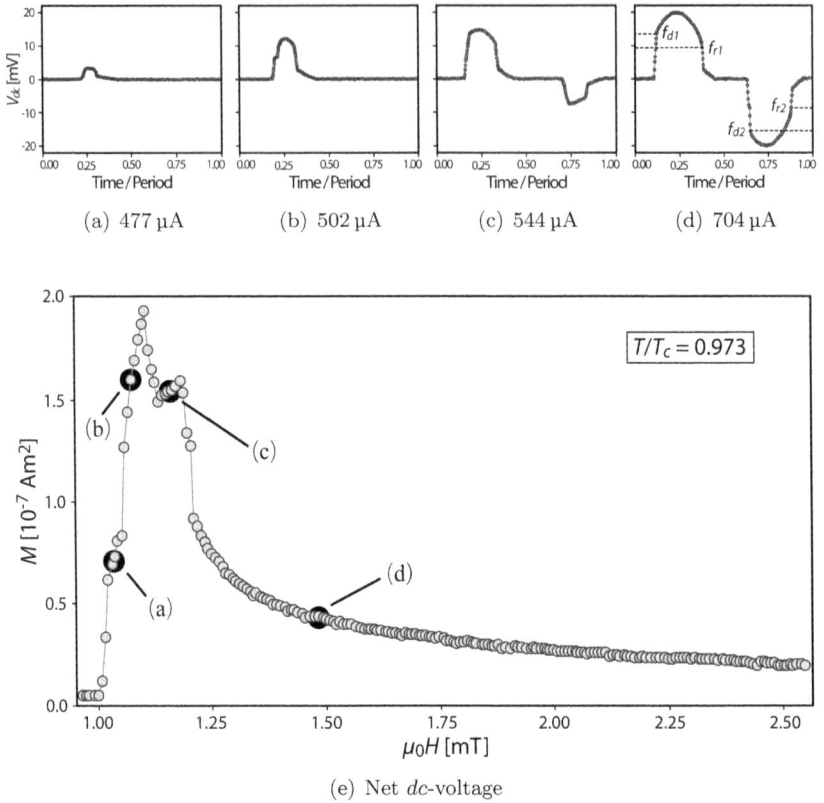

Fig. 4.37   (a–d) Voltage output at $I_{ac} = 477\,\mu A$, $502\,\mu A$, $544\,\mu A$ and $704\,\mu A$. (e) The net *dc*-voltage as a function of *ac*-amplitude, normalized by $I_{d1} = 473\,\mu A$, at $H = 0.96 \times H_1$ and $T/T_c = 0.973$ (after [477]).

tonically when the excitation amplitude A is between the weaker, f_{d1}, and the stronger, f_{d2}, depinning forces of the asymmetric potential (Fig. 4.37). We shall refer to this region as the *rectification window*. Since in this amplitude range the force cannot overcome the potential barrier for negative force direction, vortex motion occurs only during the positive half loops of the *ac*-excitation, that is, the system behaves as a half-wave rectifier. For amplitudes $A > f_{d2}$, the particles (vortices) are allowed to travel back and forth, thus leading to a decrease in the rectification signal, which vanishes slowly at high drives as the ratchet potential is gradually averaged out by the fast-moving particles. It is worth noting that these general properties of the ratchet effect also depend on the shape of the *ac*-excitation [403]. For instance, square-wave excitations lead to sharper rectification effect and a shorter tail in the $\langle v \rangle$ versus A characteristics as compared to sinusoidal excitations. Since the regime of the *dc*-measurements is equivalent to irradiating the system with a square wave pulse, the *dc*-data presented in Fig. 4.36(a) is significantly sharper than the *ac*-data shown in Fig. 4.36(b).

A closer look at the *ac*-amplitude dependence of the net *dc*-voltage reveals additional and important features of the ratchet dynamics in the considered system. Figures 4.37(e) and 4.38(a) show detailed *dc*-voltage versus *ac*-amplitude characteristics. For fields below the first matching field, some of the $V_{dc}(I_{ac})$ curves have two peaks, as for example the one at $H = 0.96H_{c1}$ shown in Fig. 4.37(e). Detailed measurements of the time evolution of the output voltage $V(t)$ suggest a superposition of two ratchet effects. Rectification is first triggered just above $I_{ac} = 473\,\mu A$ where the output signal is only present at the positive half-loop of the *ac*-current (see Fig. 4.37(a)). At $I_{ac} \simeq 495\,\mu A$, a stronger ratchet is triggered, which can be identified in Fig. 4.37(b) as a second jump in the signal in the positive half-loop. Right above $I_{ac} = 520\,\mu A$ a small signal in the negative half loop appears (see Fig. 4.37(c)), indicating that the weaker ratchet is already outside its rectification window. Finally, in Fig. 4.37(d), the observed signal is strong in both directions, meaning that the driven vortex lattice moves back and forth as a whole.

A possible reason for the presence of two ratchets can be a complicated plastic dynamics in which vortex channels flow along specific rows of pinning centers for currents considerably lower than the depinning current of the whole vortex lattice. Plastic dynamics was thoroughly studied by numerical simulations in periodic arrays of symmetrical pinning centers [400]. The local decrease in pinning efficiency in a given pinning row is caused by discommensurations, i.e. vacant pinning sites for fields $H < H_1$, distributed along the row. The extension of these ideas to arrays of asymmetric pinning sites seems to be rather straightforward. At fields very close or at H_1 this effect is minimized and, accordingly, the $V_{dc}(I_{ac})$ curve at $H = 0.98H_1$ has a single well-defined rectification peak (see Fig. 4.38(a)). The presence of the symmetry breaking pinning sites also strongly affects flux penetration and thermomagnetic avalanches, as observed in magneto-optical experiments [315].

Another remarkable feature disclosed in the $V(t)$ data is that the repinning force f_r necessary to stop vortex motion in either direction is always smaller than the corresponding depinning force f_d (see Figs. 4.37(a–d)), meaning that $V_{dc}(I_{ac})$ is hysteretic. This is a very robust phenomenon

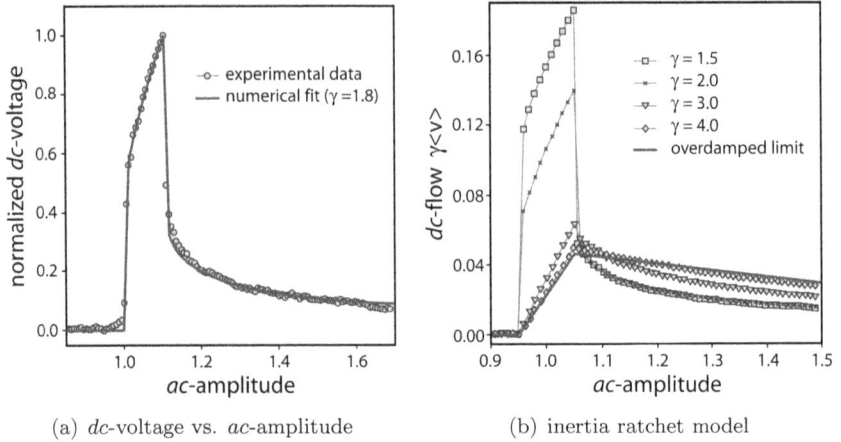

(a) *dc*-voltage vs. *ac*-amplitude   (b) inertia ratchet model

Fig. 4.38   (a) Normalized *dc*-voltage as a function of *ac*-amplitude. Circles show the experimental data at $H/H_1 = 0.98$ and $T/T_c = 0.973$, and a line the numerical fit using the inertia ratchet model (Eq. (4.14)) with only one fitting parameter, $\gamma = 1.8$. (b) Numerical integration of the inertia ratchet model for different values of γ (symbols) and the overdamped limit solution (line, $\gamma = 0$) (after [477]).

which has been observed in the entire investigated field, frequency and temperature ranges. For a deeper understanding of the above observations, the considered system can be modelled by molecular dynamics simulations of the overdamped motion of vortices in a square array of asymmetric pinning potentials at fields close to H_1, subjected to a sinusoidal Lorentz force of very low frequency [517, 518]. In the whole range of excitation amplitudes, no hysteresis was observed ($f_r = f_d$) since the $v(t)$ curves at each half period are symmetrical, in contrast with the clearly asymmetrical shape of the $V(t)$ experimental data. In addition, the $\langle v \rangle (A)$ characteristics look qualitatively different from the experimental data.

A possible reason for the misfit between the calculations and the experimental results is that the depinning process of an overdamped particle is reversible since it carries no significative inertia and cannot be deformed. However, vortices are soft objects that can be deformed under the action of competing forces and they also carry a small mass which, in the dirty limit, can be several times smaller than the electron mass. In this case, inertia is only relevant at frequencies typically above 10^{12} Hz [452]. Since our sample is in the dirty limit, the contribution of conventional vortex mass can be assumed to be negligible in the present case, meaning that the apparent inertia is more likely due to vortex deformation. Indeed, recent numerical simulations of the Ginzburg–Landau equations [388] suggested that the current-induced depinning of vortices in a periodic array of antidots is dominated by strong elongation of the vortex cores towards a neighboring antidot. Interestingly, the authors observed hysteresis in the pinning-depinning process resulting from this behavior.

To illustrate how this hysteresis influences the overall ratchet dynamics, vortices can be still modelled as rigid particles but carrying an apparent mass M. For simplicity, we consider that the vortex lattice is perfectly commensurate with the pinning array and therefore we shall analyze the experimental results measured very close to the first matching field ($H = 0.978H_1$), see Fig. 4.38(a). In this case, due to the cancelation of the vortex-vortex interactions, the vortex dynamics can be reduced to that of one particle moving in a one-dimensional potential. The corresponding equation of motion is

$$M\dot{v} = -\eta v - U_p'(x) + A\sin(\omega t), \tag{4.14}$$

where η is the viscous drag coefficient, v is the instantaneous vortex velocity and $U_p(x)$ is the one-dimensional pinning potential, which can be chosen in a simple form as

$$U_p(x) = -U\left[\sin\left(\frac{2\pi x}{a_p}\right) + \frac{1}{40}\sin\left(\frac{4\pi x}{a_p}\right)\right]. \tag{4.15}$$

This gives an asymmetry ratio $f_{d1}/f_{d2} = 0.925$ similar to that observed in our data shown in Fig. 4.38(a) ($I_{d1} = 480\,\mu A$ and $I_{d2} = 519\,\mu A$). In order to be closer to the experimental conditions, we will restrict ourselves to the adiabatic regime, that is, we shall consider only frequencies much lower than the libration frequency of the pinning potential $\omega_p \simeq 2\pi\sqrt{U/Ma_p^2}$ and the viscous drag to mass ratio $\gamma = \eta/M$. In this way, when normalizing the *ac*-amplitude by the first depinning force and the *dc*-response by the maximum response, the only fitting parameter left for comparing the calculations with the experimental data is γ.

The overdamped limit corresponds to $M \to 0$, or, equivalently, to $2\pi\gamma \gg \omega_p$. In this limit a semi-analytical approach may be used: The average flow of vortices in one cycle of the *ac*-excitation is given by

$$\langle v \rangle = \frac{1}{P}\int_0^P v(t)\,dt\,, \tag{4.16}$$

where $P = \omega/2\pi$ is the forcing period and

$$v(t) = \frac{a_p}{\eta}\left[\int_0^{a_p} \frac{dx}{-U_p'(x) + A\sin(\omega t)}\right]^{-1}. \tag{4.17}$$

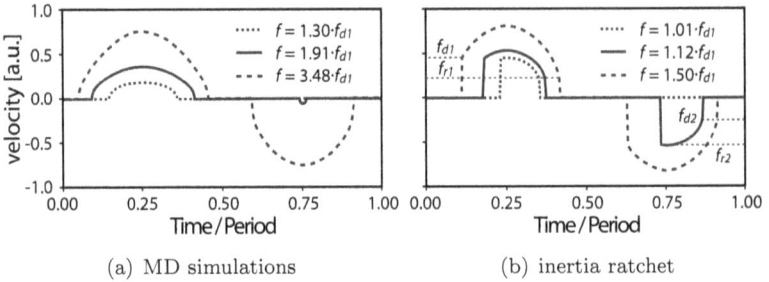

(a) MD simulations (b) inertia ratchet

Fig. 4.39 Time evolution of the vortex velocity $v(t)$ from (a) an overdamped molecular dynamics simulation and (b) calculated from the inertia ratchet model. The horizontal dashed lines indicate the instantaneous value of excitation where vortices are *de*pinned ($f(t) = f_d$) and *re*pined ($f(t) = f_r$). The subscripts 1 and 2 stand for the weaker and the stronger depinning forces, respectively (after [477]).

This model reproduces exactly the molecular dynamics simulations for $H = H_1$. The $\gamma\langle v\rangle(A)$ curve, calculated for the overdamped limit, is given by the full line in Fig. 4.38(b). The curve presents a long and slowly decaying tail qualitatively different from the curves observed in our experiments.

Integration of Eq. (4.14) for several finite values of γ (in units of $\omega_p/2\pi$) results in the curves with markers in Fig. 4.38(b). For $\gamma = 4$, the $\gamma\langle v\rangle(A)$ curve is already very close to the overdamped limit. For smaller γ values, the effect of inertia becomes increasingly more pronounced. The curves look sharper and more restricted to the rectification window. As shown in Fig. 4.38(a), the curve corresponding to $\gamma = 1.8$ provides an excellent fit to the experimental data. The time evolution of the velocity $v(t)$ for $\gamma = 1.8$ is shown in Fig. 4.39(b) for a few *ac*-drive values. Note that the curves correctly reproduce the asymmetry observed in the experimental $V(t)$ data (Figs. 4.37(a–d)), with $f_r \neq f_d$. The $v(t)$ curves calculated for $\gamma \geq 4$ resemble the overdamped limit, which gives exactly $f_d = f_r$ and no hysteresis.

4.3.2 *Controlled multiple reversals of a ratchet effect*

As discussed in the previous section, a single particle — such as a vortex — confined in an asymmetric potential demonstrates an anticipated ratchet effect by drifting along the 'easy' ratchet direction when subjected to non-equilibrium fluctuations. However, under special conditions, particles in a ratchet potential can move preferentially along the direction where the

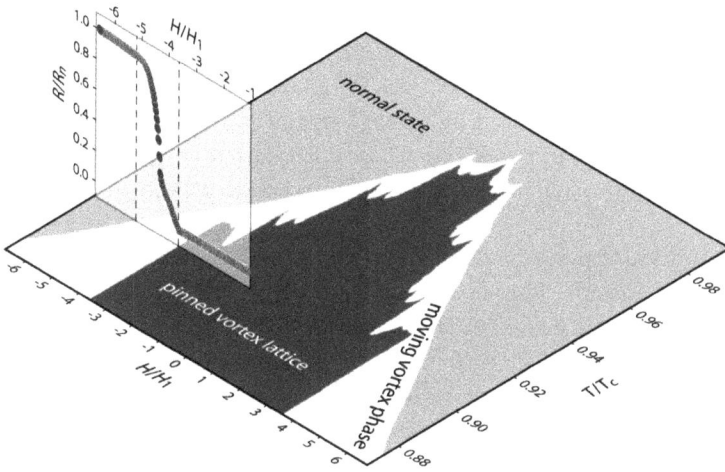

Fig. 4.40 Experimental H–T diagram (horizontal panel) of the dynamical regimes of the vortex lattice: the pinned vortex solid phase (dark gray area), the moving vortex phase (white area), bound by the criterions $R_n = 10^{-5}R_n$ and $R_n = 0.9R_n$, and the normal state (light gray area). The vertical panel shows the magnetoresistance curve at $T/T_c = 0.9$ in detail (after [122]).

potential barriers are steeper, that is, along the 'hard' direction.[6] In theory, an inversion of the drift direction of a single-particle Brownian ratchet is predicted to occur for non-zero thermal noise when the excitation frequency exceeds a certain critical value, which is usually high and very sensitive to the model parameters [39]. In a system of many weakly interacting particles this effect can, however, be strongly reduced when increasing the particle density [125].

Since the vortex density in a superconducting film can be varied continuously by changing an external magnetic field, it may be expected that the ratchet effect, discussed in the previous section, can reverse at higher field values. The vertical inset in Fig. 4.40 shows the magnetoresistance of the sample studied in the previous section at $T/T_c = 0.90$. Since vortices repel each other, the increase in occupancy of pinning centers leads to a decrease of the effective pinning force acting on the present vortices. If the vortices are under the influence of an applied *ac*-current, three different dynamical regimes can be distinguished. In the pinned vortex solid phase, the applied current is not high enough to drive vortices out of their

---

[6]This effect can be crucial in the design of artificial ratchet-based devices capable of shuttling or separating, for instance, colloidal suspensions [309] and DNA molecules [24].

equilibrium positions and, consequentially, the magnetoresistance is zero. However, when the external field is increased, a vortex density is reached at which the applied current exceeds the pinning force, hence vortices start to move and the resistance increases towards R_n. Accordingly, we define the moving vortex phase between the criterions $R = 10^{-5} R_n$ (the onset of vortex motion) and $R = 0.90 R_n$ (destruction of the superconducting state).

Temperature plays an important role in determining the pinning efficiency of an antidot. Very close to T_c, vortices are bigger then the antidots, which then become less effective pinning centers. However, at lower temperatures, vortices are getting smaller and their interaction with the antidots is stronger. In this sense, decreasing the temperature plays the role of increasing the pinning strength. This effect can be seen when plotting the three different dynamical regimes (pinned vortex solid, moving vortices and normal state) as a function of temperature and field, as it is shown in the horizontal panel of Fig. 4.40. Clearly, a broadening of the pinned vortex solid phase can be seen for decreasing temperature, indicating an increase of pinning efficiency. Still this dependence is not smooth, since the interplay between the vortex lattice and the array of pinning sites is an important factor. As a result, at some rational multiples of the first matching field, vortices assemble in a very stable lattice commensurate with the pinning array. These special configurations enhance the critical current, producing the sharp re-entrances of the pinned vortex solid phase at integer and half-integer matching fields. The presence of these matching peaks is an important experimental result, stating that even for high vortex densities, i.e. at $H > H_1$, the vortex lattice interacts strongly with the periodic potential. Therefore, the present system is an excellent candidate to study the effect of sign reversals of the ratchet effect.[7]

In the upper panel of Fig. 4.41 the field dependence of the measured dc voltage is plotted for several temperatures. At $T/T_c = 0.974$, the result of a low vortex density, discussed in Section 4.3.1, is recovered, that is a negative dc-voltage for positive magnetic fields, stating that at low vortex densities the easy-axis lies in the positive y-direction. Note that the rectification mechanism is insensitive to the vortex polarity, since the

---

[7]Concerning the exact configuration of the vortex lattice, one can estimate $n_S < 1$ at $T/T_c = 0.88$, which is the lowest temperature used. Therefore, it is clear that interstitial vortices come into play at fields higher than H_1. Nevertheless the current density is substantially higher between the antidots compared to interstitial positions, forcing the vortices to move preferentially along the antidot rows.

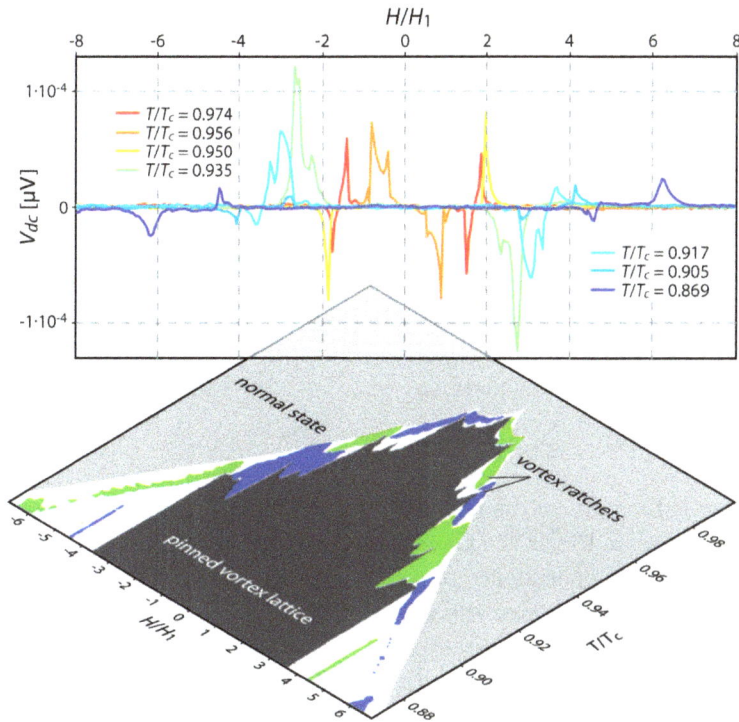

Fig. 4.41 The measured *dc*-voltage V_{dc} as a function of magnetic field for several temperatures (upper panel), and the $H–T$ diagram of the dynamical phases (lower panel). Green and blue areas correspond to regions of positive and negative values of V_{dc}, respectively. Within the white areas vortex motion is expected to be symmetric, i.e. $V_{dc} \sim 0$ (after [475]).

interaction of vortices and anti-vortices with an antidot is the same. This leads to a symmetric net *dc*-velocity $v_{dc}(H) = v_{dc}(-H)$, resulting in an antisymmetric *dc*-voltage $V_{dc}(H) = -V_{dc}(-H)$. The curves measured at lower temperatures exhibit *multiple sign reversals* of the *dc*-voltage, with maxima and minima close to integer and half integer matching fields.

The lower panel of Fig. 4.41 shows the $H–T$ diagram of the dynamical phases of the considered sample, where green and blue areas correspond to positive and negative values of V_{dc}, respectively. Within the white areas, vortex motion is presumably symmetric, i.e. $V_{dc} = 0$ within the experimental accuracy. Between the pinned vortex solid phase and the normal state, the observed voltage is dominated by vortex ratchet effects with multiple sign reversal. In order to reveal the origin of these findings, we consider a one-dimensional system of particles [122] interacting via the pair potential,

$$V_{int}(r) = -E_0 \ln(r) , \qquad (4.18)$$

with the pair separation r and the relevant energy scale E_0 in a double-well ratchet potential

$$U_p(x) = -U_{p1} \cdot \exp\left[\frac{-\sin^2(\pi x)}{2\sin^2(\pi R)}\right] - U_{p2} \cdot \exp\left[\frac{-\sin^2(\pi(x-d))}{2\sin^2(\pi R)}\right] , \qquad (4.19)$$

where U_{p1} and U_p2 determine the depth of the stronger and weaker wells, respectively. The dynamics of a chain of particles subjected to the above defined potential (Eq. (4.19)) can be studied by molecular dynamics simulations of the Langevin equation,

$$m\ddot{x}_i = -\eta\dot{x}_i - \sum_i \nabla V_{int}(x_i - x_j) - \nabla U_p(x) + F_{ext} + \Gamma_i , \qquad (4.20)$$

where m is the mass of the particles, η is the friction coefficient, F_{ext} is the external drive and Γ_i is the Gaussian thermal noise.[8]

To study the dependence of rectification on the number n of particles per unit cell, we simulated the net velocity of the chain as a function of n and $\tilde{U}_{p1} = {}^{U_{p1}}/_{E_0}$ for a constant sinusoidal ac-bias (see Fig. 4.42). In this simulation the relative pinning strength, $\tilde{U}_{p1}$, is changed by adjusting the particle interaction compared to the pinning strength of the deep potential well. Again, three different dynamic regimes can be identified, depending on the pinning force of the effective potential and the applied ac-bias. If the ac bias cannot overcome the effective potential, particles are pinned (phase 1). If more particles are inserted into the system or if the particle interaction is increased (i.e. $\tilde{U}_{p1}$ becomes smaller) the effective potential is lowered and particles start to move. The present asymmetry will be probed from the moment F_{ext} exceeds a certain threshold (phase 2). However, once F_{ext} is much larger than this specific threshold, particles will be moved over a large distance in both directions and the symmetric movement is recovered again (phase 3).

The results of the simulations clearly demonstrate that the inter-particle interaction in a chain of particles, captured by a ratchet potential, can lead in a controllable way to induce multiple drift reversals. Particulary, there is a large region in the phase diagram where the drift changes alternating sign as the number of particles per ratchet period switches from odd to even. A detailed look into the dynamics of the system [122] reveals the

---

[8]Hereafter, we use $m = 1$ and $\eta = 16$, which corresponds to strongly overdamped dynamics.

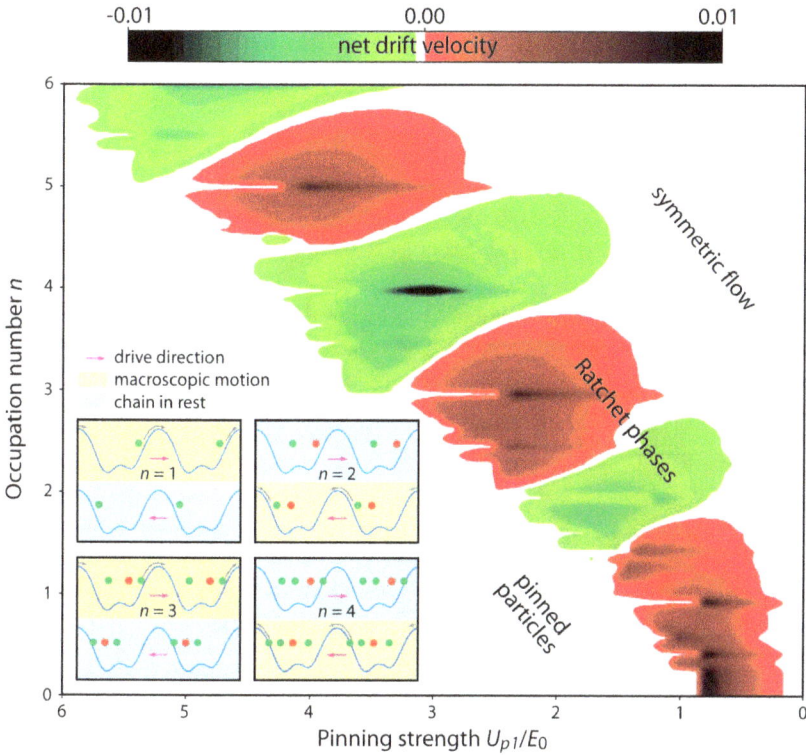

Fig. 4.42 Net drift velocity of the chain as a function of occupation number n and pinning strength U_{p1}/E_0. The chain was adiabatically excited with a sinusoidal forcing of amplitude $F_{ac} = 3E_0/a_p$ with the ratchet period a_p and frequency $f = 5 \times 10^{-7} m/\eta$ at zero thermal noise. The inset demonstrates schematically the ratchet mechanism when the chain is excited by an *ac*-square-wave force with an amplitude just above the threshold force. Purple arrows indicate the drive direction. Yellow backgrounds highlight macroscopic motion of the chain to the corresponding drive direction, whereas blue backgrounds indicate that the chain is at rest. The macroscopic drift is triggered by a transition of the most weakly pinned particle (marked in red) to the next available pinning site. In sequence, one particle in this site is knocked out to the next ratchet period (as indicated by the gray arrows) starting up motion of the whole chain (after [475]).

underlying mechanism for the particular case of moderate pinning strengths and $U_{p2}/U_{p1} > 0.56$. The effective energy, following a brick-tiling pattern, causes additional particles to be located alternatingly in the strong and weak well. This weakly pinned particle triggers the motion of other particles when it is pulled over the inner potential barrier. As a result, it will trigger the motion of the complete chain of particles in a preferential direction, which depends alternatingly on the number of particles in the potential well.

Coming back to the experimental results, the situation shown in Fig 4.41 is similar to the one in Fig 4.42, since in both cases there are three different dynamical regimes that can be distinguished, depending on the pinning force of the effective potential and the applied *ac*-bias. If the Lorentz force cannot overcome the effective pinning potential, vortices are pinned (phase 1). However, if more vortices are inserted into the system by increasing the magnetic field, or if the vortex interaction is increased by raising the temperature, the effective potential is lowered and vortices start to move. Consequentially, the present asymmetry of the pinning potential is probed as soon as F_{ac} exceeds a certain threshold (phase 2). Finally, if F_{ac} is further increased, vortices will be moved over a large distance in both directions and the symmetric movent is again recovered (phase 3).

4.3.3   *The origin of reversed vortex ratchet motion*

In Section 4.3.1 we showed that the ratchet effect can be used to manipulate the motion of flux quanta in superconductors with asymmetric pinning landscapes [280, 488]. The fact that the flux lines cannot be regarded as independent entities leads to a far richer ratchet motion which can even reverse the direction of the average motion of vortices, as demonstrated in the previous section, where the sign reversal of the ratchet signal was described as a result of trapping several vortices in each pinning site. Another possibility for the sign reversal was considered by Lu *et al.* [294] who predicted that reversed ratchet motion should also be present in a simpler system consisting of a one-dimensional asymmetric pinning potential. In the description of Section 4.3.2 the penetration depth λ was larger than the period of the ratchet potential d, whereas $\lambda = d$ in the model of Lu *et al.* [294].

Similarly, this effect has been observed experimentally [432] and modelled theoretically [299] for arrays of Josephson junctions with asymmetric periodic pinning, where $\lambda = \infty$. These studies have unambiguously identified a strong vortex-vortex interaction as the basic ingredient for producing ratchet reversals. This situation is achieved when the inter-vortex distance a_0 becomes smaller than λ. Although the inequality $a_0 < \lambda$ seems to represent a *necessary* condition, it cannot be sufficient since the natural scale where the inversion symmetry is broken, d, is not explicitly taken into account. An important question is what are the ultimate conditions for

(a) $d = 162\,\mu$m

(b) $d = 34\,\mu$m

(c) $d = 18\,\mu$m

(d) $d = 10\,\mu$m

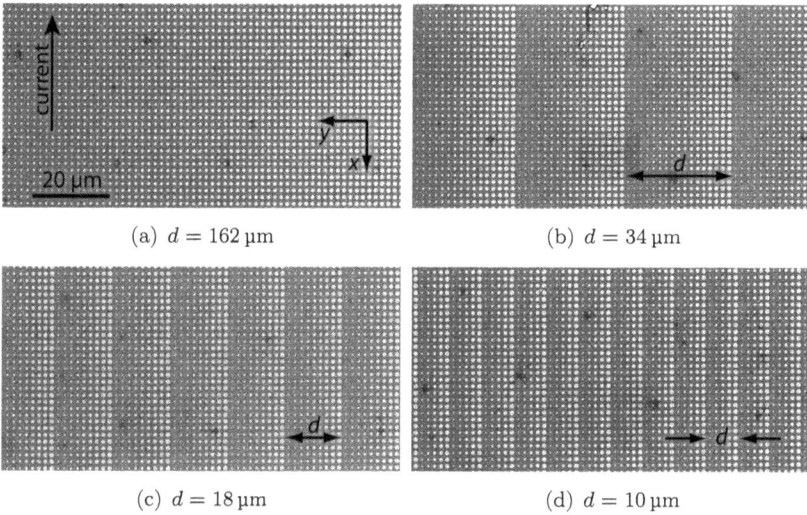

Fig. 4.43 Scanning electron microscopy images of square arrays of magnetic dots with a size-graded period of (a) $d = 162\,\mu$m, (b) $d = 34\,\mu$m, (c) $d = 18\,\mu$m and (d) $d = 10\,\mu$m. In panel (a) the positive current orientation to obtain a positive ratchet signal with positive field is indicated, see Fig. 4.44(a) (after [182]).

obtaining a reversed ratchet motion and how the number of sign reversals depends on the characteristic scales of the system.

In this section, these questions are directly addressed by considering the vortex ratchet motion in samples with different periods of the asymmetric pinning potential d. As we will show below, no sign reversal is detected for $d > \lambda$, meaning that $\lambda > d$ is a prerequisite to obtain reversed ratchet motion. In addition, it will be also demonstrated that the number of sign reversals increases with decreasing d — an effect that can be qualitatively described within a collective pinning scenario.

The pinning potential is created by a square array with a period a of $2\,\mu$m, containing ferromagnetic dots[9] of a size that changes linearly from $1.2\,\mu$m to $0.4\,\mu$m over a certain distance d (see Fig. 4.43). Four different periods $d = 10$,18, 34 and $162\,\mu$m were used. Since the typical field amplitudes used to explore the superconducting state are much lower than the coercive field of the used Co/Pt multilayers, the magnetic dots remain in

---

[9]The magnetic dots are [Co(0.4 nm)/Pt(1 nm)]$_n$ multilayers with $n = 10$, grown on top of a 2.5 nm Pt buffer layer. The dot arrays are covered with a 5 nm thick Si layer followed by an aluminium bridge of 50 nm thickness. Therefore, the aluminium is insulated from the ferromagnetic template and proximity effects are suppressed.

Table 4.2   Characteristics of the aluminium film, covering the magnetic dots

Superconducting Property	Method of Investigation
$\xi(0) = 114\,\text{nm}$	slope of $T_c(H)$
$\ell = 13\,\text{nm}$	dirty limit expression[a] $\xi(0) \sim 0.855\sqrt{\xi_0 \ell}$

[a] $\xi_0 \sim 1400$ nm is the BCS coherence length for pure aluminium.

the as-grown state during the measurements. The virgin state of such dots is a multidomain structure with zero net magnetization [181,436]. However, the stray field generated by these domains is still strong enough to deplete the superconducting condensate locally on top of the dots, thus making them effective pinning centers.

All these samples showed a featureless and nearly linear $T_c(H)$ phase boundary. Taking into account the characteristics of the aluminium film given in Table 4.2 together with the finite thickness of the film and the reduced mean free path, an effective penetration depth $\lambda(0) \sim 1.46\,\mu\text{m}$ is obtained. On the right axes of Fig. 4.44 we have indicated the variation of λ with temperature. Within the temperature window of the measurements, λ is always bigger than the vortex-vortex separation $a_0 = \sqrt{\phi_0/H}$.[10] This indicates that the vortex-vortex interaction is in all cases significant in the considered experimental conditions.

Rectification effects were investigated by applying a sinusoidal excitation of $500\,\mu\text{A}$ amplitude and frequency $f = 1\,\text{kHz}$, while simultaneously recording the average *dc*-voltage with a nanovoltmeter. When the current is parallel to the gradient and, therefore, the Lorentz force F_L is parallel to the iso-size lines (x-axis in Fig. 4.43(a)), the acquired *dc*-signal was below the experimental resolution. This is consistent with the fact that the system is fully symmetric for this direction of the Lorentz force. However, when the external *ac*-current shakes the vortex lattice with F_L in the same direction as the gradient (y-axis in Fig. 4.43), a flux line feels an effective pinning potential U_p with asymmetric forces $f = -\partial U_p/\partial y$ for motion along $+y$ and $-y$ orientations (see the schematics in Fig. 4.45). The lack of inversion symmetry of $U_p(y)$ is a natural consequence of the fact that the pinning strength increases with the radius of the dots [180,181]. Under these circumstances an external *ac*-excitation should induce a net motion of

---

[10] For instance, the earth magnetic field ($\sim$0.04 mT), if not properly shielded, would give rise to an inter-vortex distance of about $7\,\mu\text{m}$.

vortices in the direction of the smaller slope of U_p, i.e. towards $+y$. This is indeed confirmed by the positive dc-voltage V_{dc} at positive fields obtained for the largest period of size-gradient $d/a = 81$, shown in Fig. 4.44(a).

For this sample, the ratchet signal is strong only when the density of dots outnumbers the density of vortices, i.e. for fields below the first matching field of the square array $H_1^{2D} = 0.51$ mT. Near H_1^{2D}, the hopping of vortices is suppressed since most of the pinning centers are occupied, and therefore the ratchet signal decreases. The most important feature of this figure is the lack of reversal ratchet in the whole range of fields studied. This behavior, present even though the vortex-vortex interaction is important, is due to the large period of the asymmetric pinning landscape d in comparison with the penetration depth λ. For the sake of clarity horizontal dashed lines are added in Fig. 4.44 at the temperatures where $\lambda \sim d$.

The most conclusive evidence for the ratio d/λ being indeed the relevant parameter accounting for the appearance of reversed ratchet motion can be seen in Figs. 4.44(b–d) where the ratchet period is progressively reduced to meet the condition $d < \lambda$. In contrast to the previous sample, where d/λ, clear multiple sign reversals of the ratchet signal are observed. This finding shows that the existence of a strong vortex-vortex interaction is not a sufficient condition to reverse the easy ratchet direction, but that it is rather the relative size of λ in comparison with the period of the ratchet potential d which determines the occurrence of this effect. Furthermore, the number of sign reversals in the same window of temperature and field ranges from none for $d/a = 81$, 2 for $d/a = 17$ and 3 for $d/a = 9$, to 5 for $d/a = 5$, illustrating the influence of this ratio on the number of observed reversals of sign of the vortex rectification effect.

Lu *et al.* [294] recently showed that a two dimensional vortex lattice interacting with a one-dimensional ratchet potential produces a conventional ratchet when the vortex lattice is highly ordered in such a way that the vortex-vortex interactions cancel out. In contrast, disordered configurations of the vortex lattice lead to reversed ratchet motion. The vortex lattice disordering occurred due to buckling transitions of the vortices confined in each row of the ratchet in the high field regime, but additional reversals are expected at lower fields. As shown in Refs. 282 and 498, a two-dimensional diluted vortex lattice, interacting with a one-dimensional periodic pinning array, undergoes a series of continuous phase transitions as a function of field provided that the vortex-vortex interaction remains strong enough. The number of possible vortex lattice orderings that appear for changing vortex density increases when the lattice constant of the pin-

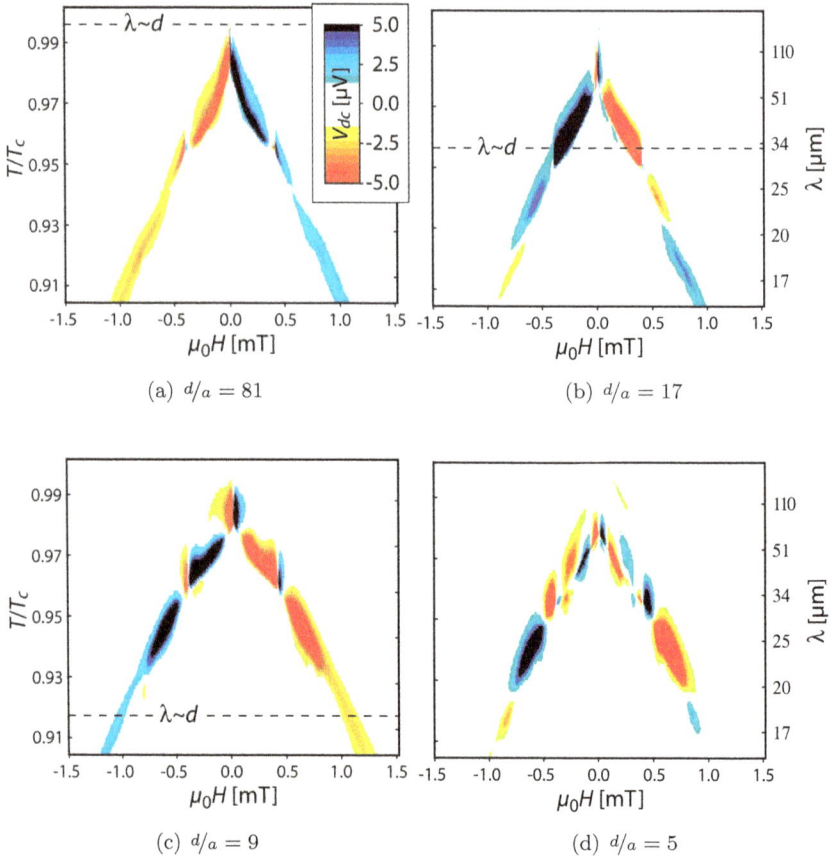

(a) $d/a = 81$

(b) $d/a = 17$

(c) $d/a = 9$

(d) $d/a = 5$

Fig. 4.44   Rectified signal V_{dc} as a function of field H and temperature T for samples with a period d of the asymmetric pinning potential of (a) 162 μm, (b) 34 μm, (c) 18 μm and (d) 10 μm, corresponding to the images shown in Fig. 4.43. In all cases the external ac-current is applied along the x-axis. The horizontal dashed lines indicate the temperatures at which $\lambda \sim d$ (after [182]).

ning array decreases, so we expect to find more order-disorder transitions, and thus more ratchet reversals, for smaller lattice periods d.

This can be illustrated with molecular dynamics simulations of the system described in Ref. 294 for pinning strength $A_p = 0.025 f_0$, ac-field $F_{ac} = 0.065 f_0$ and ratchet periods $d = 8\lambda$, 3λ, λ and 0.5λ, where $f_0 = 6\phi_0^2/2\pi\mu_0\lambda^3$. Figure 4.46 shows the simulated velocity $\langle V \rangle$ versus vortex density n_v, indicating that the number of ratchet reversals for $n_v < 1.1/\lambda^2$ changes from no reversals at $d = 8\lambda$ to two reversals for $d = 3\lambda$,

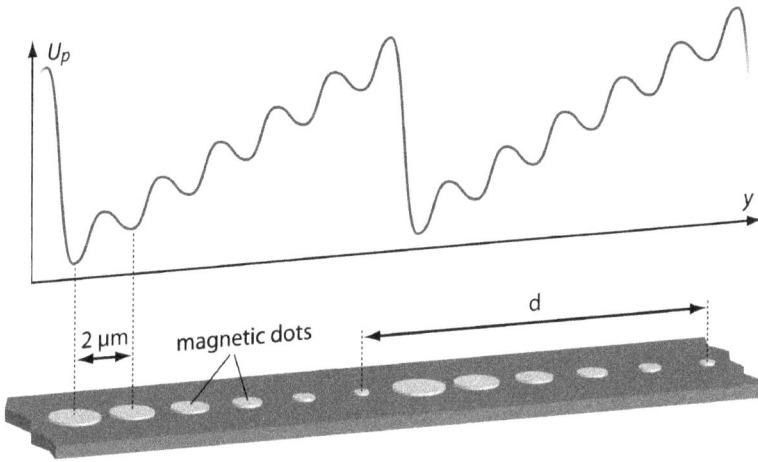

Fig. 4.45 Schematic representation of the asymmetric ratchet potential U_p along the y-axis (see also Fig. 4.43). The smaller average slope when vortices move towards $+y$ is the easy direction for the net vortex drift under oscillatory excitations.

four at $d = \lambda$, and ten reversals at $d = 0.5\lambda$. In agreement with the experimental results, this demonstrates that more ratchet reversals occur when the ratchet period d is decreased. We note that our experimental system represents a discrete two dimensional version of the situation described by these simulations.

In conclusion, an inversion of the effective ratchet potential for vortex motion can occur not only due to the vortex-vortex interactions [122], but also when the characteristic vortex-vortex interaction scale length λ is of the order of the period of the asymmetric pinning landscape d . Although the inequality $d < \lambda$ represents a sufficient condition for the occurrence of multiple reversal in vortex systems, it should also remain valid for any ratchet system with repulsively interacting particles such as colloids in an optically created ratchet potential. In this case the strength of the ratchet potential can be adjusted by controlling the laser intensity whereas the colloid–colloid interaction can be modified by, for example, using a non-polar solvent.

Nanostructured Superconductors

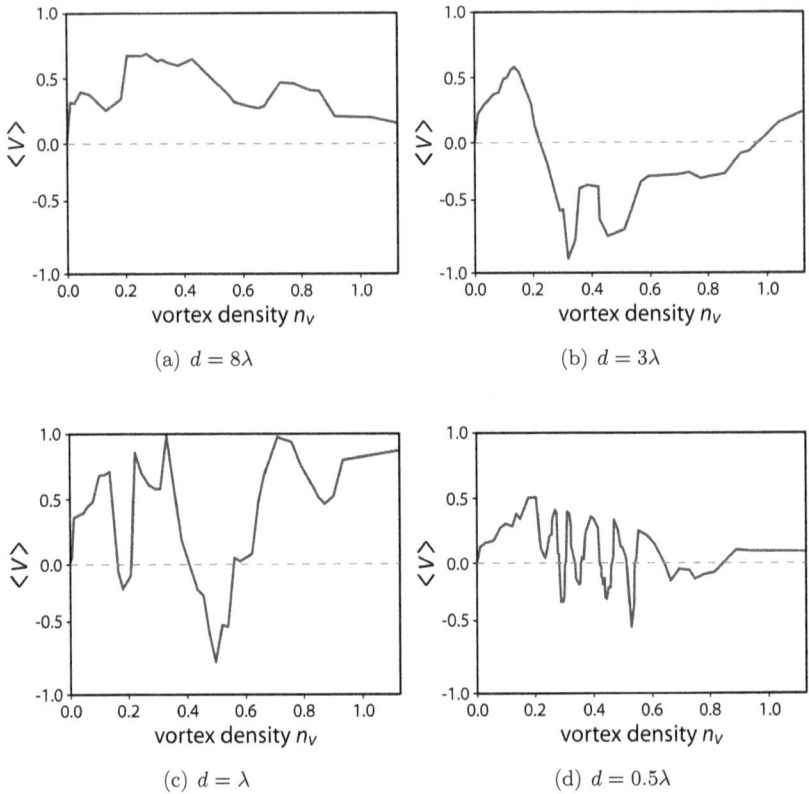

(a) $d = 8\lambda$

(b) $d = 3\lambda$

(c) $d = \lambda$

(d) $d = 0.5\lambda$

Fig. 4.46 Simulated ratchet response $\langle V \rangle$ versus vortex density n_v for systems with different pinning periods d of (a) 8λ, (b) 3λ, (c) λ and (d) 0.5λ (after [182]).

Chapter 5

Superconductor–Ferromagnet Hybrid Systems

According to the classical Bardeen–Cooper–Schrieffer theory of superconductivity, the ground state of the superconducting condensate is formed by electron pairs with opposite spins — the so-called spin-singlets — bounded via phonon interactions [36, 37]. As early as 1956, Ginzburg [183] pointed out that this fragile state of matter could be destroyed by the formation of a homogeneous ferromagnetic ordering of spins, if its corresponding magnetic field exceeds the thermodynamic critical field of the superconductor. Later on, Matthias *et al.* [306–308] demonstrated that besides the orbital effect, i.e. a pure electromagnetic interaction between the ferromagnetic and superconducting subsystems, there is also a strong suppression of superconductivity arising from the exchange interaction that tends to align the spins of the electrons in detriment of Cooper-pair formation. Anderson and Suhl [9] predicted that a compromise between these antagonistic states can be achieved if the ferromagnetic phase is allowed to break into domains of size much smaller than the superconducting coherence length ξ in such a way that from the superconductivity point of view, the net magnetic moment averages to zero. Alternatively, Larkin and Ovchinnikov [276] and Fulde and Ferrel [164], theoretically predicted that superconductivity can survive in a uniform ferromagnetic state if the superconducting order parameter is spatially modulated.

In general terms, the effective polarization of the conduction electrons, either due to the external field H (orbital effect) or the exchange field H_{ex} (paramagnetic effect), leads to a modification (suppression and modulation) of the superconducting order parameter. Typically, in ferromagnetic metals, the exchange field is considerably higher than the internal magnetic field and it dominates the properties of the system. However, in some cases where both fields can have opposite directions, an effective compensation of the conduction electrons polarization can occur and superconductivity can consequently be recovered at high fields $H \simeq -H_{\mathrm{ex}}$, where field-induced superconductivity can take place (Jaccarino and Peter [236]). Bulaevskii *et al.* [76] gave an excellent overview of both experimental and theoretical aspects of coexistence of superconductivity and ferromagnetism where both orbital and exchange effects are taken into account.

The progressive development of material deposition techniques and the advent of refined lithographic methods have made it possible to fabricate different nanoscale superconductor–ferromagnet (S/F) structures at nanometer scales. Unlike the investigations dealing with the coexistence of superconductivity and ferromagnetism in ferromagnetic superconductors (for review see Ref. 153), the ferromagnetic and supercondicting subsystems in the artificial heterostructures can be also physically separated by an insulating buffer. As a consequence, the strong exchange interaction is limited to a certain distance around the S/F interface, whereas the weaker electromagnetic interaction can persist to longer distances into each subsystem. In recent reviews, Izyumov *et al.* [234], Buzdin [80] and Bergeret *et al.* [53] discussed in detail the role of proximity effects in S/F heterostructures dominated by exchange interactions.

In order to unveil the effects of electromagnetic coupling, it is important to suppress proximity effects by introducing an insulating buffer material between the S and F components (see also the reviews by Lyuksyutov and Pokrovsky [298] and Aladyshkin *et al.* [6]). In this chapter we will discuss the thermodynamic and transport phenomena in the S/F hybrid structures dominated by electromagnetic interactions, focussing only on the S/F hybrids that consist of conventional low-T_c superconductors. These S/F heterostructures with pure electromagnetic coupling can be described phenomenologically using Ginzburg–Landau and London formalisms rather than sophisticated microscopical models.

5.1   Field polarity dependent vortex pinning in laterally nanostructured S/F systems

Already in 1972, Autler suggested that an array of magnetic particles, prepared by Bitter decoration, could result in an enhancement of the critical current density of an underlying superconductor [17] — a configuration that was realized and studied later on experimentally by Fasano *et al.* [142]. Further investigations on magnetic pinning centers were focussed initially on dots with in-plane magnetization: Geoffroy *et al.* [170] and Otani *et al.* [372] reported on oscillations in the magnetoresistance of niobium films, covered with periodic arrays of micrometer sized RE-cobalt[1] ferromagnetic particles. The observed effect was attributed to fluxoid quantization effects and the modulation of the superconducting state by the stray field of the particles. Niobium films on top of submicron-sized nickel dots were investigated with transport measurements [219, 235, 301–303, 448, 485]. The magnetoresistance of these films showed pronounced minima at integer multiples of the first matching field, revealing that the dots act as strong vortex pinning centers. Since the observed matching effects are much less pronounced when reference non-magnetic silver dots are used [219], the strong pinning must have its origin in the ferromagnetic state of the nickel dots. The pinning of rectangular dots, which are in the single domain state, is stronger compared to the case of dots in a multidomain state [474]. Moreover, local vortex visualization revealed that vortices are attracted to one specific side of in-plane magnetized dots in the single domain state [472], leading to the conclusion that the strong pinning by arrays of dots with in-plane magnetization originates from a magnetic interaction between the dots and vortices.

The magnetization of the dots used in [142, 170, 219, 235, 302, 303, 372, 448, 472, 474, 485] was preferentially parallel to the surface of the sample, due to the shape anisotropy of the dots. Morgan and Ketterson [339] reported on pinning experiments using nickel dots with an aspect ratio of almost one, due to which the major component of the magnetization is expected to be perpendicular to the surface of the sample after saturation with an out-of-plane field. With this magnetization procedure, critical current measurements revealed that significantly stronger pinning is obtained if the applied field and the magnetic moments m_z of the dots have the same polarity, compared to the case of opposite polarity ('field-polarity dependent pinning').

In this section, we shall analyze the vortex pinning properties of magnetic nanostructures that consist of cobalt-platinum (Co/Pt) or cobalt-palladium (Co/Pd) multilayers with perpendicular magnetic anisotropy. Especially, we will focus on two different S/F nanosystems, containing regular arrays of magnetic dots with out-of-plane magnetization and regular arrays of antidots in Co/Pt multilayers. The used dots have a clear perpendicular magnetic anisotropy and a small aspect ratio, with the result

---

[1]In this case, RE represents rare-earth samarium (Sm) or gadolinium (Gd) elements.

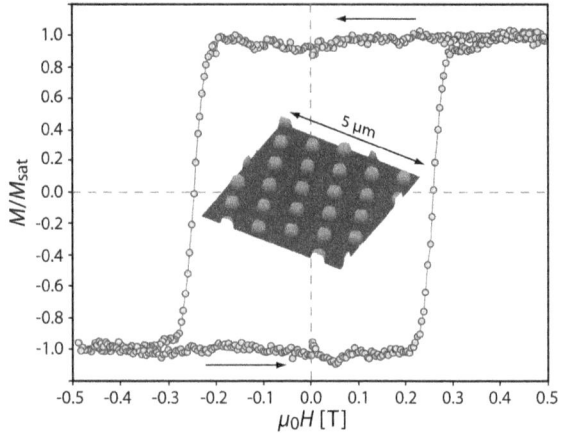

Fig. 5.1 Magnetic hysteresis loop of a Co/Pt dot array, measured with the magneto-optical Kerr effect at room temperature. The inset shows an atomic force micrograph of a typical array of used magnetic dots, consisting in this case of a $[Co(0.4\,nm)/Pt(1.0\,nm)]_{10}$ multilayer on top of a 2.5 nm Pt base layer (after [273]).

that the superconducting films cover the dots completely without being perforated. Moreover, scanning Hall probe microscopy makes it possible to investigate the vortex pinning on a microscopic scale and reveal the origin of the field polarity dependent pinning in these S/F-nanosystems.

5.1.1   *Vortex pinning by magnetic dots*

For the present flux pinning experiments it is of crucial importance that the magnetic state of the dots remains unchanged when applying a small magnetic field perpendicular to the surface of sample. Therefore, the prepared structures were controlled by measuring their magnetic hysteresis loops using the magneto-optical Kerr effect, which is a useful tool for the characterization of magnetic nanostructures [256].

Figure 5.1 shows a typical example of such loop, when the external magnetic field is oriented perpendicular to the surface of the sample. Apparently, the Co/Pt dots have 100 % magnetic remanence, implying that their magnetic state is very stable once they are magnetically saturated. Compared to reference continuous Co/Pt multilayers, the dots usually have an enhanced coercive field H_{coe}, which might be due to their smaller demagnetization factor. However, in the case of the particular sample shown in Fig. 5.1, application of magnetic fields $|\mu_0 H| < \pm 200\,mT$, which is much higher than typically used fields for measurements of flux pinning properties, cannot change the stable magnetic state of the dots.[2]

---

[2]At low temperatures, the coercive fields are even higher due to the decrease of thermal fluctuations of the magnetic moments.

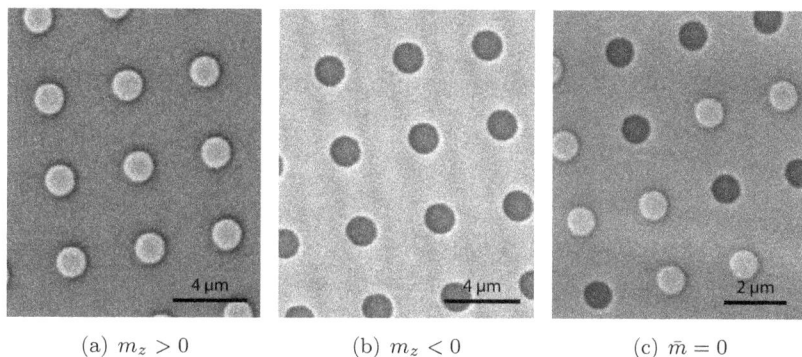

(a) $m_z > 0$ (b) $m_z < 0$ (c) $\bar{m} = 0$

Fig. 5.2   Magnetic force microscope images of a square array of circular magnetic dots, consisting of Co/Pd multilayers. Diameter and periodicity of the dots are 1.5 μm and 4 μm, respectively. The images were taken (a) after magnetically saturating the sample in a field of $+1\,\mathrm{T}$ and (b) $-1\,\mathrm{T}$, and (c) after demagnetizing the dots (after [273]).

Figure. 5.2 shows magnetic force micrographs of typical samples after saturation in positive and negative fields, where bright and dark shades can be associated with magnetic moments of the dots pointing in positive and negative z-direction, respectively[3]. The uniform signals obtained from each dot indicate that, after magnetic saturation, they all are in the single domain state. This is consistent with the observed remanence of 100% of the saturated magnetization M_{sat}, shown in Fig. 5.1.

When demagnetized[4], the dots appear either as bright or as dark spots in the magnetic force micrograph (Fig 5.2(c)), meaning that the dots are still in the single domain state, but the direction of their magnetic moments is along either the positive or the negative z-direction. Within the statistical error, there is an equal number of bright and dark colored dots, meaning that the average magnetic moment is zero.

As an example, in the scanned area shown in Fig. 5.3, there are 58 bright and 70 dark shaded dots. An interesting question that arises is how dots with different sense of magnetization should be distributed. For non-interacting dots one can expect a totally random distribution of dots with up and down magnetization. However, there are some areas in

---

[3]For convenience, we will refer to the state in which the magnetic moments of the dots point exclusively in positive (negative) z-direction as to $m_z > 0$ ($m_z < 0$). When demagnetized, the average magnetic moment is zero and, accordingly, we call this state $\bar{m} = 0$.

[4]The demagnetization was carried out by oscillating a perpendicularly applied field around zero with decreasing amplitude, especially taking care of applying small field steps near the coercive field of the dots.

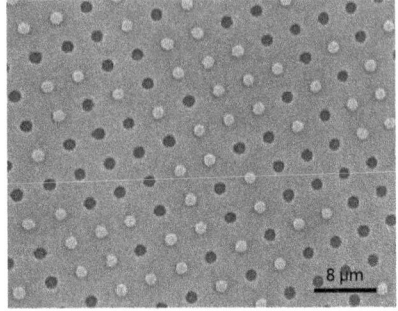

Fig. 5.3 Magnetic force micrograph of a larger area of the sample shown in Fig. 5.2 in the demagnetized state (after [273]).

Fig. 5.2 where dots of one color form clusters. Hyndman *et al.* have studied this problem in a similar system [229], finding that patterns with a strong dipolar interaction preferentially have an antiferromagnetic ordering, while weakly coupled dots show patterns comparable to the one shown in Fig. 5.2(c). Therefore, one can assume that, in the present case, relatively weak dipolar interactions occur between the dots. Thin magnetic discs with perpendicular magnetization can be modelled by current carrying loops coinciding with the disc circumference. Such currents, among other systems, can be present in π-rings [217], and order and clustering effects can be seen in the square arrays of the π-rings [367].

In the vortex pinning experiments presented in the following, the magnetic moments m of the dots and the applied field H are both perpendicular to the surface of the sample. This leads to magnetic interactions between the magnetic dots and vortices in a superconducting film covering the dots, depending on the mutual orientation of m_z and H. Additionally, the regular arrangement of the dots is expected to give rise to the appearance of matching effects in the magnetization curves. A crossover in the pinning behavior of magnetic dots can be observed when increasing their size. However, in this section, we will focus mostly on the pinning properties of small dots whose stray magnetic fields are not strong enough to induce non-zero fluxoids in the superconducting film.

The discussed sample consists of magnetic dots that are composed of a $[Co(0.3\,nm)/Pt(1.1\,nm)]_{10}$ multilayer on a 3 nm platinum buffer layer, covered with a $[Ge(10\,nm)/Pb(50\,nm)/Ge(25\,nm)/Au(50\,nm)]$ superconducting layer.[5] Having a square shape with side length $a = 0.4\,\mu m$, the magnetic dots are arranged in a square lattice with a period of 1 μm. Once

---

[5]The two germanium layers are used as an electrically insulating buffer, while the top layer of gold allows for approaching the surface with the sensor of a scanning Hall probe microscope in tunnelling controlled mode.

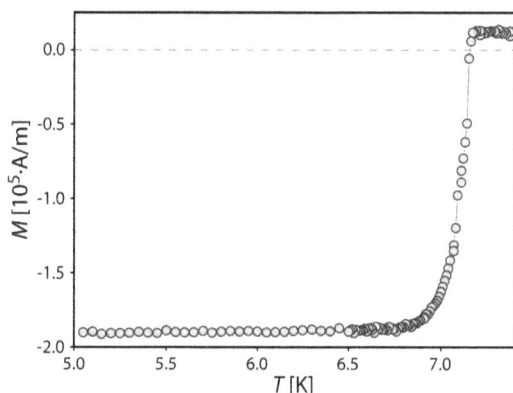

Fig. 5.4 The field-cooled magnetization of a 50 nm lead film at $\mu_0 H = 0.5\,\text{mT}$, on top of an array of Co/Pt dots in the state of $m_z > 0$. Above the critical temperature, the magnetization of the sample shows a small offset from zero, caused by the contribution of the positively magnetized dots (after [273]).

magnetically saturated, their coercive field of $\mu_0 H_{\text{coe}} = 0.23\,\text{T}$, together with a remanence of 100%, allows to sweep $\mu_0 H$ between $\pm 0.1\,\text{T}$ without changing the magnetic response of the array, indicating that the magnetic state of the dots remains virtually unchanged.

Figure 5.4 shows the $M(T)$ curve of the sample in the state of $m_z > 0$ at $\mu_0 H = 0.5\,\text{mT}$. From this graph, a critical temperature of $T_c = 7.17\,\text{K}$ can be read off, with a transition width of about 0.4 K. The same critical temperature is obtained when the dot array is in the state of $m_z < 0$ or $\bar{m} = 0$. Note that at temperatures higher than T_c, the magnetization of the sample shows a small offset from zero, which is caused by the contribution from the positively magnetized dots.

Magnetization curves of the sample at $T/T_c = 0.983$ are shown in Fig. 5.5, corresponding to the two states $m_z \lessgtr 0$ of magnetized dots. In these diagrams, the field axes are normalized to the first matching field $\mu_0 H_1 = \Phi_0 \cdot (1\mu\text{m})^{-2} = 2.068\ \text{mT}$, at which the external field generates exactly one flux quantum Φ_0 per unit cell of the dot array. The most obvious features of the curves shown in Fig. 5.5 is the asymmetry of $M(H)$ with respect to the polarity of the applied field, and a number of matching effects at rational and integer multiples of the first matching field. The asymmetry manifests itself in a much larger width of the magnetization loop ΔM — which is proportional to j_c in the Bean model — when H and m_z have the same polarity, and a smaller ΔM for mutual opposite polarity of H and m_z. In the first case, pronounced matching effects appear at integer (1, 2, and 3) and rational multiples ($1/2$ and $3/2$) of H_1. Moreover, weaker features can be seen in the $M(H)$ curve at $H/H_1 = 1/4$, $3/4$, $5/4$ and $7/4$. By contrast, when H and m_z have opposite polarity, only weak

matching features are observed at $H/H_1 = 1/2$ and 1. These magnetization loops show some similarities with those of thin superconducting films
containing arrays of antidots (see Sec.4.2). For example, when m_z and H
have the same polarity, the magnetization has a plateau between $H = 0$
and $H = H_1$. This is also observed for antidot samples close to the critical
temperature (Fig. 4.8), when there are more pinning sites than vortices in
the superconducting film. The magnetization drops rapidly as interstitial
vortices appear, as it is observed for $H > H_1$ in Fig. 5.5(a).

To demonstrate how the different matching effects depend on temperature, Fig. 5.6 shows magnetization curves of the S/F structure for
$T/T_c = 0.997$, 0.990, 0.976 and 0.962, obtained in the $m_z < 0$ state. Apparently, close to T_c, rational matching effects become more pronounced
when H and m_z have the same polarity. Additional kinks and peaks appear at $H/H_1 = -1/3$ and $-2/3$ in the $M(H)$ curve at $T/T_c = 0.997$, and
at $H/H_1 = -4/3$ and $-5/3$ at $T/T_c = 0.990$ (see Fig. 5.6(a) and (b), respectively). These rational matching effects disappear when the temperature
is lowered, and instead, integer matching effects become more distinct. At
even lower temperatures, all matching effects are smeared out, similar to
the case of superconducting films with antidot lattices [30], and the field-

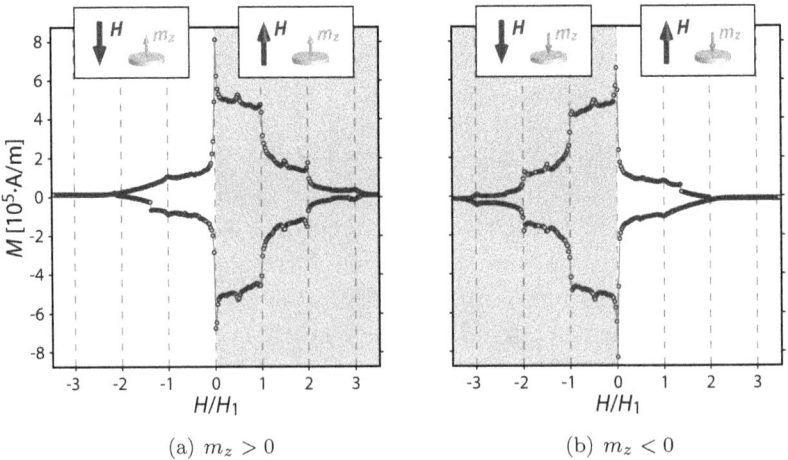

(a) $m_z > 0$ (b) $m_z < 0$

Fig. 5.5   Magnetization curves at $T/T_c = 0.983$ of a lead film on top of an array of
Co/Pt dots, which were saturated (a) in a positive and (b) in a negative field prior to the
measurement. Note that the offset of the magnetization, caused by the ferromagnetic
contribution of the dots, has not been subtracted. The insets indicate the respective
orientation of the magnetization of the dots and $\mathbf{H}$ (after [273]).

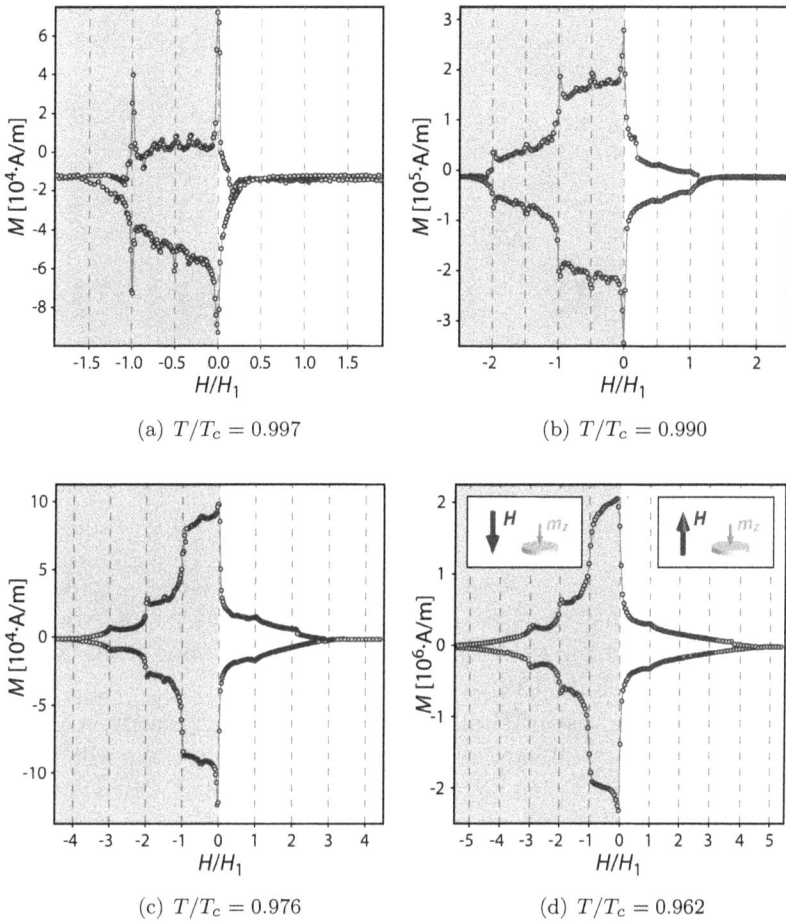

(a) $T/T_c = 0.997$

(b) $T/T_c = 0.990$

(c) $T/T_c = 0.976$

(d) $T/T_c = 0.962$

Fig. 5.6   Magnetization curves of a superconducting lead film on top of an array of Co/Pt dots at different temperatures, after magnetically saturating the dots in a negative field, i.e. in the state of $m_z < 0$: (a) $T/T_c = 0.997$, (b) $T/T_c = 0.990$, (c) $T/T_c = 0.976$ and (d) $T/T_c = 0.962$ (after [273]).

polarity dependence of the magnetization curve becomes less pronounced. For m_z and H having opposite polarity, matching effects are observed only at $H/H_1 = 1/2$ and 1 at all temperatures.

As already discussed in Sec. 4.2, matching effects indicate the presence of commensurate vortex configurations in a strong periodic pinning potential. Which specific vortex configuration is realized is determined by the vortex-dot interaction and by the repulsive vortex-vortex interaction. Assuming

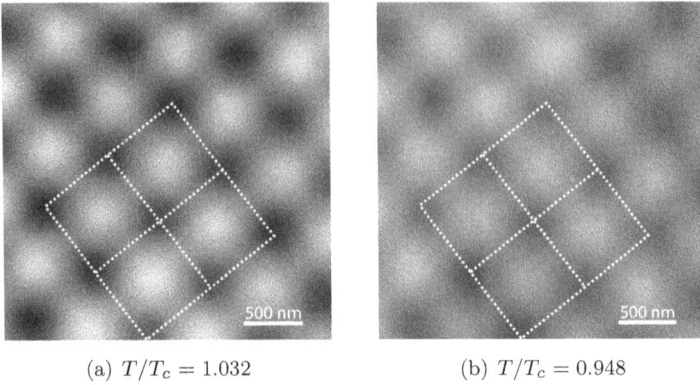

(a) $T/T_c = 1.032$ (b) $T/T_c = 0.948$

Fig. 5.7   Scanning Hall-probe microscopy images of the z-component of the magnetic stray field above the sample surface in the state of $m_z < 0$ at two different temperatures: (a) slightly above and (b) below T_c. The dots are positioned at the crossings of the dotted lines (after [473]).

an attractive interaction between dots and vortices for *the same* polarity[6] of H and m_z and that the dots are not able to pin multiquanta vortices, the same vortex configurations will be realized as those seen in the case of antidot lattices (see Figs. 4.12 and 4.15).

However, for the case of H and m_z having *opposite* polarity, it is less obvious whether the total interaction between dots and vortices will be attractive or repulsive. To find an answer to this question, the z-component of the magnetic flux density $B_z(x,y)$ was locally imaged above the surface of the sample, using scanning Hall-probe microscopy at low temperatures. Figure 5.7 shows $B_z(x,y)$, recorded without an external field at temperatures slightly above and below T_c. Magnetized in a large negative perpendicular field, the dots appear as a square array of dark spots at $T/T_c = 1.032$ (see Fig. 5.7(a)). As a guide to the eye, dotted lines are added in Fig. 5.7 to indicate the positions of the dots, which are located at the points where the lines cross. The returning stray field causes the bright contrast between the dots, resulting in a peak-to-valley contrast ΔB of the detected stray field of $30\,\mu$T in Fig. 5.7(a). After zero field cooling to $T/T_c = 0.948$, ΔB decreases by about 20% to $25\,\mu$T, as can be seen from the weaker contrast in Fig. 5.7(b). This effect can be attributed to the response of the superconductor to the local magnetic stray field of the

---

[6]This assumption is reasonable, since both possible interactions should be attractive: the magnetic interaction between dots and vortices on the one hand, and the pinning caused by the thickness modulation of the lead film on the other hand.

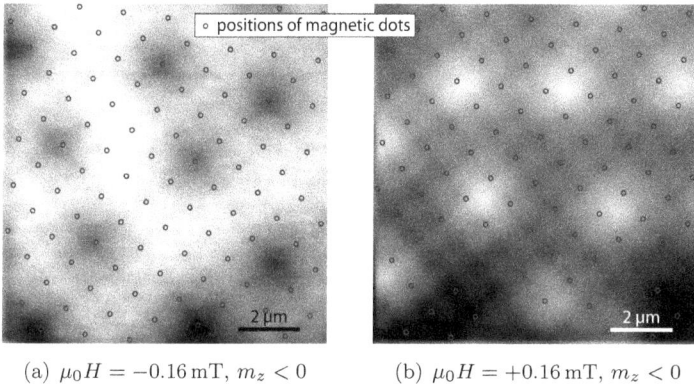

(a) $\mu_0 H = -0.16\,\mathrm{mT}$, $m_z < 0$ (b) $\mu_0 H = +0.16\,\mathrm{mT}$, $m_z < 0$

Fig. 5.8   The magnetic stray field above the sample surface in the state of $m_z < 0$, after field cooling in two different fields at $T/T_c = 0.948$: (a) $\mu_0 H = -0.16\,\mathrm{mT}$ (external field and magnetization of the dots are parallel) and (b) $\mu_0 H = +0.16\,\mathrm{mT}$ ($\boldsymbol{H}$ and $\boldsymbol{m}$ are antiparallel). Small circles indicate the positions of the dots (after [473]).

dots [136,300,325,326,472]: depending on its strength, the stray field will be screened or non-zero fluxoids will be induced in the superconductor. Below T_c, supercurrents appear in the lead film, encircling the dots, depending on the amount of flux generated by each dot [128,129,300,326]. In the present case, the stray field of the dots is not sufficiently high to induce additional fluxoids in the superconductor. Consequently, local supercurrents screen the flux of the dots and reduce the measured field B_z above the sample surface.[7]

Figures 5.8(a) and (b) show images of $B_z(x,y)$ in the state of $m_z < 0$, obtained after field-cooling the sample in small negative and positive fields of $\mu_0 H = \mp 0.16\,\mathrm{mT}$, respectively, corresponding to about 8.5 flux quanta in the scanned area of $10.5 \times 10.5\,\mathrm{\mu m^2}$. Clearly, one can identify 9 negative (dark) vortices in Fig. 5.8(a) and 8 positive (bright) vortices in Fig. 5.8(b), in perfect agreement with the expected number of flux quanta. Since the magnetic dots produce a much weaker contrast than the vortices, we indicated their positions by small circles for clarity. Apparently, the location of vortices depends on the field polarity: if the field of the vortices points in the same direction as the magnetic moments of the dots — as it is the case in Fig. 5.8(a) where both m_z and H are negative — vortices

---

[7]The reason for the still non-zero field, detected above the dots with the Hall probe, is likely to be related to incomplete screening, since at temperatures close to T_c the penetration depth is of the order of the spacing between the dots. Moreover, the field B_z is measured at a certain distance of $\sim 200\,\mathrm{nm}$ above the surface of the superconductor.

are positioned at the dot sites. In contrast, vortices with opposite field polarity are located at interstitial positions (see Fig. 5.8(b) where $m_z < 0$ and $H > 0$).

These local visualization experiments can be correlated with the global magnetization experiments presented above in Figs. 5.5 and 5.6. When H and m_z are parallel, the flux lines are pinned by the dots and high critical currents and pronounced matching effects are observed. On the other hand, for the antiparallel alignment of H and m_z, the vortices are caged at interstices, where they have a higher mobility and the pinning is substantially reduced.

The gradual disappearance of rational matching effects upon the decrease of temperature (Fig. 5.6) can be explained by the temperature dependence of the penetration depth and the field gradient in the superconductor [471]. The above observation that matching anomalies are seen in the magnetization loops only close to T_c, can be understood by the requirement of having long-range interactions between the vortices over at least several periods of the dot array. The range over which vortices are interacting is determined by the temperature dependent penetration depth $\lambda(T)$. Since $\lambda(T)$ diverges for $T \to T_c$, rational matching effects disappear first upon decreasing T, for they are based on vortex lattices with larger unit cells (see Fig. 4.15) compared to those of integer matching effects (Fig. 4.12), which vanish, accordingly, at lower temperatures.

Moreover, the flux gradient inside the sample plays an important role in the considered systems. As already mentioned in Sec. 4.2.4, the Bean model assumes a linear flux gradient in the sample, which is not realistic in a superconducting film with arrays of magnetic dots or antidots. Rather, it was suggested that in films with such periodic pinning sites, the flux gradient could be better described by a multi-terrace [110,398] or a single terrace flux profile [343] (see. Fig. 4.18). However, matching effects will only appear in the sample when the flux gradient in the sample is small, because larger gradients prohibit the formation of large regions with commensurate vortex arrangements. At temperatures close to T_c, the vortex-vortex interaction becomes stronger, since the overlap of the magnetic fields of adjacent vortices increases with temperature. This results in a smaller flux gradient close to T_c, in agreement with the observed pronounced matching effects. At lower temperatures, the matching effects vanish due to the large flux gradient. Finally, at very low temperatures, the coherence length $\xi(0)$ becomes of the order of the grain size in the used lead films. As a consequence, core pinning at grain boundaries can become stronger than the magnetic pinning of the dot array, making the observation of matching effects practically impossible. Periodic arrays of magnetic dots with perpendicular anisotropy also strongly affect vortex channeling and flux patterns as revealed in magneto-optical imaging experiments [176].

5.1.2 *Commensurate vortex domain formation*

An important feature that distinguishes fractional matching states from integer ones is that two or more possible degenerate structures always exist under the symmetry operations of the pinning array. Vortex imaging experiments using scanning Hall-probe microscopy have revealed the spontaneous formation of complex composite structures of degenerate domains, separated by domain walls in vortex arrays near rational fractional filling [144, 198]. This phenomenon was entirely unanticipated since, in contrast to ferromagnetic materials, the energies and magnetization of different domains are identical. In this section, we analyze the evolution of the domain structure with magnetic field, showing that domain formation is a direct consequence of long-range vortex-vortex interactions. Under these conditions, corners in domain walls prove to be efficient 'sinks' for unmatched excess vortices or vacancies, and allow commensurate states to exist over a broad range[8] of applied magnetic fields [197].

Figure 5.9 shows scanning Hall-probe images of the local induction of the S/F system discussed above in Sec. 5.1.1, visualizing the vortex structures that are formed after field cooling to low temperatures at fields close to $-1/2H_1$. At exactly $-1/2H_1$ (Fig. 5.9(d)) a very well ordered 'checkerboard' structure is observed where every second pinning site is occupied by a vortex. Remarkably, cooling the sample at fields slightly above or below $-1/2H_1$ did not result in excess vortices or vacancies in the commensurate lattice, as is the case at field values close to integer matching. Instead coexisting domains of the two degenerate commensurate states — related by a translation of one lattice vector of the pinning array — are observed, separated by domain walls. The latter appear to move smoothly as slightly different cooling fields are employed (see the transition Fig. 5.9(b → c)) and their presence does not seem to be related to the possible existence of inhomogeneities in the pinning array. A crucial difference between domain structures at fields slightly above and slightly below the half matching field

---

[8]Theoretically, the probability of observing 'pure' matching structures on a macroscopic scale is proportional to the measure of the parameter space at which they exist and, strictly speaking, is supposed to be zero when interactions can be neglected.

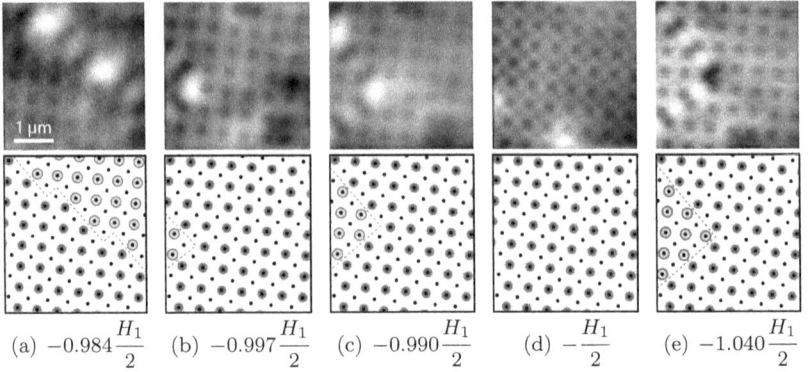

(a) $-0.984\dfrac{H_1}{2}$ (b) $-0.997\dfrac{H_1}{2}$ (c) $-0.990\dfrac{H_1}{2}$ (d) $-\dfrac{H_1}{2}$ (e) $-1.040\dfrac{H_1}{2}$

Fig. 5.9 Scanning Hall-probe micrographs of commensurate vortex configurations after field cooling close to the half-matching field at (a) $H = -0.984H_1/2$ and $T/T_c = 0.767$, (b) $H = -0.997H_1/2$ and $T/T_c = 0.767$, (c) $H = -0.990H_1/2$ and $T/T_c = 0.767$, (d) the 'checkerboard' configuration at $H = -H_1/2$ and $T/T_c = 0.948$, and (e) $H = -1.040H_1/2$ and $T/T_c = 0.837$. The grayscale spans from (a) to (e) are 0.28 mT, 0.24 mT, 0.27 mT, 0.20 mT and 0.20 mT. The lower panels show clearer sketches of the domain structure in each image (after [197]).

can be observed: for $H < -1/2H_1$, when vacancies should exist in the checkerboard structure, the corners where domain walls bend by $90°$ comprise a 'low density' square cluster of three unoccupied pins and one occupied one. In contrast, for $H > -1/2H_1$, when excess vortices should exist, the corners are composed of a 'high density' cluster of three occupied pins and one unoccupied one. Hence, despite the increase in energy associated with the straight segments of domain wall, domain formation is favored because the incorporation of the mismatched vortices and vacancies into domain wall corners lowers the overall energy.

For a short-range vortex-vortex interaction one would expect the energy of the straight domain walls to be prohibitively large and the ground state should consist of 'checkerboard-interstitials' or on-site 'vacancies'. However, in thin films at temperatures close to T_c, when $\lambda(T)$ is much larger than the thickness τ, Pearl [382] has shown that vortex-vortex interactions are long range, decaying as $\sim \ln(\Lambda/r)$ for $r \ll \Lambda(T)$ and $\sim 1/r$ for $r \gg \Lambda(T)$, and domains can become favorable. Figure 5.10 illustrates some possible domain structures. Simple 'bookkeeping' reveals that the square domain in Fig. 5.10(a) with four high-density corners can accommodate exactly one excess vortex (i.e. $1/4\Phi_0$ per corner). Translating the square domain one lattice site vertically (or horizontally), as shown in Fig. 5.10(b), generates a domain with low-density corners which can accommodate exactly one

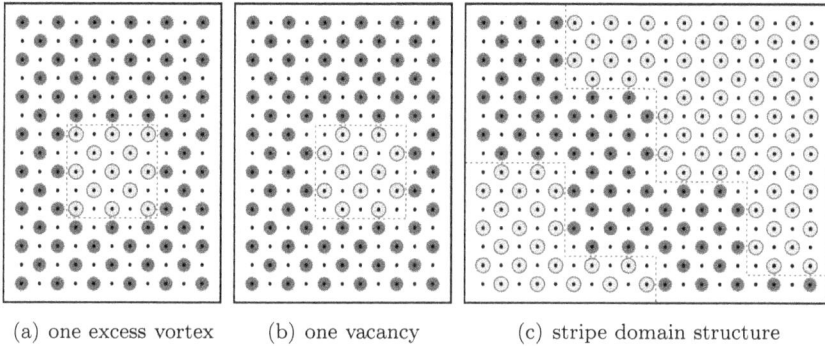

(a) one excess vortex   (b) one vacancy   (c) stripe domain structure

Fig. 5.10   Sketches of possible domain structures of commensurate vortex lattices: A square domain accommodating (a) one excess vortex and (b) one vacancy, and (c) a possible stripe domain structure (after [197]).

vacancy. Alternatively, stripe-like domains could form, as illustrated in Fig. 5.10(c) and observed in Fig. 5.9(a), with either high- or low-density domain wall corners.

To demonstrate that long-range vortex-vortex interactions promote domain formation, molecular dynamics simulations were performed with a generic pinning potential. Considering vortices as stiff massless lines, overdamped Langevin dynamics simulations of an effectively two-dimensional vortex system were carried out [397]. The long-range vortex-vortex interaction energy, i.e. for the case of $\lambda(T) \gg \tau$, was described by [382]

$$U_{vv} = \frac{\Phi_0^2 \tau}{2\pi\mu_0\lambda(T)^2} \left[H_0\left(\frac{r}{\Lambda}\right) - Y_0\left(\frac{r}{\Lambda}\right) \right] , \tag{5.1}$$

where H_0 and Y_0 are second kind Struve and Bessel functions, respectively. The short-range vortex-vortex interaction energy, which is appropriate in thick films with $\tau \gg \lambda(T)$, is described according to Eq. (1.26) by

$$U_{vv}(r) = \frac{\Phi_0^2 \tau}{2\pi\mu_0\lambda(T)^2} K_0\left(\frac{r}{\lambda}\right) , \tag{5.2}$$

where K_0 is the modified Bessel function of the second kind.

The corresponding differential equation was solved numerically, subject to periodic boundary conditions in order to compute the time progression of the system. Starting from a high temperature molten vortex configuration, the system was slowly annealed in the presence of a pinning potential with $d = 1\,\mu m$ periodicity to zero temperature. The interaction of a vortex with

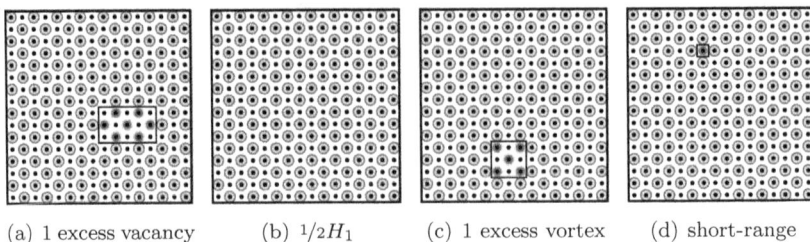

(a) 1 excess vacancy   (b) ¹/₂H₁   (c) 1 excess vortex   (d) short-range

Fig. 5.11   Results of molecular dynamics simulations with a long-range vortex-vortex interaction for (a) half-matching field plus one vacancy, (b) half-matching field, and (c) half-matching field plus one excess vortex. (d) The same simulation as (c) but with a short-range vortex-vortex interaction (after [197]).

each pinning site was assumed to be proportional to $-1/(\rho^2 + c)$, where ρ was the distance to the center of the pin, and c^{-1} a measure of the pins depth.[9] Figures 5.11(a–c) show typical results of the simulation for 400 pinning sites, using the long-range interaction of Eq. 5.1 with $\Lambda = 2\,\mu$m. For exactly half-matching conditions, the checkerboard structure is reproduced (Fig. 5.11(b)). With one vacancy present, a rectangular domain with four low-density corners is created (Fig. 5.11(a)), while with one excess vortex a square domain with four high-density corners is found (Fig. 5.11(c)), in agreement with the images in Fig. 5.9. Repeating the calculation with one excess vortex but using the short-range vortex-vortex potential of Eq. (5.2) results in the configuration shown in Fig. 5.11(d). As expected, the excess vortex is located in this case at an unoccupied pin in the checkerboard lattice.

A simple 'mean-field' analytic model of the system, where the flux associated with each lattice site is averaged over the four adjacent unit cells, confirms that Λ sets a characteristic length scale for the size of domains. Within this picture, the domain walls vanish and the only long-range interactions are between the fractional flux quanta $(1/4\Phi_0)$ associated with the domain corners. Adding the domain wall energy σ per lattice site and assuming $w > \Lambda > d$, the total energy associated with a domain of side length w is approximately

$$E(w) \sim \frac{4\sigma w}{d} + \frac{(4+\sqrt{2})\Phi_0^2}{8\mu_0\pi^2 w}.$$ (5.3)

[9]Note that modeling results are not very sensitive to the details of the chosen pinning potential.

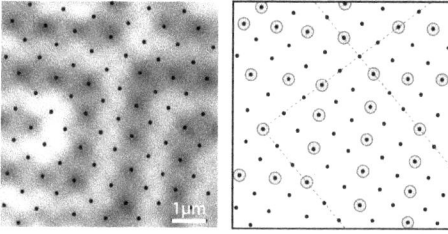

Fig. 5.12 Scanning Hall-probe image (left) of commensurate vortex domains after field cooling at $|H| = -1/3H_1$ and $T/T_c = 0.767$ (gray scale ~ 0.26 mT). The right panel shows a clearer sketch of the observed domain structure (after [273]).

For typical sample parameters, one can calculate numerically that $\sigma \sim U_{vv}(d)/50 \sim \Phi_0^2/(50\pi\mu_0\Lambda)$ and the domain energy has a minimum at $w \sim 2\sqrt{d\Lambda}$. The effective penetration depth $\Lambda = 2\,\mu\text{m}$, used in simulations, represents a much higher reduced temperature ($T/T_c = 0.954$) than that corresponding to our field-cooled scanning Hall-probe images. We assume, however, that the latter represent metastable states which become 'frozen in' close to T_c where the penetration depth is much longer.

The checkerboard structure at $|H| = 1/2H_1$ preserves the fourfold rotational symmetry of the pinning array and appears rather robust with respect to the specific form of vortex-vortex interactions. This is not true for the structure at $|H| = 1/3H_1$ where the ground state was originally predicted to be one of six degenerate symmetry-breaking chain states as sketched in the upper right panel of Fig. 4.15 [497]. These represent an inefficient way of packing vortices, and one might anticipate that a multidomain state would have lower energy, even exactly at the matching field [397]. The multidomain scanning Hall-probe image at $|H| = 1/2H_1$ (Fig. 5.12) shows this to be the case.

In summary, vortex domain formation near rational fractional matching is a consequence of domain wall topology and long-range vortex-vortex interactions in thin films close to T_c. Domain corners can incorporate fractions of excess vortices and vacancies and the overall energy is lowered because the system is able to lock to the symmetry of the pinning array within a domain, yet the magnetic induction averaged over several domains can vary around the exact matching field. This explains the absence of some fractional magnetization peaks even when matching occurs (see Fig. 5.6). Matching patterns, which strongly break the symmetry of the pinning array, can even be unstable with respect to domain formation exactly at the matching field. The number of domains and domain sizes are generally governed by $\Lambda(T)$ at which the domain structure becomes frozen, the periodicity d, and the number of mismatched vortices.

(a) $H = 0$

(b) $H > 0$

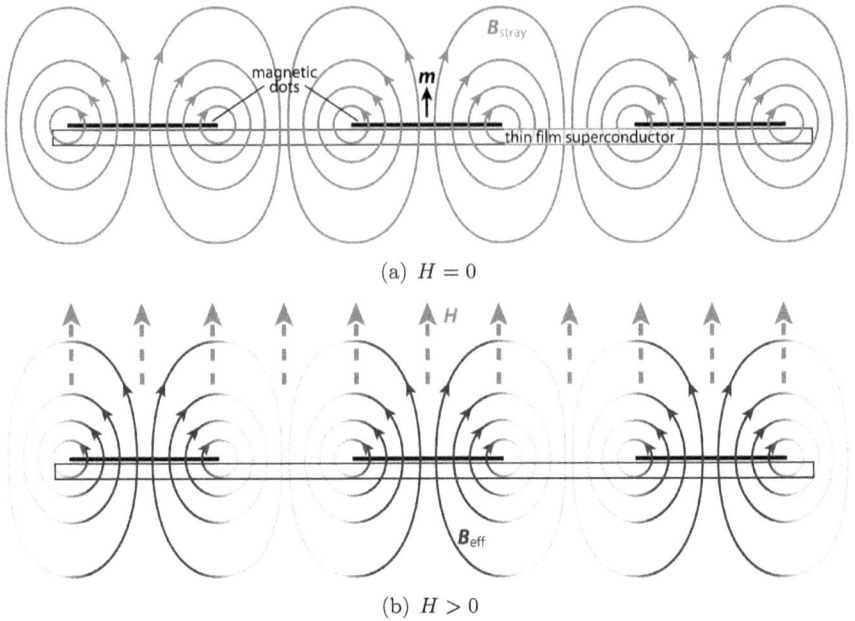

Fig. 5.13 Schematic drawing of the appearance of magnetic-field-induced superconductivity (FIS) in a nanostructured S/F hybrid system. Without an external magnetic field (a), the z-component of the stray magnetic field B^z_{stray} of the dots is such that superconductivity is suppressed in the entire underlying superconductor. However, for $H > 0$, B^z_{stray} can be compensated between the dots, resulting in the conditions necessary for the observation of FIS. Note that the aspect ratios of dots and S-film are those used in the experiment described below.

5.2   Field induced superconductivity

A quite general property of the superconducting state — namely its suppression by a magnetic field which exceeds a certain critical value — sets limitations for practical applications of superconductors, since a current that flows through a superconductor generates a magnetic field, which can lead to the loss of zero resistance. However, in this section we will demonstrate that it is possible to *induce* superconductivity in special S/F hybrid systems by *applying* an external magnetic field.

   This rather counterintuitive phenomenon, typically referred to as '*magnetic-field-induced superconductivity*' (FIS), has first been observed in three distinct homogeneous bulk materials. Wolf *et. al.* [502] and later on Meul *et. al.* [322] found that the compound $(\text{EuSn})\text{Mo}_6\text{S}_8$ is superconducting at zero field below a certain temperature, entering the normal state at low fields. Surprisingly, at higher fields of the order of 10 T, the material turns superconducting again. This appearance of FIS was interpreted in terms of the Jaccarino–Peter effect [236], where the exchange fields from the paramagnetic europium (Eu) ions compensate an applied magnetic field, so that the destructive influence of the applied field is neutralized. Another system that exhibits FIS is $HoMo_6S_8$ [184], although at significantly lower fields around 0.1 T. Simultaneous measurements of magnetization and resistance of this compound, as a function of the applied field, revealed that jumps in the resistance and in the magnetization are going hand in hand. This indicates that a purely electromagnetic field compensation effect between the applied field and internal magnetic fields are responsible for the occurrence of FIS in this material. Recently, FIS was also discovered in organic λ-$(BETS)_2FeCl_4$ materials [33, 470] in ultra high fields between 18 and 41 T. In these organic compounds, the exchange field of iron (Fe) ions is responsible for the compensation of the applied field, and thus for the occurrence of the superconducting state. Interestingly, the formation of the nonuniform Fulde–Ferrell–Larkin–Ovchinnikov (FFLO) state is predicted within a certain temperature and field region of the phase diagram of λ-$(BETS)_2FeCl_4$ [223].

For the above three cases of $(EuSn)Mo_6S_8$, $HoMo_6S_8$ and λ-$(BETS)_2FeCl_4$, the occurrence of FIS is explained by a compensation of the applied field by an internal field produced by the polarization of magnetic ions. Consequentially, for all these materials, the induced moment is parallel to the applied field. However, also a rather 'true' coexistence of ferromagnetism and superconductivity has been observed in the bulk materials UGe_2 [419], URhGe [13], UIr [4] and UCoGe [228]. In the specific case of URhGe, for example, superconductivity occurs when the ordered magnetic moment is tilted by 30 to 55° to $\boldsymbol{H}$, making it unlikely that an exchange field due to magnetic moments can compensate the effect of H over such broad field range. Lévy *et al.* observed in high quality URhGe crystals a superconducting re-entrance between 8 and 13 T that is correlated with the vicinity to a magnetic transition [283], suggesting that superconductivity is driven by the exchange of magnetic excitations [284]. Their results seem to indicate a triplet order parameter [143], i.e. pairs of electrons with parallel spins, with the maximum gap directed along the magnetically hard a-axis, which may result from pairing due to magnetic fluctuations propagating along the a-axis in connection with a rotation of the magnetic moments in the bc-plane.

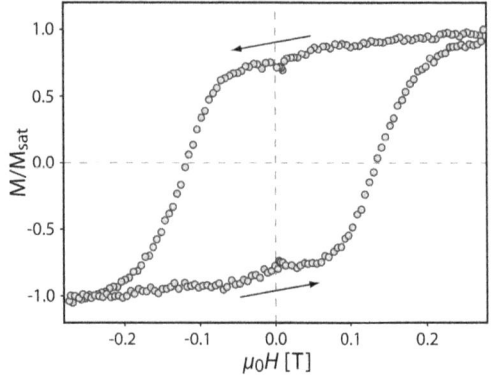

Fig. 5.14 Magnetic hysteresis loop of the considered array of magnetic Co/Pd dots, obtained by the magneto-optical Kerr effect at room temperature. The applied field $\boldsymbol{H}$ is perpendicular to the surface of the sample (after [273]).

In the following we will present in details an example for the realization of FIS in nanostructured hybrid S/F systems. A schematics of the underlying idea is given in Fig. 5.13, showing a lattice of magnetic dots with magnetic moments $\boldsymbol{m}$ aligned in the positive z-direction, on top of a thin film superconductor. The z-component of the stray magnetic field B_{stray}^z of the dot array is positive and negative underneath and between the dots, respectively. Close to the phase boundary of the superconductor, B_{stray}^z exceeds the critical field and, as a result, the superconducting film is in the normal state. However, when added to a homogeneous magnetic field H (see Fig. 5.13(b)), the stray magnetic field of the dots *enhances* the z-component of the effective magnetic field $\mu_0 H_{eff} = \mu_0 H + B_{stray}^z$ in the regions underneath the dots and, at the expense of that, *reduces* H_{eff} everywhere else in the S-film. Therefore, in a certain range of temperature, superconductivity can appear in the S-film at the interstitial regions upon application of a magnetic field. Note that the described effect is a general property of any superconducting film with an array of ferromagnetic dots of perpendicular magnetic anisotropy, and that the field range for the FIS observation can be controlled by varying the distance between the dots or the amount of magnetic material in each dot.

In the present case we consider a nanostructured S/F hybrid system similar to the one of the previous section, i.e. a superconducting lead film that is covered by an array of magnetic dots, but with slightly different dimensions.[10] A room-temperature hysteresis loop of the structure, obtained by the magneto-optical Kerr effect, is shown in Fig. 5.14 for $\boldsymbol{H}$ perpendic-

---

[10]In particular, a lead film of 85 nm thickness and $0.8 \times 0.8 \, \mu m^2$ magnetic dots consisting of a $[Co(0.4nm)/Pd(1.4nm)]_{10}$ multilayer, forming a square array with period $d = 1.5 \, \mu m$, were used.

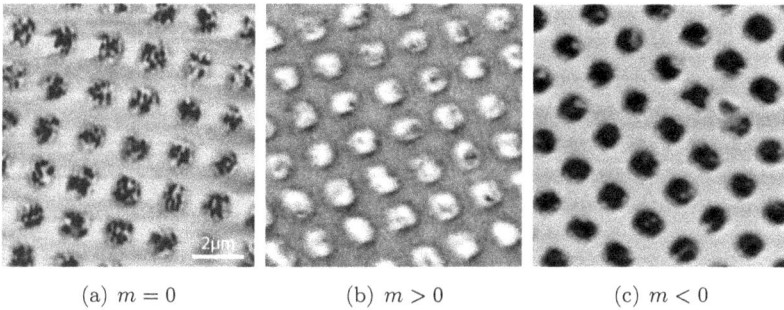

    (a) $m = 0$ (b) $m > 0$ (c) $m < 0$

Fig. 5.15 Room temperature magnetic force micrographs of the used magnetic dots at $H = 0$ in (a) the demagnetized state $m = 0$, and (b, c) the magnetized states after saturation at $\pm 1\,\mathrm{T}$ (after [273]).

ular to the surface of the sample. The $M(H)$ loop reveals a high magnetic remanence of $\sim 80\%$ of the saturation magnetization M_{sat}, as well as a large coercive field $\mu_0 H_{\mathrm{coe}} \simeq 150\,\mathrm{mT}$. It is therefore possible to produce quite stable remanent domain states in the ferromagnetic subsystem by using different magnetization procedures.

After demagnetization, the dots are in a multidomain state, as can be seen in the corresponding magnetic force image of Fig. 5.15(a), which is in contrast to the small Co/Pt dots used in the studies discussed in the previous section. In this state, the net magnetic moment m of each dot is approximately zero, and we will thus refer to it as to the state of $m = 0$. By contrast, after magnetization in high fields, the dots are either magnetized up or down, see Figs. 5.15(b, c), depending on the polarization of the saturating field. Note that the spots, visible in some of the dots in Figs. 5.15(b, c), could be correlated directly to inhomogeneities of the dots topography [273], meaning they have no magnetic origin and that the magnetized dots are in a single domain state.

Figure 5.16 shows the $T_c(H)$ phase boundary of the superconducting lead film for the three different magnetic states of the dot array (see Fig. 5.15), obtained from resistivity measurements as a function of temperature. For $m = 0$, a conventional phase boundary is obtained, which is symmetric with respect to H, see Fig. 5.16(a). The only features that indicate the presence of the dot array are two weak kinks at $H = \pm H_1$, meaning that in this state, the dots have only minor influence on the superconducting phase diagram. The phase boundary is clearly altered when magnetizing the dot array, leading to a strongly asymmetric $T_c(H)$ with

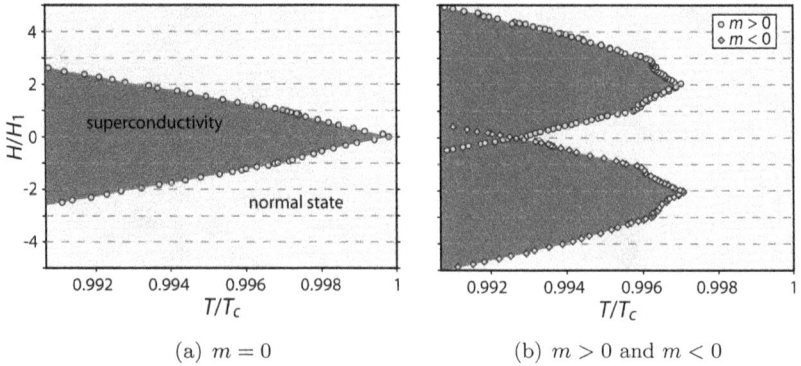

(a) $m = 0$ (b) $m > 0$ and $m < 0$

Fig. 5.16 H–T phase diagrams of the considered lead film in (a) the demagnetized state and (b) the magnetized states $m \gtrless 0$. Dark and bright areas correspond to regions of superconductivity and normal state, respectively. Note that the phase boundaries for $m > 0$ and $m < 0$ are combined in (b) and indicated with circles and diamonds, respectively (after [273]).

respect to H, see Figs. 5.16(b). This shift is a sort of 'magnetic bias' of the $T_c(H)$ boundary, as described in Sec. 1.4. The maximum critical temperature is shifted to $\pm 2H_1$ for $m > 0$ and $m < 0$, respectively, equivalent to the observation of FIS. In addition, kinks can be observed in the $T_c(H)$ curves at $H/H_1 = 0, +1, +2$ and $+3$ for $m > 0$, and at $H/H_1 = 0, -1, -2$ and -3 for $m < 0$.

These kinks are a result of fluxoid quantization effects [290], confirming that superconductivity indeed nucleates in multiply connected regions of the film, similar to the case of superconducting wire networks or thin films with periodic arrays of antidots (see Chap. 4). The maximum T_c at exactly $H = +2H_1$ can therefore be understood in terms of fluxoid quantization: the flux created by the stray field between the dots can be estimated from magnetostatical calculations [273] to be about $-2.1\Phi_0$ per unit cell of the dot array. This makes $H = +2H_1$ a favorable field for fulfilling the fluxoid quantization constraint. Similar arguments can be applied for the dots in the $m < 0$ state to explain the shift of the maximum T_c to $H = -2H_1$. In the state of $m = 0$, B_{stray} is strongly reduced due to the presence of the domain structure in the dots, and as a result, the stray field influences the superconductor only slightly, leading to a phase boundary without peculiarities except the weak kinks at $H = \pm H_1$. Detailed modeling of these effects in the framework of the Ginzburg–Landau theory was carried out in Ref. 324.

The field region in which FIS is observed can be tuned by changing the period of the dot array or the magnitude of the stray field. For instance, an increase of the fields emanating from the dots could be achieved by using materials with larger magnetic moments, shifting the maximum of T_c to even higher applied fields. Good candidates for this are arrays of nanodots [410] and nanopillars [99], including ferromagnetic nanowires electro-deposited in porous alumina templates [209]. For instance, dot arrays with a period of 70 nm have been fabricated [255], corresponding to $H_1 \approx 0.4\,\mathrm{T}$, which is already a remarkably high field. However, one should bear in mind that decreasing the dot size could result in a too weak stray field, so that the Meissner state is realized in the superconductor and the phase boundary is not field-shifted. This can be prevented by the use of nanopillars as mentioned above, consisting of magnetic material with a large magnetic moment like rare-earth ferromagnets such as gadolinium.

Finally, we would like to point out that the dipole array field compensator could possibly be used in applications. A lot of attention is currently paid to materials showing giant [31] and colossal magnetoresistance effects [493], and the implementation of such materials as field sensors in magnetic recording. Besides improving and shifting the critical fields, the nanoengineered FIS could also be used in these kind of applications, since tunable huge magnetoresistance effects, as presented in Figs. 5.16(b, c), are ideal for logical switches or field sensors.

5.3 Dipole-induced vortex ratchet effects

If a periodic pinning potential for vortices lacks inversion symmetry in a certain direction, any correlated fluctuating force can induce a net vortex motion based on the ratchet effect [403] (see Secs. 2.7 and 4.3). As proposed by Carneiro [86], a new way to create vortex ratchets can be realized by using in-plane magnetized dots, breaking the spatial inversion symmetry rather by the interaction between vortices and magnetic dipoles than by the shape of the pinning sites as in the case of antidot lattices (Sec. 4.3).

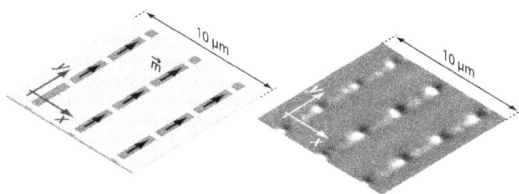

Fig. 5.17 Schematic sketch of the used array of magnetic micro-bars (left), and a MFM-image of the bars after magnetization along the y-axis (right), where dark and bright regions indicate high H^z_{stray} (after [121]).

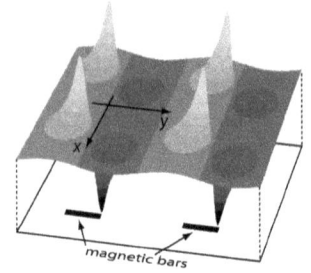

Fig. 5.18   Calculated pinning potential for the used array of magnetic micro-bars, following a standard procedure described in Refs. 86 and 323 (after [475]).

This dipole-induced ratchet motion depends on the orientation and strength of the local magnetic moments, thus allowing for the control of the direction of the vortex drift, which makes this kind of pinning potentials attractive for practical applications.

On the basis of a series of transport experiments, it was demonstrated [121] that in-plane magnetized dipoles can indeed rectify vortex motion. Here, we shall discuss rectification effects in two different structures: a square array of magnetic cobalt bars on top of, and a close-packed array of permalloy square loops covered by a thin superconducting aluminium film (see Tab. 5.1 for some characteristics of the considered structure).

5.3.1   *Generation of vortex-antivortex pairs*

Figure 5.17 shows a room-temperature[11] magnetic force micrograph of an array of magnetic microbars, after magnetization along the y-direction.

Table 5.1   Characteristics of the studied S/F systems with micromagnets

Property	Cobalt Bars	Square Permalloy Rings
periodicity d [μm]	3	1.07
dimensions of magnets [μm^2]	2.6×0.5	1×1 (150 nm line width)
thickness τ_{mag} of magnets [nm]	47	25
thickness τ of superconductor [nm]	50	50
critical temperature T_c [K]	1.325	1.365
coherence length $\xi(0)$ [nm]	87	140
first matching field H_1 [mT]	0.230	1.806

[11]Note that the high Curie temperature of cobalt ($\sim$1400 K) guarantees that the magnetic states shown in Fig. 5.17 remain essentially the same at low temperatures. Moreover, due to a coercive field of $\sim$40 mT, small perpendicularly applied magnetic fields, less than 3 mT, used in the present experiments to generate specific vortex distributions, are not able to change the magnetic state of the micromagnets.

Owing to their large aspect ratio, the remnant state of all bars is single domain [431], with the magnetic poles lying very close to the extremities of the bars. As a consequence, the small separation between neighboring bars works as a dipole with an effective magnetic moment pointing against the magnetization of the bars.

The pinning potential for a singly quantized vortex U_{bar}, generated by the array of magnetic bars, is given in Fig. 5.18, calculated with a standard procedure [86, 323]. This potential is antisymmetric along the y-direction with respect to the origin of the midpoint between consecutive bars, thus producing the broken inversion symmetry required for the ratchet effect. Inducing a Lorentz force on a vortex along the $+y$-direction, the asymmetry of this potential may be probed by applying an oscillating current along the x-axis. For antivortices, however, U_{bar} changes sign, that is while vortices drift more easily in the $+y$-direction, the easy drive for antivortices is the $-y$-direction, which is in contrast with ratchet potentials generated by asymmetric antidots (see Sec.4.3). Because $V_{dc} \propto v_d H$, where v_d is the vortex drift velocity, the measured dc-voltage V_{dc} is insensitive to the polarity of the external field. Therefore, the presence of mirror-like regions in $V_{dc}(H)$ (with respect to $H = 0$) is a fingerprint of the dipole-induced ratchet effect.

In order to test whether this situation describes the studied system properly, the output dc-voltage V_{dc}, generated by the flow of vortices, was recorded while applying a sinusoidal ac-current. For a symmetric pinning landscape like the one provided by a plain film, one should expect a very low response V_{dc}, as it is indeed observed in our experiment. Note that it is very difficult to realize the $V_{dc} = 0$ situation due to the inevitable deviation of the real sample shape from a perfect symmetrical shape [389]. For the polarized state, a well-defined ratchet drift (i.e. $V_{dc} \neq 0$) can be seen in Fig. 5.19, where the corresponding H–T diagram clearly demonstrate a symmetrical $V_{dc}(H)$. Interestingly, at lower matching fields and even at $H = 0$, narrow regions of positive voltage are embedded in larger regions of negative V_{dc}. Such zero-field ratchet effect suggests the existence of vortex-antivortex $(v - av)$ pairs, generated by the micro magnetic array, which interact with the underlying asymmetric potential and give rise to the observed ratchet signal even at zero field.

Under these circumstances, a $v - av$ excitation with vorticity (L_v, L_{av}) introduces a change in both the superfluid kinetic energy and the condensate energy, given by

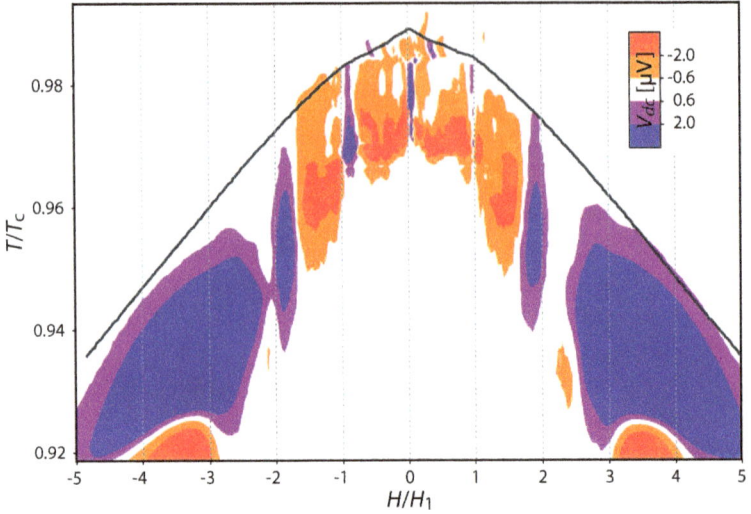

Fig. 5.19 Rectified voltage V_{dc} in the temperature–field plane for a superconducting aluminium film with in-plane magnetized micro-bars on top, subjected to a 1 kHz sinusoidal ac-current. The full line indicates the transition between superconducting and normal state (after [121]).

$$E_k(R) = L_v L_{av}(V(R) - V(0)) + \frac{L_v - L_{av}}{2} V(0) \qquad (5.4a)$$

$$E_c = \pi r_0^2 \tau H_c^2 \frac{L_v + L_{av}}{8\pi} , \qquad (5.4b)$$

respectively. In Eqs. (5.4), $V(R)$ is the attractive interaction energy of a singly quantized $v - av$ pair, $R = |\boldsymbol{r}_v - \boldsymbol{r}_{av}|$, r_0 the effective vortex core radius and H_c the thermodynamic critical field [328]. By symmetry considerations one can show that the $v - av$ pair is aligned with the y-axis and $\boldsymbol{r}_{av} = -\boldsymbol{r}_v$, from which is follows that $R = 2y_v$. The total change in the Gibbs free energy, caused by the appearance of a pair, can thus be expressed in terms of y_v only:

$$\begin{aligned} \Delta G_{L_v, L_{av}}(y_v) &= (L_v + L_{av})U_{bar}(y_v) + E_k(2y_v) \\ &\quad + E_c - (L_v - L_{av})\Phi_0 H\tau , \end{aligned} \qquad (5.5)$$

where the last term accounts for the interaction with the external field.[12] Suppose that $R \ll \lambda^2/\tau$, $V(R)$ and r_0 can be calculated within Clems

---

[12]Note that in Eq. (5.5), the index L_v, L_{av} in E_k and E_c is omitted for clarity.

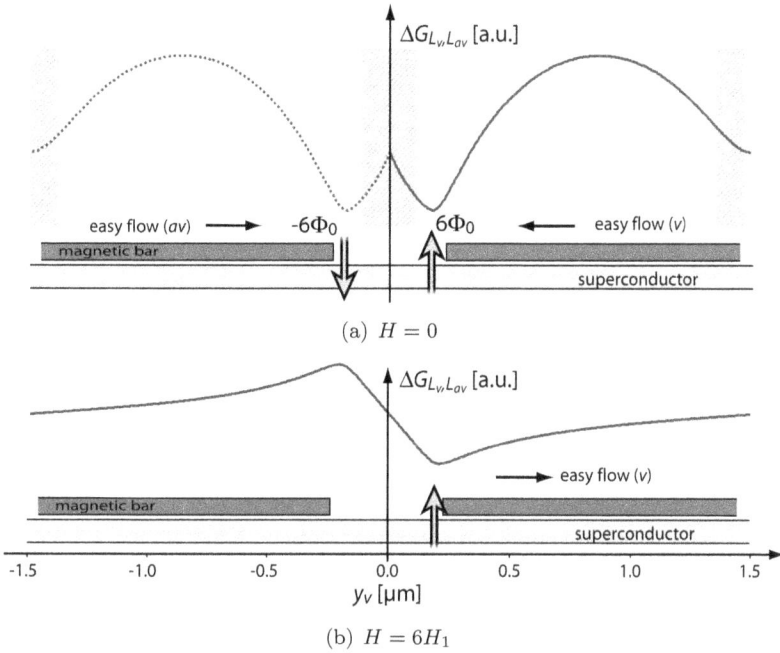

Fig. 5.20 Total energy change of stable $v - av$ pair configurations as a function of pair position y_v and $y_{av} = -y_v$ at (a) $H = 0$ and (b) $6H - 1$, calculated for the parameters of the present sample at $T/T_c = 0.9$. The curves are shown from the point of view of vortices (full lines) and antivortices (dashed lines), in one unit cell of the array of micro-magnets. The cartoons illustrate the equilibrium configuration for vortices (upward arrows) and antivortices (downward arrows). Small arrows indicate the direction of easy motion of vortices and antivortices. Hatched areas roughly correspond to a pair separation of $R < \xi(T)$, at which $v - av$ annihilation takes place (after [121]).

variational approach [105] for the Ginzburg–Landau equations. Let us assume that at $H = 0$ each magnet stabilizes a (L_0, L_0) pair, thus generating a lattice of neutral $v - av$ pairs. Accordingly, $E_k(R)$ must account for the interaction of a vortex with the whole array of antivortices and vice versa. The zero-field vorticity L_0 and the equilibrium pair position y_0 can be estimated by minimizing $G_{L_v,L_{av}}(y_v)$ at $H = 0$. The solution is thermodynamically stable if $G_{L_0,L_0}(y_0) < 0$. For the parameters of the present structure, we find $L_0 = 6$ at $T/T_c = 0.9$. Although this is a rough estimate, it is a good indication that the magnetic template indeed generates $v - av$ pairs.

We can now go a step further in this analysis and determine the zero-field ratchet motion. Figure 5.20(a) shows the $\Delta G_{L_v,L_{av}}(y_v)$ curve at $H = 0$ in a unit cell of the array of micro-magnets, where hatched areas correspond to $v - av$ annihilation zones, at which the present model is no longer valid. A current in $+x$-direction (i.e. out of the page) forces both vortex and antivortex to approach each other, while a current in $-x$-direction tears them away from each other. Notice that the symmetry constraint $y_{av} = -y_v$ is implicit here. The resulting energy profile exhibits a steeper slope for motion of vortices (antivortices) to the right (left) and, therefore, compression rather than expansion of the $v - av$ pair is favored. As a consequence, an oscillating current induces more vortex motion during the positive half-cycles, which results in a positive dc-voltage as observed at $H = 0$ in the T–H diagram of Fig. 5.19. However, if the external field exceeds a certain value (namely $H > L_0 H_1$), all antivortices are annihilated. In this case, the $v - av$ attraction term is absent and the vortex energy profile is determined by the bare vortex-magnetic-dipole potential, which has a smaller slope for vortex motion along the $+y$-direction (see Fig. 5.20(b)). Consequentally, in such high fields, an oscillating current induces $V_{dc} < 0$.

The ratchet motion at zero field relies on a very symmetric $v - av$ configuration and, in general, it cannot be expected to hold for small incommensurate fields. Indeed, changing the field from $H = 0$ leads to a negative voltage sign as observed in Fig. 5.19. One possibility to explain these findings is schematically sketched in Fig. 5.21, where at low incommensurate fields the extra vortices or antivortices are loosely attached to the pinning site, and thus are able to move at lower critical forces. Let us consider, for instance, the situation at a slightly positive field where only a few vortices are distributed along the sample. At high temperatures, extra vortices will — dependent on the magnetization of the bar — either be attracted by an antivortex with subsequent annihilation, or merge into another vortex. In both cases there will be a distribution of defects in the $v - av$ pair lattice, namely $(L_v, L_{av} - 1)$ or $(L_v + 1, L_{av})$, breaking the high discrete symmetry of the system.

The case of a $(L_v, L_{av} - 1)$ defect is illustrated in Fig. 5.21, where we took $L_v = L_{av} = 1$ for simplicity. The situation for zero current is depicted in Fig. 5.21(b) — c.f. the equilibrium configuration at $H = 0$ shown in panel (a). For a current in $+x$-direction, the defect (in this case a singly quantized vortex) will be pushed to the left, experiencing the steeper direction of the magnetic pinning potential while being pulled back by the nearest antivortex (Fig. 5.21(c)). However, for a current in $-x$-direction, the vortex experiences the weaker pinning potential while being pulled in the same direction by the nearest antivortex. On its way to the next avail-

(a) Undisturbed symmetric $v - av$ configuration at $H = 0$

(b) A defect in the vortex lattice due to $H > 0$

(c) Vortex dynamics for positive currents $\boldsymbol{J}$ at $H > 0$

(d) Vortex dynamics for negative currents at $H > 0$

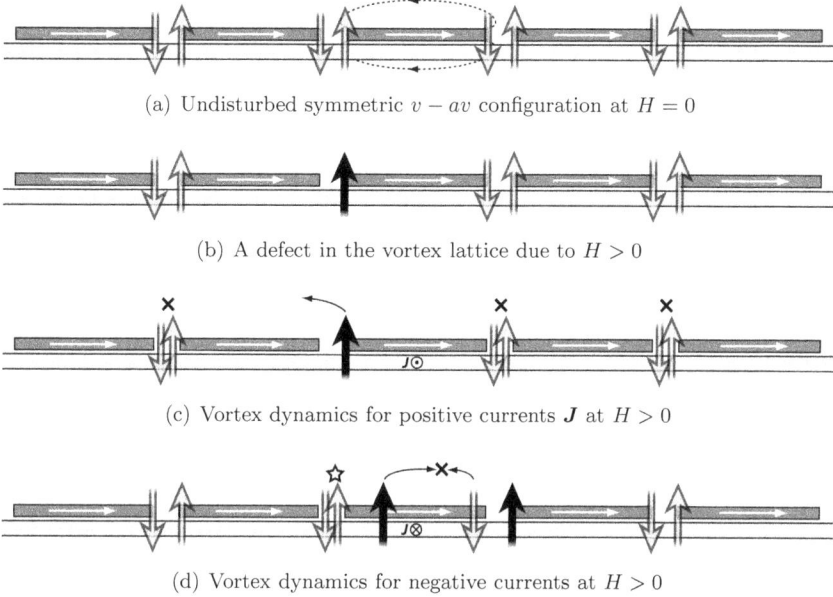

Fig. 5.21 (a) Configuration of $(1,1)$ $v - av$ pairs at zero field close to T_c, assuming $L_v = L_{av} = 1$. (b) An extra vortex annihilates an antivortex of a $(1,1)$ pair at a small positive field, giving rise to a $(1,0)$ defect (dark upward arrow). The dynamics of the system when subjected to a positive (c) and a negative (d) transport current. Crosses and stars indicate $v - av$ annihilation and creation, respectively.

able pinning site, the vortex annihilates with the antivortex on the right, leaving space for creation of a new $v - av$ pair (Fig. 5.21(d)). The resulting situation is the same as in Fig. 5.21(b), but translated to the right by exactly one period of the array of micromagnets. By alternating between positive and negative current, the vortex will be rectified step-by-step to the right, moving further during the positive current half-cycles and thus causing a negative V_{dc}. This is in agreement with the experimental observations at high temperatures in Fig. 5.19, where narrow bands of positive voltage at some lower matching fields and at $H = 0$ are embedded into larger regions of negative voltage.

5.3.2 Switching rectification properties

As a second example for systems showing magnetic dipole induced vortex motion rectification, we shall now consider a ring-like magnet and thus multiply connected structure, which is a necessary characteristic for ob-

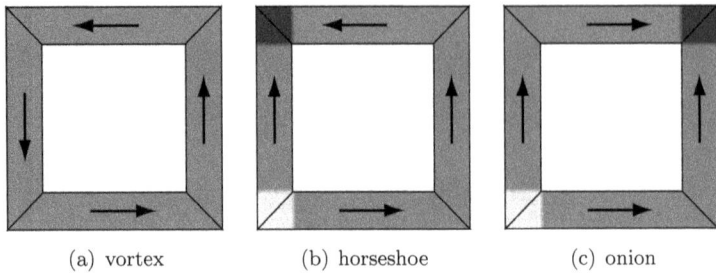

(a) vortex   (b) horseshoe   (c) onion

Fig. 5.22   The different magnetic states of magnetic square loops with in-plane magnetization: vortex (a), horseshoe (b), and 'onion' state (c). Black arrows indicate the
direction of the magnetic moment in each domain. In (b, c), positive and negative out-
of-plane components of the magnetic stray field are indicated by light and dark spots,
respectively.

taining a switchable 'ON/OFF' pinning landscape [440]. Vortex pinning
by magnetic loops in different states has been visualized by using Bitter
decoration and scanning Hall-probe microscopy [314]. These systems exhibit two clearly distinct possible magnetic states, a flux-closure state with
negligible out-of-plane stray field component B_z, which induces weak pinning in a superconducting film, and dipolar states (horse shoe and onion
states) with strong B_z, acting as effective pinning sites. The procedure
necessary to switch between these states is quite simple. In the following
we shall demonstrate with transport data that different sorts of ratchet
motion take place, depending on the magnetic state of the rings. In particular, it will become clear that the net direction of the vortex motion can
be reversed by changing the magnetic state of the micromagnets.

It has been experimentally demonstrated [232, 484] that a two-
dimensional multiply-connected magnetic structure with shape of group
symmetry C_n can be set in two fluxclosure states of opposite chirality (see
Fig. 5.22(a)) and $n\,(n-1)$ different polarized states (see Fig. 5.22(b, c)). For
a square loop $n = 4$ and, therefore, 12 states corresponding to six different
dipole directions with two opposite dipole orientations are expected. If the
net dipolar moment is parallel to the side of the square, the final state is
named 'horseshoe' state (see Fig. 5.22(b)) whereas we refer to as the 'onion'
state if the dipole is along the diagonal of the square (see Fig. 5.22(c)).

In an array of microscopic magnetic loops, separated by more than
50 nm, most of the spectrum of possible magnetic states coexist in a very
disordered distribution in the as-grown state, thus suggesting that there
is little interaction between neighboring elements [437]. A different sit-

uation emerges when the rings are magnetized by applying a relatively high ($H \sim 800\,\text{mT}$) in-plane magnetic field along their diagonal and subsequently turning the field off (Fig. 5.22(c)). In this case, the effective magnetic field, felt by the superconducting layer, exhibits a highly ordered dipole-like structure of alternating positive and negative stray field. However, if starting from the single domain fully polarized state, reversing H to about $-35\,\text{mT}$ and then reducing the field to zero, domain walls are formed and at the corners of the square the magnetic moment gradually rotates by $90°$ [484], thus forming a flux-closure state with little stray field (Fig. 5.22(a)). Object oriented micromagnetic simulations suggest that all four segments of the square shaped rings are nearly single domain states. Different orientation of these single domain states form different domain patterns. The type and orientation of the elements can be controlled by an external in-plane magnetic field.

If the micro-loops are set in the magnetized state (in the present case $+45°$ off x-axis), they form a square array of in-plane dipoles with the same periodicity as the array of squares. The interaction between the resulting field component B_z and the flux lines produces a minimum in the pinning potential at the maxima of B_z. This potential U_{loop} mirrors the shape of the B_z profile, inverting the asymmetry and hence producing an also asymmetric pinning force $F = -\partial_y U_{\text{loop}}$. Under these circumstances, if the system is submitted to an oscillatory current in x-direction with zero average, a net motion of the flux line lattice will be induced towards $\pm y$ with the sign depending on the orientation of the dipoles. This is indeed confirmed by the dc-output voltage V_{dc} recorded under a sinusoidal ac-current I_{ac} as shown in Fig. 5.23 for different scenarios.

For a completely symmetric pinning landscape, like the one provided by a plain film, V_{dc} is expected to be negligible. In contrast to that, plain reference samples displayed a clear $V_{dc} \neq 0$, which can be attributed, for example, to inevitable asymmetries between the edge barriers for vortex penetration [389, 490]. For both the patterned sample with the magnetic square loops in a disordered state (see upper panel in Fig. 5.23) and for the plain film, V_{dc} is an antisymmetric function of field, indicating that an undesirable ratchet potential dominates the effective vortex motion. As in the previous example of the magnetic micro-bars, it is precisely the imperturbable nature of this standard ratchet motion which allows us to unambiguously separate the contribution coming from the dipole-induced ratchet motion.

The dc-response shown in the middle panel of Fig. 5.23 corresponds to the case of magnetic square loops in the 'onion' state. Since for $H > 0$

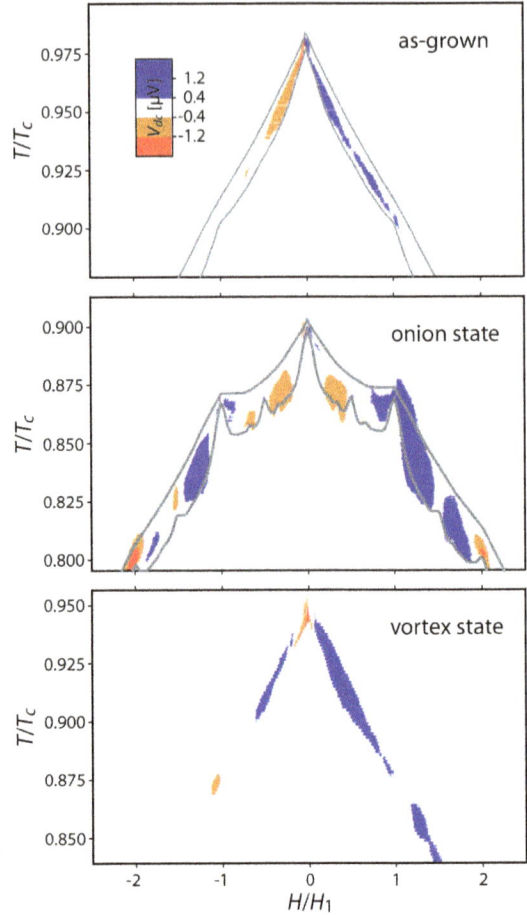

Fig. 5.23 Rectified voltage V_{dc} as a function of field and temperature for a sinusoidal excitation of 1 kHz frequency for the square micromagnets in the as-grown state (upper panel), the 'onion' state (middle panel) and vortex state (lower panel). Solid lines indicated the transition between superconductivity and normal state as determined by a 10% and 90% normal state resistance criterion (after [440]).

vortices are located at different positions than for $H < 0$, v_d turns out to be an antisymmetric function of H and, therefore, positive and negative fields give rise to the same signs of V_{dc}. In other words, vortices and antivortices have an opposite easy direction of motion, leading to mirror-like $V_{dc}(H)$ regions. Interestingly, unlike for the as grown state, in the 'onion' state the ratchet signal changes sign as $|H|$ is increased. This behavior can be in principle attributed to the fact that several dipolar moments coexist in the array of square rings as shown in Fig. 5.24. The relevance of each of these possible dipolar moments depends on the size of the coherence length and the vortex density [122]. Thus, as $|H|$ increases (as so does the vortex density), noting also that lower temperatures are needed to obtain

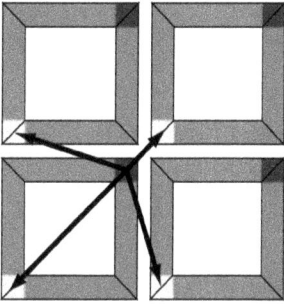

Fig. 5.24 Schematic representation of the possible nearest neighbor dipolar moments that can be formed for the close packed array of square rings (after [440]).

an optimum ratchet signal, the coherence length decreases and therefore a crossover between different regimes can be expected.

If starting from the 'onion' state the rings are set in a magnetic vortex state (see lower panel in Fig. 5.23), although the shape of the ratchet signal resembles the as-grown state, its field independent polarity indicates that it is magnetic in origin. This unusual behavior (sharing common features of the as-grown and 'onion'-state) results from the incomplete formation of the flux-closure state in all rings simultaneously. Indeed, following the procedure indicated above for building a flux-closure state, the final distribution at remanence consists of a majority of squares in the vortex state coexisting with a minority in the 'onion' state. This reduced density of rings in the 'onion' state suppresses the superconducting condensate locally but does not affect the minimum resistance percolation path which defines the superconducting phase boundary. However, since the remaining rings in the 'onion' state have a well defined dipolar moment, they still influence the ratchet signal.

5.4 Superconductivity in stray fields of magnetic domains

So far in this chapter we have concentrated on S/F hybrid systems that contain arrays of magnetic nanostructures, used to manipulate the super-conducting state in thin film superconductors close by. However, as already discussed in Sec. 1.4, it is possible to induce and/or suppress superconductivity by the stray magnetic fields of magnetic domains or domain-walls, which offer an alternative for the manipulation of the superconducting state at the nanoscale. In the present section, we shall discuss two phenomena that emerge from the interaction of a thin film superconductor with the stray magnetic field of underlying magnetic domains, namely the states of

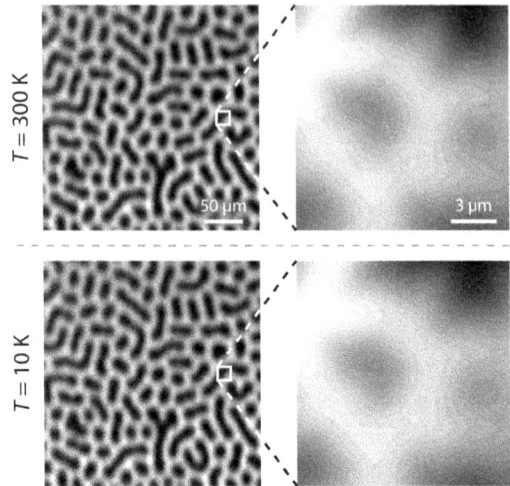

Fig. 5.25 Polarized light microscopical (left) and scanning-Hall
probe images (right) of the magnetic domain pattern at the
basal plane of a barium hexaferrite single crystal at different
temperatures. White squares in
the optical images indicate the
size of the scan areas of the scanning Hall-probe images.

domain-wall superconductivity (DWS) and reverse-domain superconductivity (RDS), see Sec. 1.4. In particular, we shall demonstrate the existence
of these two states by analyzing local imaging and transport data.

The used S/F hybrid systems are composites of thin film superconductors, prepared on top of (and electrically isolated from) bulk single crystals
of the group of the Hexagonal Ferrites. These specific magnetic materials are convenient for two reasons: on the one hand, their magnetization is
relatively high ($\sim 10^5 \, \mathrm{Am^{-1}T^{-1}}$), resulting in stray magnetic fields of magnetic domains of some hundreds of millitesla [162], which can have a severe
influence on the superconducting state in adjacent S-films. On the other
hand, they exhibit very well defined magnetic domain patterns, which are
basically independent of changes in temperature below $300 \, \mathrm{K}$, see Fig. 5.25.
This means that the magnetic properties of the considered S/F systems can
be studied conveniently at ambient temperature.

5.4.1   *Domain superconductivity and domain-wall supercon-*
ductivity

The microscopic domain structure of a single crystalline barium hexaferrite
($\mathrm{BaFe_{12}O_{19}}$) substrate is shown in Fig. 5.26 as a function of an external
magnetic field, following the major hysteresis loop from remanence to saturation and back to remanence. In the initial state, the magnetic force
micrograph displays a branched domain pattern with a domain width of

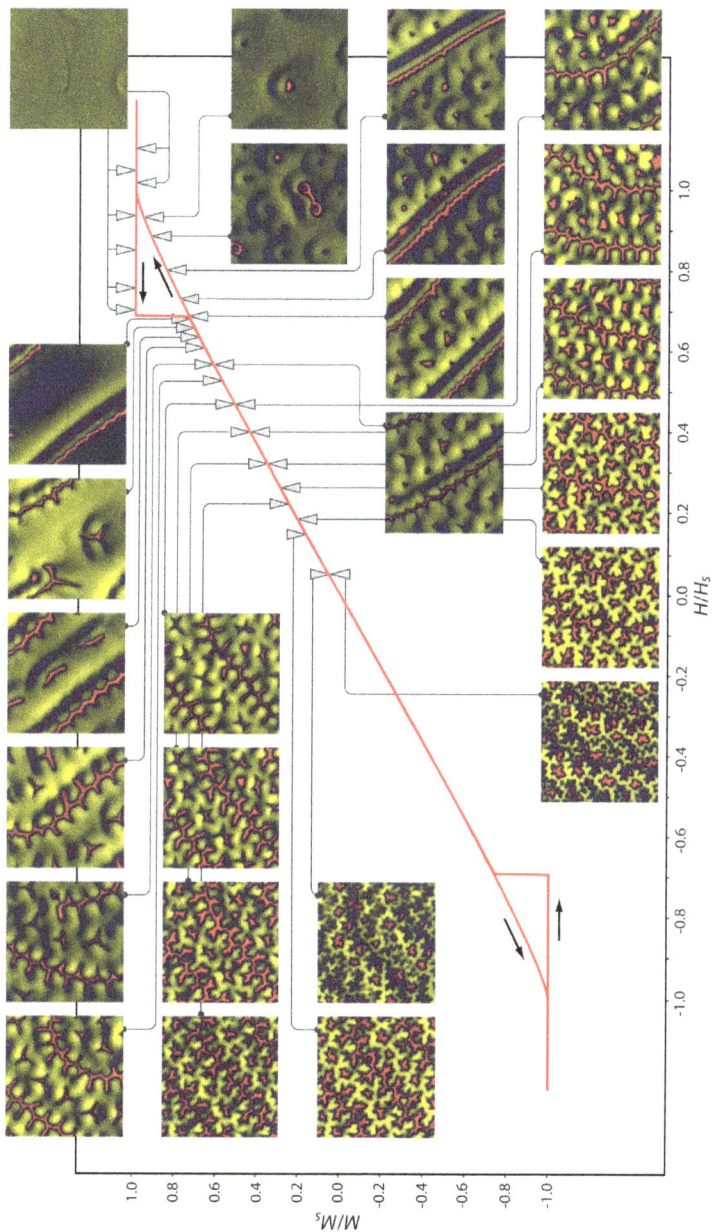

Fig. 5.26 Magnetic force micrographs (60×60 μm) of the magnetic domains at the basal plane of a barium hexaferrite single crystal, measured at room temperature with an external magnetic field applied perpendicular to the basal plane. When the field is reduced from saturation, the evolution of the domain structure displays hysteresis around the saturation field H_s, which is in agreement with the $M(H)$ loop, see the red line (after [508]).

Fig. 5.27 The normalized resistance of 50 nm niobium on top of a barium hexaferrite single crystal, as a function of temperature in the field range of 0 to 1 T. The field is ramped from -1.5 T to $+1.5$ T. Note that the measured $R(T)$ curves depend strongly on the magnetic history of the substrate, as indicated by the dashed line, corresponding to a field of 0.5 T when ramping the field in the opposite direction (after [508]).

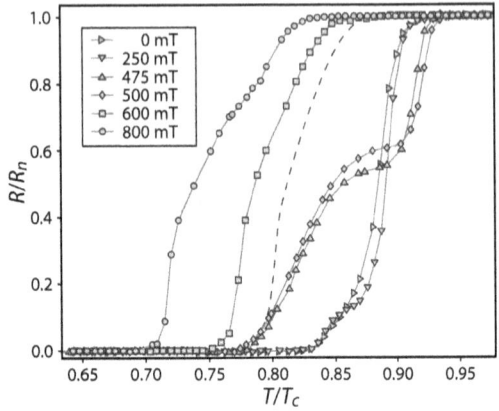

about 1.9 µm. This specific kind of domain pattern has its origin in domain branching towards the surface of the crystal and is typical for materials with uniaxial magnetic anisotropy [195, 224]. In the first approximation, dark yellow and red regions can be attributed to magnetic domains with up and down magnetization, respectively. The width of the domain walls is about 200 nm.

With increasing field, domains that are polarized parallel to $\boldsymbol{H}$ ('parallel-domains' — dark yellow areas) grow at the expense of their counterparts, the 'reverse-domains' (red regions), and the domain pattern becomes less branched. At 188 mT, elongated domains coexist with interstitial, isolated domains, while the domain structure disappears completely at a saturation field of about 354 mT. When the field is ramped down after saturation, reverse-domains start to nucleate at 299 mT. The evolution of the domain structure displays clear hysteresis around the saturation field, in agreement with the magnetization loop $M(H)$ in Fig. 5.26. No evident hysteresis is observed at low fields, suggesting that the domain walls move freely in response to the variation of the external field [512].

Deposited on top of such magnetic domain pattern, a thin niobium film shows a substantial broadening of the resistive superconducting transition even at zero field, with crossing $R(T)$ curves for $H < 600$ mT (Fig. 5.27). In contrast, transitions of a reference sample on a non-magnetic substrate are sharp, occurring within 0.1 K, and T_c shifts monotonically to lower temperature as H is increased. An H–T phase diagram of this $Nb/BaFe_{12}O_{19}$ hybrid is given in Fig. 5.28 for three different resistance criteri. For the criterion of $0.9 R_n$, the critical temperature *increases* with increasing field until $H \approx 525$ mT, and the phase boundary is almost symmetric with respect to H except near the saturation field. On increasing the field further, T_c decreases abruptly, and the $T_c(H)$ phase boundary passes into a conventional linear regime for $|H| > 600\,\mathrm{mT}$.

This very unusual effect— the increase of T_c with increasing magnetic field — is in accordance with theoretical predictions of the $T_c(H)$ behavior for domain-wall superconductivity [5, 82]. To apply this model to our system, let us further consider the present domain structure and its changes with H. In our case, the width of domain-walls is around 200 nm, which is much larger than the coherence length $\xi \sim 40\,\mathrm{nm}$ of the niobium film at $T/T_c = 0.975$. As the temperature is decreased at zero applied field, it is more favorable for superconductivity to nucleate above the domain walls owing to the lower stray fields at those locations [454]. The overlap of the superconducting areas above different domain walls can be excluded because of the relatively large domain size $w \sim 1.9\,\mu\mathrm{m}$, which is much higher than ξ. Therefore, the present system corresponds well to the limit of isolated domain-wall superconductivity,[13] and its phase boundary is predicted to behave as [5]:

$$T_c(h) = \Delta T_c^{\mathrm{orb}}(0.5 - E_{\min})h^4 + \Delta T_c^{\mathrm{orb}}(2E_{\min} - 0.5)h^2 + T_c(0) , \quad (5.6)$$

where $h = H/H_{\mathrm{stray}}$ and the superscript 'orb' stands for orbital suppression of T_c. Fitting the experimental phase boundary with Eq. (5.6), using $E_{\min}$ as the only variable parameter, we find that the experimental data are in good agreement with the theoretical model for $E_{\min} = 0.37$. Although Eq. (5.6) — strictly speaking — was derived for domain-wall width less than the coherence length[14], the good fit displayed in the inset of Fig. 5.28 suggests that in the present system, domain-wall superconductivity can still be described by that model.

The appearance of these localized superconducting nuclei results in a substantial broadening of the superconducting transition as observed in

---

[13]The critical parameter $\pi H_{\mathrm{stray}}^{\max} w^2/\Phi_0 \approx 3000 \gg 1$, where $H_{\mathrm{stray}}^{\max}$ is the maximum absolute value of the stray field above the magnetic domain, in our case $\sim 540\,\mathrm{mT}$ [5].
[14]To deduce Eq. (5.6), Aladyshkin *et. al.* assumed that the width of a domain wall is much less than the superconducting coherence length [5], using a distribution of the magnetic field of $4\pi M_s \mathrm{sgn}(x) + H$ near the surface, with the saturation magnetization M_s. With this assumption, the nucleation of superconductivity at zero field is analogous to surface superconductivity below the surface critical field, resulting in a $E_{\min} = 0.59$. However, in the present case of a $BaFe_{12}O_{19}$ single crystal, the stray field changes smoothly over the domain wall, so that the assumption of infinitely thin domain walls is no longer justified, and the direct comparison with surface superconductivity is no longer valid. As a result, one can expect that $E_{\min}$ will deviate from the predicted value of 0.59.

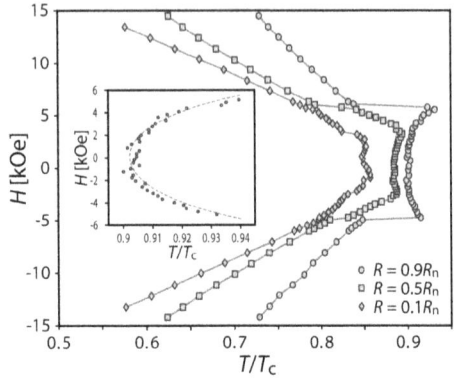

Fig. 5.28 *H–T* phase diagram of a Nb/BaFe$_{12}$O$_{19}$ hybrid for three different resistance criteri. The inset shows an enlarged view for the resistance criterion of $0.9R_n$ with a fit to Eq. 5.6, using $E_{min} = 0.37$, $H_{stray} = 540\,\mathrm{mT}$ (corresponding to the field at which $R(H)$ shows a minimum at $T/T_c = 0.94$) and $T_c(0) = 7.84$ (after [508]).

the corresponding $R(T)$-curves in Fig. 5.27. The onset of the resistance decrease with temperature corresponds to the appearance of the domain-wall superconductivity, and the crossover to the linear $T_c(H)$ behavior indicates the presence of bulk superconductivity. When increasing H, the superconducting areas shift away from the domain walls towards the regions at which the absolute value of the total magnetic field is minimal due to compensation effects [5]. However, one should bear in mind that an increase of the field also leads to an extension of parallel-domains, accompanying the motion of the domain walls.

In order to understand the evolution of superconductivity with magnetic field at different fixed temperatures, we shall have a closer look at the variation of the local field across different magnetic domains, as it is schematically illustrated in Fig. 5.29. If the applied field exceeds the saturation field H_s, the magnetic substrate is in the single domain state and magnetized along $\boldsymbol{H}$, see Fig. 5.29(a). At fields slightly below $+H_s$, first domains of negative polarity appear (if the difference between the nucleation field and H_s is ignored). Their local fields H_{stray} are opposite to the applied field and, eventually, compensate the applied field (Fig. 5.29(b)). As a result, superconductivity will appear in the areas just above domains of the opposite polarity in the ferromagnet. As the field is ramped down towards zero, the difference in amplitudes between $|H_{stray}|$ and $|H|$ grows and the compensation effect becomes less efficient above the domains themselves. Remarkably, within the area of domain walls, the stray fields are substantially lower, making a weak field compensation effect at these locations still possible (Fig. 5.29(c)). For negative applied fields, reverse-domain superconductivity will appear again near $H \approx -H_s$, see Fig. 5.29(d).

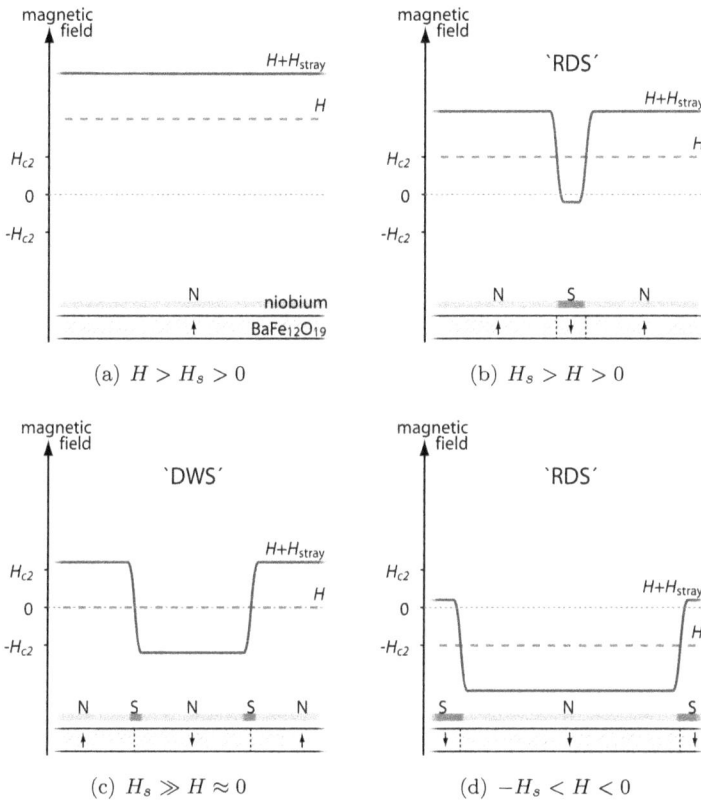

Fig. 5.29 Schematic illustration of reverse-domain (RDS) and domain-wall superconductivity (DWS) in S/F hybrids: (a) The F-component ($BaFe_{12}O_{19}$) is magnetically
saturated by the external field, resulting in a homogeneous total field $H + H_{stray} > H_{c2}$
that suppresses superconductivity in the entire S-film (niobium). (b) Appearance of
reverse-domain superconductivity at $H < H_s$, when the stray field of domains polarized opposite to $\boldsymbol{H}$ compensate the external field. (c) Domain-wall superconductivity at
$H \approx 0$, and (c) a similar scenario as in (b) but for a negative external field.

On the way from $H \approx 0$ to $H \approx -H_s$, the superconducting area, initially located only at the domain walls, progressively captures the whole
set of domains with polarity opposite to that of the applied field. In other
words, at $H \approx 0$, superconductivity survives only above the domain walls,
but when the field is ramped up towards $\pm H_s$, the superconducting state
spreads from the circumference of the domain to the whole domain area,
provided that the latter has polarity opposite to that of the applied field.

5.4.2   *Direct visualization of reverse-domain superconductivity*

The experimental results of the previous section demonstrate rather unambiguously the existence of the two states of domain and domain-wall superconductivity. However, in the spirit of 'seeing is believing' [104], we shall show in the following how the nucleation of superconductivity in the presence of magnetic domains can be visualized directly. In order to do so, a $Nb/PbFe_{12}O_{19}$ hybrid system, similar to the one discussed above in Sec. 5.4.1, was studied with low-temperature scanning laser microscopy[15] (LTSLM) [163]. This method is based on the principle of scanning a sample with a laser beam (probe) while simultaneously recording a response of the sample [444]. Thereby, a change in any characteristics of the sample, arising as a result of its local interaction with the probe, can serve as a response signal. For studying thin superconductors with this technique, one possibility is to apply a bias current to the film under investigation, and to measure the voltage drop across the sample, caused by changes in resistance at the illuminated spot due to local destruction of superconductivity by heating.[16]

In any case, imaging the local strength of superconductivity is, by itself, not sufficient for visualizing states such as domain superconductivity. It is rather the simultaneous knowledge about the actual domain configuration in the magnetic substrate, that allows for a conclusive observation of this phenomenon. Therefore, as parallel probing of both superconductivity and magnetization via LTSLM has not been realized yet, the magnetic domain pattern in the F-substrate (a single crystal of magnetoplumbite) and the distribution of superconductivity in the S-film (niobium) were visualized consecutively with magnetic force microscopy and LTSLM, respectively.

Being closely related to barium hexaferrite, magnetoplumbite ($PbFe_{12}O_{19}$) is also a well known representative of uniaxial materials with high magnetic anisotropy [202, 355], exhibiting beautiful branched domain patterns in a single crystal. The evolution of magnetic domains at the basal plane of a $PbFe_{12}O_{19}$ single crystal is shown in Fig. 5.30 as a function of

---

[15]Although being probably the most straightforward method for direct imaging of superconductivity, the well-known alternative of scanning-tunneling spectroscopy requires extremely flat surfaces [242] — an obstacle that is rather difficult to overcome in the fabrication of S/F hybrid systems that are composed of thin film superconductors prepared on underlying magnetic substrates.

[16]Seeming intuitive in the case of a current loaded superconductor, changes in voltage due to local illumination are caused by various mechanisms, such as lattice heating via low-energy phonons or pair-breaking processes induced by photons [159].

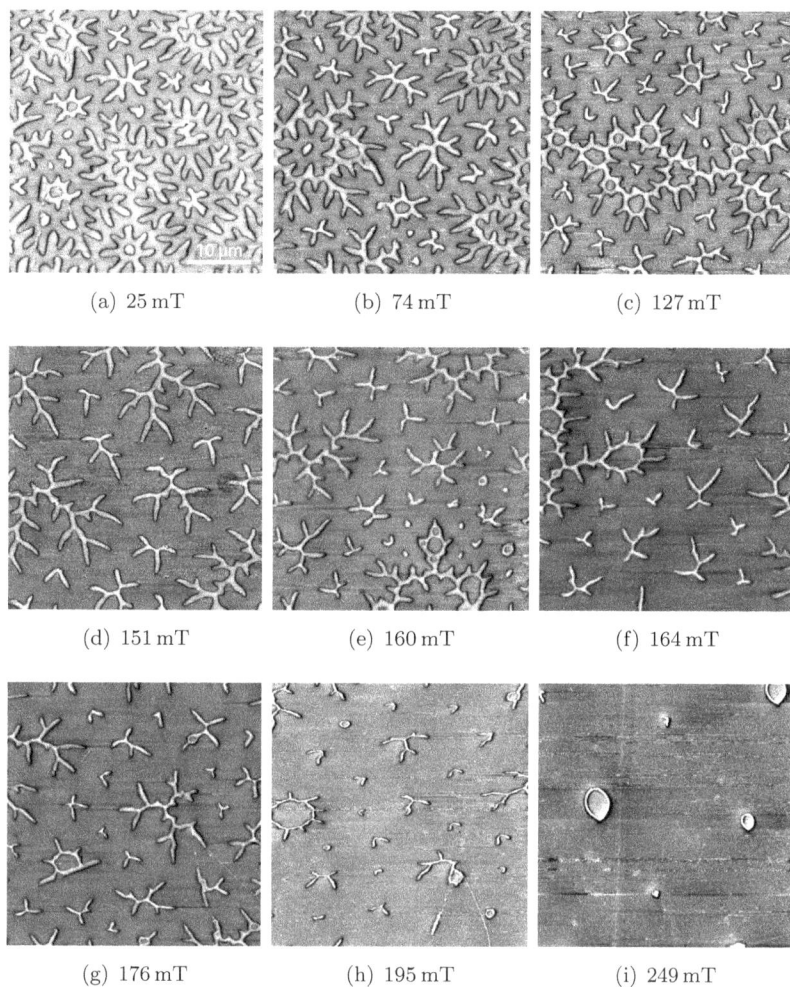

(a) 25 mT   (b) 74 mT   (c) 127 mT

(d) 151 mT   (e) 160 mT   (f) 164 mT

(g) 176 mT   (h) 195 mT   (i) 249 mT

Fig. 5.30   Magnetic force micrographs of the magnetic domain pattern at the basal plane of a single crystal of magnetoplumbite, at various external magnetic fields (after [163]).

the external field, ramped up from remanence to almost saturation. At the lowest field of 25 mT shown in Fig. 5.30(a), an intertwined maple leaf like pattern is formed by the domains, making it rather confusing to figure out their direction of polarization. Contrary to that, when H is slightly increased to 74 mT (Fig. 5.30(b)), the reverse domains become thinner and can be distinguished clearly from the parallel domains. The trend of the domain thinning continues at higher fields together with a transition of the

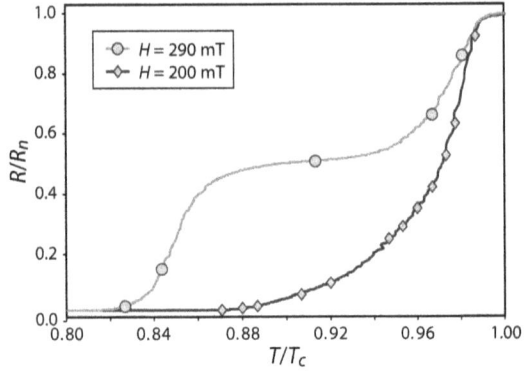

Fig. 5.31 Typical examples for different transitions of a niobium transport bridge from the superconducting to the normal state at 200 mT (dark line) and 290 mT (bright line). Markers indicate those temperatures at which the LTSLM-images of Fig. 5.31 were taken (after [163]).

reverse domains from chains of spiked rings (127 mT — Fig. 5.30(c)) over isolated branches (151 to 195 mT — Fig. 5.30(d–h)) to single hollow ovals patterns at 249 mT (Fig. 5.30(i)).

Prepared on top of a single crystal of magnetoplumbite, a niobium transport bridge of $40 \times 135 \, \mu m^2$ showed two kinds of transitions from the superconducting to the normal state in external magnetic fields, both strikingly different in nature. For each of these kinds, a typical example is shown in Fig. 5.31 where the normalized resistance of the bridge is plotted versus temperature. Mostly, the transitions were characterized by pronounced plateaus similar to the case of 290 mT, but sometimes at lower fields, smooth transitions were observed alike the $R(T)$-curve corresponding to 200 mT. To get to the origin of this remarkable difference in transition, LTSLM images were taken at those temperatures marked in Fig. 5.31 with gray circles and diamonds.

Turning first to the case of 200 mT of which corresponding LTSLM images are shown in the left series in Fig. 5.31, several observations can be made. To begin with, the signal is dominated by a feature 'Λ' that develops continuously with increasing temperature. At $T/T_c = 0.871$, first signs of 'Λ' appeared which became clearer when T was raised. All details of 'Λ' were visible at $T/T_c = 0.920$ whereas at higher temperatures, the feature became less defined in shape ($T/T_c = 0.947$), weakened ($T/T_c = 0.960$), smeared out ($T/T_c = 0.967$ to 0.973), and finally faded away. This continuous development is in agreement with the smooth increase of the corresponding resistance, see the $R(T)$-curve in Fig. 5.31. Second, the feature 'Λ' is extended along the whole bridge and is mostly a single path. However, at some places it branches out and forms rings as can be seen in the picture corresponding to $T/T_c = 0.920$. Finally, the normalized

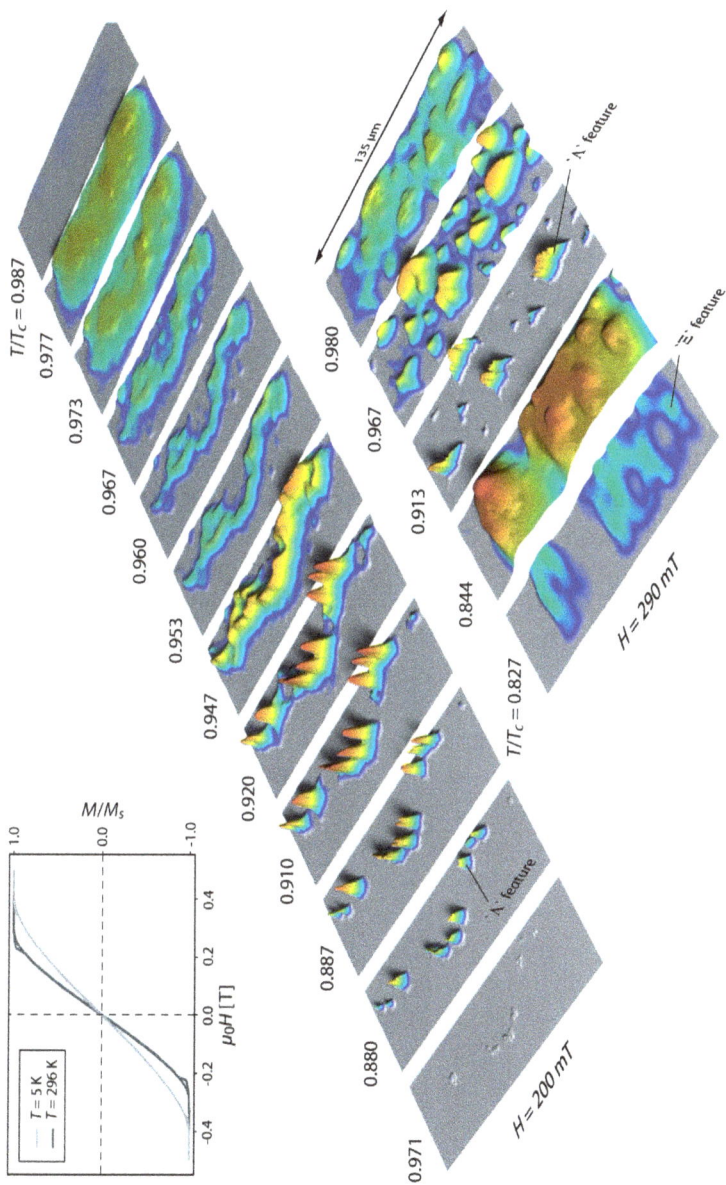

Fig. 5.32 Scanning laser microscopical images of a 40 nm thick niobium transport bridge on top of a magnetoplumbite single crystal as a function of temperature. Left and right series show the response $\partial R/\partial T$ of the bridge to the local perturbation by the laser at two different external fields (after [163]).

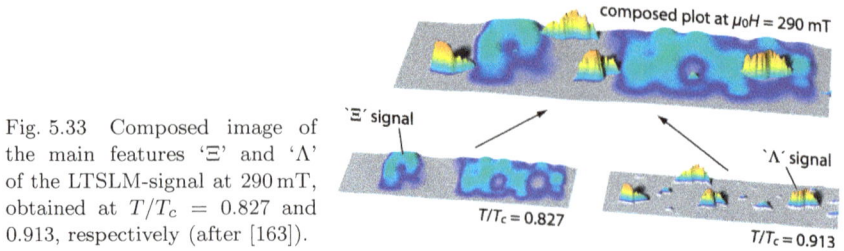

Fig. 5.33 Composed image of the main features 'Ξ' and 'Λ' of the LTSLM-signal at 290 mT, obtained at $T/T_c = 0.827$ and 0.913, respectively (after [163]).

magnetization of the substrate is the same at 200 mT and 5 K and at 130 mT and 296 K, see the inset in Fig.5.32.

Therefore, at 200 mT and 5 K, a magnetic domain pattern is expected that is similar to that at the room temperature image taken at 127 mT (Fig. 5.30(c)). In that case, an extended reverse domain exists, forming a chain of spiked rings. The size of these rings is comparable to the size of the rings formed by 'Λ' in the LTSLM-images. It is obvious to conclude from these observations that an extended reverse domain connects the left and right side of the bridge, causing an uninterrupted path of reverse-domain superconductivity in the niobium film. However, only main features of that reverse domain are mirrored in the LTSLM image of the reverse-domain superconductivity, as those parts of the RDS which did not contribute to current transport could not be detected.

In addition to the main features of the LTSLM images, several finer details can be observed, which can be described by the same interpretation used above. One example is the appearance of small dots next to 'Λ' at temperatures higher than $0.920\,T_c$. In this temperature range, not all the current flows inside the extended path of reverse-domain superconductivity, because the critical current density is too low. Instead, areas of RDS next to 'Λ' which are induced by small isolated reverse domains, see Fig. 5.30(c), also contribute to the current transport and can therefore be detected. Another example is the difference in temperatures at which rings and single paths of 'Λ' appear in the LTSLM signal (at $T/T_c = 0.920$ and 0.880, respectively). Since each ring provides two parallel superconducting paths, heating one branch of the ring causes the current to redistribute to the other branch. As a result, the signals obtained from rings are smaller than those from single paths of 'Λ'. However, this is only true as long as current redistribution is possible, i.e., as long as the critical current density is not reached in the rings. Accordingly, at higher temperatures ($T/T_c > 0.920$), rings and single paths of 'Λ' are of the same height in the LTSLM signal.

Switching to the case of 290 mT, the LTSLM signal (right series of images in Fig. 5.32) develops in a strikingly different way with temperature, compared to the LTSLM signal for the case of 200 mT. At low temperatures ($T/T_c = 0.827$ and 0.844), a broad feature 'Ξ' was observed which was not very well defined in shape. At $T/T_c = 0.913$, 'Ξ' vanished completely and instead, a new feature 'Λ' with small sharp features was detected. This abrupt change in the LTSLM signal goes along with the appearance of the plateau in the corresponding $R(T)$-curve in Fig. 5.31. While at temperatures below the plateau region the feature 'Ξ' was seen, in the plateau region and at higher temperatures the feature 'Λ' was observed.

In order to compare these two signals, we plotted them on top of each other in Fig 5.33. This composed picture shows clearly that 'Ξ' and 'Λ' are complementary in nature, because in the vicinity of each part of 'Λ', 'Ξ' tends to vanish. Again, interpretation of this observation is possible when taking the domain images of Fig. 5.30 into account. At 290 mT and low temperatures, the appearance of elongated reverse domains is to be expected. This suggests that 'Λ' is an image of the RDS, while 'Ξ' originates from superconductivity above domains polarized along $\boldsymbol{H}$, i.e. above the parallel-domains. Accordingly, 'Ξ' was obtained at a lower temperature than 'Λ' since T_c on top of parallel-domains must be lower than above reverse-domains. At $T/T_c = 0.827$, the signal 'Ξ' does not appear in areas that are shorted by strongly superconducting paths along the RDS regions showing up as signal 'Λ' at $T/T_c = 0.913$. The same explanation holds for the absence of 'Ξ' in the case of 200 mT, where the whole bridge was shorted by one strongly superconducting path of reverse-domain superconductivity.

5.4.3 *Superconducting – normal-state junctions induced by stray magnetic fields*

From the feasibility of creating superconducting and normal-state regions in thin superconductors by use of underlying magnetic domains, a new aspect emerges when considering such states as a network of S/N junctions. From this point of view, one may ask what qualities these junctions possess and whether they might be opening new perspectives. While it will be the subject of the present section to address the first of these questions, the answer to the latter is 'yes' – for the following two reasons.

Firstly, since induced by the stray field of magnetic domains which, for example, can be manipulated via external fields [459] or spin polarized currents [252,507], such S/N junctions are flexible themselves. This property is

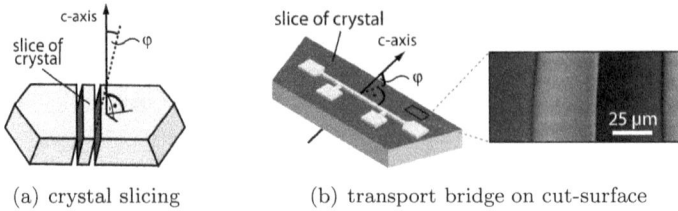

(a) crystal slicing

(b) transport bridge on cut-surface

Fig. 5.34   (a) Schematics of a $BaFe_{12}O_{19}$ single crystal after slicing under a cut angle φ with respect to its c-axis. (b) Sketch of a transport bridge processed perpendicular to the c-axis on the cut surface of a slice of the crystal from (a). The magnified area shows a magnetic force micrograph of the domain pattern at the cut surface ($\varphi = 10°$) of a $BaFe_{12}O_{19}$ single crystal (after [161]).

in clear contrast with commonly fabricated junctions, which are predefined, static structures, not allowing for any modification after their fabrication is completed.

For another thing, in modern research and applications, junctions between superconductors and other materials are often created by processing the different components into contact via state of the art nanotechnology. Naturally, in such systems, interfaces between the individual substances exhibit imperfections resulting from oxidation, lattices mismatch etc. [201, 455]. By contrast, in the considered case, superconducting and normal state are created *in the same material*, thus completely avoiding in series preparation of various materials together with its inherent difficulties. Accordingly, S/N interfaces, induced inside one single material, should be of high quality by nature, potentially leading to new devices of superior functionality.[17]

The investigation by transport experiments of stray field induced S/N interfaces requires specially designed S/F hybrid systems with the following two qualities: (i) the opportunity to realize superconductivity either above the magnetic domain walls or above the reverse-domains of a magnetic substrate, and (ii) the possibility to force transport currents to flow across the induced S/N interfaces. The preparation of such system is challenging as several strict requirements need to be fulfilled. First of all, formation of magnetic stripe domains in the magnetic substrate is desirable [44]. Align-

---

[17]Such devices, based on superconducting junctions, are today of great importance in both research and life, offering measuring techniques with unbeatable precision. It is thanks to these devices that the activity of a human brain can be detected [108], and magnetic fields ten trillion times smaller than the field of our Earth became accessible for precise measurements [396].

ment of a transport bridge perpendicular to such domains guarantees a bias current to cross them successively. Second, the magnetic domain pattern of the template must not change significantly when subjected to external fields, required for setting up the different states, such as RDS, in the thin film superconductor. Finally, the out-of plane component of the stray field above magnetic domains has to reach the upper critical field of the superconductor in order to allow for the realization of DWS down to temperatures well below T_c (see Fig. 5.29).

One option to fulfill the above three requirements is to prepare an aluminum transport bridge in a specific direction on the surface of a barium hexaferrite single crystal that is cut under a small tilt to its c–axis, as illustrated in Fig 5.34. When cut along the proper crystallographic axis, single crystals of $BaFe_{12}O_{19}$ exhibit a one-dimensional stripe-type domain structure, see Fig. 5.34(b), with dominant in-plane magnetization [224]. To demonstrate that these magnetic domains do not change significantly in perpendicular external magnetic fields, Fig. 5.35 shows the influence of H on the size and position of the magnetic domains at low temperatures (77 K), visualized with a scanning Hall-probe microscope. Apparently, the width of the parallel-domains increases linearly for $H < 150$ mT with a rate of approximately 43 nm/mT. In accordance with the very small coercivity of these ferromagnets, the domain walls return to their initial positions within the experimental resolution of 1 µm, each time H is reduced to zero. Moreover, the z–component H_{stray}^z of the stray field of these magnetic domains is on the order of $\mu_0 H_{stray}^z \sim 50$ mT, which is more than twice as high as typical critical fields of the used aluminum films. Therefore, by applying compensation fields of $H \sim \pm H_{stray}^z$, it is possible to completely suppress superconductivity above parallel-domains while allowing for it above reverse-domains (see Fig. 5.29).

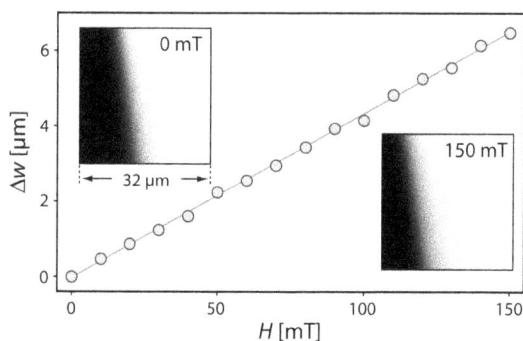

Fig. 5.35 The increase of the width w of a parallel domain (bright areas in the inserts) in a $BaFe_{12}O_{19}$ substrate (cutting angle $\varphi = 10°$) as a function of H. The presented data is the average of the results obtained for an increasing and a decreasing external magnetic field. Insets show two examples of the position of the same domain-wall at 0 mT and 150 mT (after [161]).

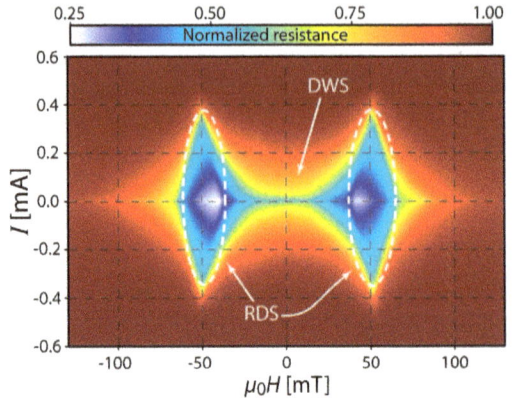

Fig. 5.36 The normalized resistance $R_n = R/R(1.1T_c)$ of the transport bridge shown in Fig. 5.37 at $T/T_c = 0.28$, as a function of bias current and external field. Regions where reverse-domain and domain-wall superconductivity occur are indicated (after [161]).

Figure 5.36 shows the normalized $dc-$resistance of a $35 \times 250\,\mu\text{m}^2$ aluminum transport bridge on top of such magnetic stripe domains (see Fig. 5.37 for an illustration of the entire S/F system) well below T_c, crossing several domain walls of the substrate. As can be expected from the above results, two pronounced minima in resistivity are seen around $\pm 53\,\text{mT}$, indicating that stray fields above magnetic domains are compensated by H. Application of these compensation fields thus induces the state of reverse-domain superconductivity in the S/F hybrid system. Moreover, while beyond the compensation fields, the resistance rises quickly towards its value in the normal state, parts of the bridge remain superconducting when subjected to external fields lower than the compensation fields. Particularly, in the case of zero applied field, when superconductivity is likewise suppressed above domains of opposite magnetization, the reduced resistance is a clear fingerprint of domain-wall superconductivity.

While the minima in Fig. 5.36 together with the reduction in resistance at zero applied field can be understood in terms of occurrence of domain and domain-wall superconductivity, respectively, the *values* of the resistance reached at these points is surprising. As can be seen from Fig. 5.37, approximately half of the area of the bridge is covered by each kind of domains. Nucleation of superconductivity above one type of domains should thus cause the bridge to loose roughly half of its resistance in the normal state. By contrast, for compensation fields of both polarities, only half of the expected resistance is seen. A similar observation can be made at zero field when superconductivity survives above domain walls only. In that case, the drop in resistance is, a priori, expected to be equal to the ratio between the width of magnetic domains and domain walls. But from the detailed magnetic force micrograph in the lower part of Fig. 5.37, it becomes clear that all changes in stray fields are confined to approximately 1 μm around domain walls, whereas the domains are typically 25 μm wide. Therefore, in absence of external fields, the observed reduction in resistance by ~ 45% is surprising.

The normalized differential conductance $G_N = {}^{dI}/{}_{dV} R_n$ of the considered bridge in the state of reverse-domain superconductivity is shown in Fig. 5.38 as a function of temperature and voltage. At the lowest temperatures, G_N is sharply peaked at zero voltage (note also that $G_N = 2.8$ at zero bias exceeds considerably the theoretical limit $G_N = 2.0$), declining symmetrically to its minima at $\pm V_1$ before recovering to its normal value at higher voltages. Together with some smaller local minima, these features gradually collapse with increasing temperature. In order to interpret these conductance spectra, several aspects must be taken into account.

(i) The entire transport bridge is in the normal state for $V > 2\,\text{mV}$ even at the lowest temperature of $0.28\,T_c$. The reason for this is that the critical current density j_c is exceeded due to the low resistance of the bridge. Therefore, in a certain low-voltage region where $j < j_c$, a higher value for G_N is expected since the parts of the bridge above reverse-domains are superconducting.

(ii) The normalized differential conductance reaches 2.8 at $0.28\,T_c$ and zero voltage. Assuming that this increase in G_N was solely caused by the N → S transition mentioned under point (i), approximately 64% of the transport bridge had to become superconducting. However, the hybrid sys-

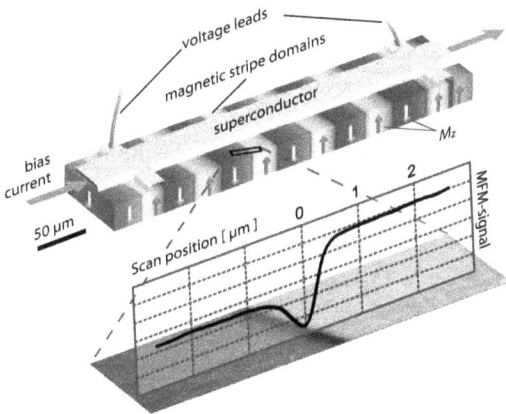

Fig. 5.37 An optical image of the considered aluminum transport bridge, illustrated in 3D, with its underlying magnetic domains, obtained from magnetic force microscopy. Note that the MFM-image is vertically extended to illustrate the domains. Arrows indicate the z–component M_z of the magnetization in the substrate. The enlarged region shows a detailed MFM-image of a typical domain-wall in the substrate (after [161]).

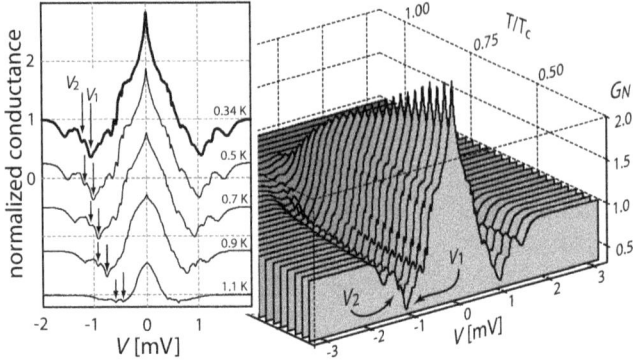

Fig. 5.38   Differential conductance spectra of the considered aluminum bridge as illustrated in Fig. 5.37 at 53 mT in the state of reverse-domain superconductivity. Results are shown twice for clarity: the left 2D-panel displays a few conductance curves that are vertically shifted, whereas all obtained curves are given in a 3D-representation at the right. Arrows mark two minima whose positions can be traced. In the left panel, curves are shifted and the vertical scale corresponds to the top curve (after [161]).

tem behaves similar for both polarities of H (see Fig. 5.36), meaning that an unequal distribution of parallel- and reverse-domains cannot be the reason for such high conductances observed at positive and negative compensation fields. Moreover, as discussed above, the external field increases the width of the parallel-domains by 43 nm/mT (see Fig. 5.35). Accordingly, at 53 mT, a normalized conductance of only 1.8 instead of 2 should be expected, provided that parallel- and reverse-domains are equally distributed at $H = 0$. Finally, the characteristic length $\xi_N = \sqrt{\hbar D/k_B T}$, over which the Cooper pair amplitude decays exponentially with the distance from an S/N interface, is in the present case on the order of 400 nm.[18] Due to this proximity effect, the superconducting state extends into the normal regions and vice versa, but the corresponding increase in G_N at $V = 0$ is only minor for this effect.

Taking account of the above considerations, the observed conductance of 2.8 cannot be explained by a corresponding expansion of the superconducting state along the transport bridge. However, below the superconducting

---

[18] An estimation of the diffusion coefficient is $D = v_F l_p/3 \approx 6.6 \cdot 10^{-3}$ m^2/s, with the Fermi velocity $v_F = 2.03 \cdot 10^6$ m/s [15] and the electron mean free path $l_p = mv_F/\varrho_i ne^2 \approx 9.8$ nm (m and e are electronic mass and charge, $n = 18.1 \cdot 10^{28}$ m^{-3} is the density of conduction electrons [15]). The residual resistance ϱ_i was estimated according to $\varrho_p/\varrho_i + 1 \approx R(295\,\mathrm{K})/R(1.5\,\mathrm{K}) = 1.68$, with the resistivity $\varrho_p = 2.74 \cdot 10^{-8}$ Ωm of aluminum at 295 K [251].

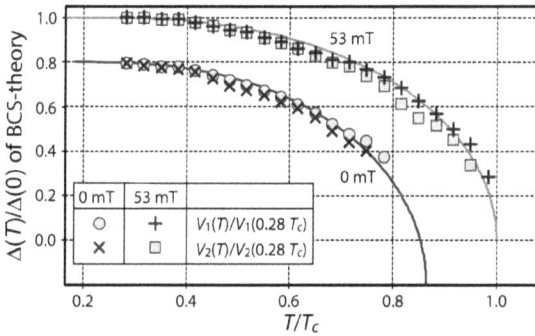

Fig. 5.39 The traced positions of characteristic minima in the conductance spectra of Fig. 5.38 and Fig. 5.40 (markers), compared with $\Delta(T)$ of the BCS-theory (solid lines). For clarity, the results obtained at 0 mT are shifted by -0.2 (after [161]).

gap Δ, i.e. for $V < \Delta/e$, an excess of the conductance can generally result from Andreev reflection [10] processes at S/N interfaces, if the latter are highly transparent for incident electrons. In the present case, normal and superconducting states are created inside the same material and, therefore, the presence of highly transparent S/N interfaces is reasonable. Accordingly, the observed excess of conductance suggests that the mechanism of charge transfer across the S/N interfaces is indeed affected by Andreev reflection.

The theory of Blonder, Tinkham and Klapwijk (BTK) [61] describes the effects of Andreev reflection on the conductance of a single S/N junction for the particular case of ballistic transport in the normal-state region. From that theory it follows that inside the gap, the conductance can be enhanced up to twice its above-gap value. In the present case, as discussed above, the above-gap conductance in the state of reverse-domain superconductivity can be estimated to be 1.8. Therefore, the observed zero-voltage conductance of 2.8 is smaller than twice the above-gap conductance (2 × 1.8), meaning that these findings are not in contradiction with the BTK-theory.

Moreover, the BTK-theory predicts for highly transparent interfaces a flat conductance below the gap, which has been verified experimentally with superconducting point contacts [446]. By contrast, the G_N-curves of Fig. 5.38 are sharply peaked at zero voltage. Such anomalies in the conductance spectra in the form of zero-bias peaks have been reported before in systems that deviate from the BTK model, such as for example planar niobium–gold contacts [506], junctions between superconductors and semiconductors [244, 253, 362] and series of S/N/S-junctions [268]. In the present case, the used sample differs also significantly from the model system of BTK, since the bridge crosses nine domain walls (see Fig. 5.37), each of them inducing one S/N interface. It is also reasonable to expect that, due

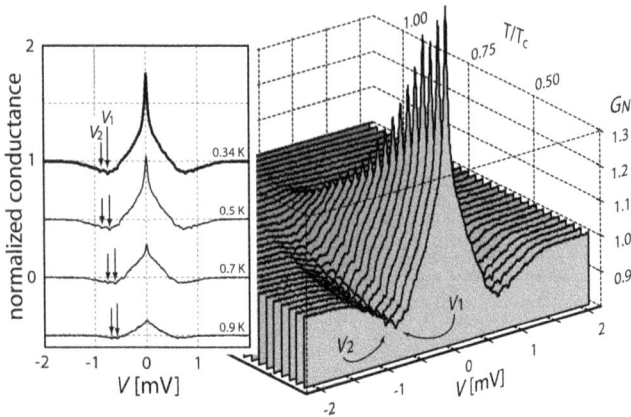

Fig. 5.40   Differential conductance spectra of the considered S/F hybrid at 0 mT in the state of domain-wall superconductivity. Similar to Fig. 5.38 two minima whose positions can be traced are marked by arrows (after [161]).

to the large size of the domains, the electric transport in the normal-state regions is not ballistic. A theoretical description of such series of diffusive S/N/S junctions goes beyond the ballistic theories [61, 366], and it should also include nonlocal coherent effects in the normal-state regions [360, 483].

(iii) Two of the local minima of the conductance spectra of Fig. 5.38, marked as V_1 and V_2, can be traced from $0.28\,T_c$ to nearly T_c. Their relative position on the V-axis was compared to the superconducting gap function

$$\ln\left(\frac{1.13E_c}{k_B T_c}\right) = \int_0^{E_c} \frac{\tanh\left(0.5 k_B^{-1} T^{-1} \sqrt{\xi^2 + \Delta^2}\right)}{\sqrt{\xi^2 + \Delta^2}}\, d\xi, \qquad (5.7)$$

of the BCS-theory [465], using a value of 423 K for the ratio between cut-off frequency E_c and Boltzmann constant k_B [203]. A solution $\Delta(T)$ of the above equation can be found by iteration, integrating numerically over energies ξ while treating T_c as a fitting parameter. As illustrated by the upper curve in Fig. 5.39, $V_{1,2}$ data follow quite closely the superconducting gap Δ in temperature.

For the ideal case of a single ballistic S/N/S junction, it is known that multiple Andreev reflection leads to minima in the conductance curves at voltages smaller than the gap ($V < \Delta/e$). Their positions follow $\Delta(T)$ in the same way as $V_{1,2}$. However, the present case is quite different in that the measured $G(V)$-curves belong to *a series* of diffusive S/N/S junctions. When dividing V by the number of S/N interfaces [40] and considering that V dropped mainly over the normal-state regions of the bridge, it could be concluded that $V_{1,2}$ lie inside the gap and originate from multiple Andreev reflection (typical values for Δ/e are $200\,\mu$V for aluminum [177]). However, it is also possible that series of Andreev reflection processes lead to multiplication effects and to different effective voltages across subsequent S/N interfaces. Therefore, even if caused by the same process, features in $G(V)$ could repeatedly appear at different voltages, and could result in the observed set of local minima. Moreover, a multiplication effect in series of junctions might also lead to an increase of the conduction by factors higher than two [433].

Intriguingly, all observations described above can be made not only in the case of reverse-domain superconductivity but also in absence of external fields when domain-wall superconductivity is realized (Fig. 5.40). In that case, local minima are less pronounced, but nonetheless, two of them can be traced up to higher temperatures (lower curve in Fig. 5.39). As before, their positions in the conductance spectra follow the collapse of Δ. It is remarkable that values of T_c obtained by fitting are significantly different in the cases of RDS $(1.20 \pm 0.02\,\mathrm{K})$ and DWS $(1.05 \pm 0.03\,\mathrm{K})$. These findings reflect directly that, due to quantum size effects, T_c values of superconducting micro-structures differ significantly from those of bulk superconductors — an effect that leads to the reduction of T_c when superconductivity is confined above the domain walls of an underlying magnet [179].

Concluding Remarks

In this book we have presented recent research activities on nanostructured superconductors. We have demonstrated that two important superconducting critical parameters, critical currents and critical fields, can be enhanced dramatically by optimizing the nanoscale flux and condensate confinement through nanostructuring. In individual nanostructures, this has led to the observation of strong topology- and geometry-dependent critical fields, as well as to the discovery of symmetry induced antivortices. In superconductors with nanoengineered periodic pinning arrays, field-induced superconductivity due to the field compensation by stray fields of magnetic dots, and a rich variety of novel flux phases and stable configurations have been investigated — providing the underlying mechanisms that lead to a practical realization of the theoretical limit for the critical current (depairing current) and for the development of fluxonics devices.

However, the situation with the most important critical parameter, the superconducting critical temperature T_c itself, turns out to be much more complex. In superconductors, the condensate of Cooper pairs is formed below the critical temperature T_c and a pronounced diamagnetic response, supported by persistent shielding currents, as well as a zero resistance state are established. In spite of a reasonably good understanding of superconductivity in bulk materials (except high-T_c and other 'exotic' materials), a consistent full description of this strongly correlated state at the nanoscale is, however, still lacking. In particular, crucial issues such as the emergence and development of superconductivity on the way from atoms to bulk solids need to be solved [11, 218]. We know that magnetism occurs already in individual atoms and that atomic shells are filled according to Hund's rules. On the other hand, it is not clear how and when superconductivity emerges in nanocells containing N_{at} atoms and having progressively growing dimensions from Ångströms through nanometers to micrometers.

In this context, questions of great fundamental importance can be asked, concerning the evolution of the superconducting state at the nanoscale:

- At which dimension does superconductivity appear for the first time? And of how many atoms is such a 'critical' nanoparticle composed?
- Will T_c and the superconducting gap remain the same as a function of N_{at}, or is the $T_c(N_{at})$ dependence really highly non-trivial, demonstrating, for instance, a multiple re-entrance or oscillations?
- Can such nanograins eventually have a critical temperature exceeding the T_c value of the reference bulk material?

Following the pioneering experimental work of Ralph *et al.* [395], only partial answers to these questions could be given. Superconductivity is a coherent quantum state with a characteristic superconducting coherence length ξ. At first sight, superconductivity should be suppressed in samples with a size smaller than ξ, but in fact it persists down to much smaller sizes. In samples that are smaller than ξ, the modulus of the order parameter remains more or less constant, implying that the superconducting state can survive the confinement in small grains. Since ξ is typically in the range of some tens of nanometers, metallic grains with that size can be perfectly superconducting. But what happens to superconductivity if the sample is made considerably smaller — a few nanometer and less? This fundamental question was addressed theoretically by Anderson and Suhl already in 1959 [9]. They argued that superconductivity is no longer possible when the average spacing d_E between discrete energy levels, which appear in ultra small metallic grains due to the confinement, becomes larger than the bulk superconducting gap Δ. In this limit, even a single pair of electrons is not available in the energy interval Δ, not to speak of an ensemble of Cooper pairs forming a Bose-condensate similar to the case of bulk superconductors. Later on, in contradiction to the theory of Anderson and Suhl, tunneling experiments on small Sn particles embedded in an oxide film did not show any lower-size limit for superconductivity — at least down to a particle radius of 2.5 nm [178].

In transport measurements on ultrathin films [212], a superconductor–insulator transition was observed and this was attributed to the disappearance of superconductivity below a certain grain size. However, recent scanning tunneling microscopy (STM) experiments have demonstrated the existence of a well-defined superconducting gap in the grains composing insulating thin films [267]. Moreover, the insulating state itself appears due to the Coulomb blockade effects between the grains, rather than due to the suppression of superconductivity in the ultra small grains.

Further progress in this field was promoted by the experiments of Ralph, Black and Tinkham, who fabricated a single-electron tunneling device containing only one aluminum particle with a diameter below 10 nm [395]. Tunneling through the aluminum nanoparticle was found to be parity dependent. The magnetic field dependence of the superconducting correlations clearly revealed the difference in tunneling through a particle containing an odd or an even number of electrons. These experiments stimulated the development of theoretical models [11, 186, 420, 435], showing some new surprising features of pairing correlations with decreasing grain size, including a parity-dependent gap in the excitation spectrum, a different behavior of odd and even grains in magnetic fields, and a smooth crossover between the bulk superconducting regime ($d_E \ll \Delta$) and the fluctuation-dominated regime ($d_E \gg \Delta$). The superconducting state in ultra small grains can also survive in very high magnetic fields [178]. Interestingly, the exact solution by Richardson of the pairing problem for discrete levels, developed many years ago in the context of nuclear physics [404, 405], has been 're-discovered' by the solid state theorists working on nanoscale superconductivity [69]. The remarkable theoretical predictions of superconductivity in nanosystems have not yet been verified in experiments. At the same time, recent impressive experiments with superconductivity in single atomic layers of Pb [510] and in just four pairs of $(BETS)_2 GaCl_4$ molecules [103] is an important step in this direction, which fully justifies further studies of the nanoscale evolution of the superconducting critical temperature in different nanoscale systems.

Bibliography

[1] Aarts, J., Geers, J. M. E., Brück, E., Golubov, A. A. and Coehoorn, R. (1997). Interface transparency of superconductor/ferromagnetic multilayers, *Phys. Rev. B* **56**, 5, pp. 2779–2787.

[2] Abrikosov, A. A. (1957). On the Magnetic Properties of Superconductors of the Second Group, *Sov. Phys. JETP* **5**, p. 1174.

[3] Abrikosov, A. A. (1988). *Fundamentals of the Theory of Metals* (North–Holland, Amsterdam).

[4] Akazawa, T., Hidaka, H., Kotegawa, H., Kobayashi, T. C., Fujiwara, T., Yamamoto, E., Haga, Y., Settai, R. and Onuki, Y. (2005). Pressure-induced superconductivity in UIr without inversion symmetry, *Physica B: Cond. Matter* **359–361**, 3, pp. 1138–1140.

[5] Aladyshkin, A. Y., Buzdin, A. I., Fraerman, A. A., Mel'nikov, A. S., Ryzhov, D. A. and Sokolov, A. V. (2003). Domain-wall superconductivity in hybrid superconductor-ferromagnet structures, *Phys. Rev. B* **68**, 18, p. 184508.

[6] Aladyshkin, A. Y., Silhanek, A. V., Gillijns, W. and Moshchalkov, V. V. (2009). Nucleation of superconductivity and vortex matter in superconductor-ferromagnet hybrids, *Supercond. Sci. Technol.* **22**, 5, p. 053001.

[7] Alexander, S. and Halevi, E. (1983). Superconductivity on networks: II. The London approach, *J. Phys. France* **44**, 7, pp. 805–817.

[8] Ammann, C., Erdös, P. and Haley, S. B. (1995). Superconducting micronets: The Wheatstone bridge, *Phys. Rev. B* **51**, 17, pp. 11739–11747.

[9] Anderson, P. W. and Suhl, H. (1959). Spin Alignment in the Superconducting State, *Phys. Rev.* **116**, 4, pp. 898–900.

[10] Andreev, A. F. (1964). The thermal conductivity of the intermediate state in superconductors, *Sov. Phys. JETP* **19**, pp. 1228–1231.

[11] Andreev, A. F. (1999). Superfluid, Superconducting, and Magnetic Ordering in Mesoscopic Quantum Dots, *J. Supercond.* **12**, 1, pp. 197–207.

[12] Andreev, A. F. and Bashkin, E. P. (1975). Three-velocity hydrodynamics of superfluid solutions, *Sov. Phys. JETP* **42**, p. 164.

[13] Aoki, D., Huxley, A., Ressouche, E., Braithwaite, D., Flouquet, J., Brison,

J. P., Lhotel, E. and Paulsen, C. (2001). Coexistence of superconductivity and ferromagnetism in URhGe, *Nature* **413**, pp. 613–616.

[14] Arutunian, R. M. and Zharkov, G. F. (1983). Behavior of a hollow superconducting cylinder in a magnetic field, *J. Low Temp. Phys.* **52**, 5, pp. 409–431.

[15] Ashcroft, N. W. and Mermin, N. D. (1976). *Solid State Physics* (Brooks/Cole).

[16] Auer, J. and Ullmaier, H. (1973). Magnetic Behavior of Type-II Superconductors with Small Ginzburg–Landau Parameters, *Phys. Rev. B* **7**, 1, pp. 136–145.

[17] Autler, S. H. (1972). Fluxoid pinning in superconductors by a periodic array of magnetic particles, *J. Low Temp. Phys.* **9**, 3–4, pp. 241–253.

[18] Avraham, N., Khaykovich, B., Myasoedov, Y., Rappaport, M., Shtrikman, H., Feldman, D. E., Tamegai, T., Kes, P. H., Li, M., Konczykowski, M., van der Beek, K. and Zeldov, E. (2001). 'Inverse' melting of a vortex lattice, *Nature* **411**, pp. 451–454.

[19] Babaev, E. (2002). Vortices with Fractional Flux in Two-Gap Superconductors and in Extended Faddeev Model, *Phys. Rev. Lett.* **89**, 6, p. 067001.

[20] Babaev, E. and Ashcroft, N. W. (2007). Violation of the London law and Onsager–Feynman quantization in multicomponent superconductors, *Nature Phys.* **3**, pp. 530–533.

[21] Babaev, E., Faddeev, L. D. and Niemi, A. J. (2002). Hidden symmetry and knot solitons in a charged two-condensate bose system, *Phys. Rev. B* **65**, 10, p. 100512.

[22] Babaev, E. and Speight, M. (2005). Semi-Meissner state and neither type-I nor type-II superconductivity in multicomponent superconductors, *Phys. Rev. B* **72**, 18, p. 180502.

[23] Babaev, E., Sudbø, A. and Ashcroft, N. W. (2004). A superconductor to superfluid phase transition in liquid metallic hydrogen, *Nature* **431**, pp. 666–668.

[24] Bader, J. S., Hammond, R. W., Henck, S. A., Deem, M. W., McDermott, G. A., Bustillo, J. M., Simpson, J. W., Mulhern, G. T. and Rothberg, J. M. (1999). DNA transport by a micromachined Brownian ratchet device, *Proc. Natl. Acad. Sci. USA* **96**, pp. 13165–13169.

[25] Baelus, B. J. and Peeters, F. M. (2002). Dependence of the vortex configuration on the geometry of mesoscopic flat samples, *Phys. Rev. B* **65**, 10, p. 104515.

[26] Baelus, B. J., Peeters, F. M. and Schweigert, V. A. (2000). Vortex states in superconducting rings, *Phys. Rev. B* **61**, 14, pp. 9734–9747.

[27] Baelus, B. J., Yampolskii, S. V. and Peeters, F. M. (2002). Coupled mesoscopic superconductors: Ginzburg–Landau theory, *Phys. Rev. B* **66**, 2, p. 024517.

[28] Baert, M. (1995). *Critical currents, pinning and matching effects in superconducting PbGe multilayers*, Ph.D. thesis, Katholieke Universiteit Leuven.

[29] Baert, M., Metlushko, V. V., Jonckheere, R., Moshchalkov, V. V. and Bruynseraede, Y. (1995a). Composite flux-line lattices stabilized in super-

conducting films by a regular array of artificial defects, *Phys. Rev. Lett.* **74**, 16, pp. 3269–3272.

[30] Baert, M., Metlushko, V. V., Jonckheere, R., Moshchalkov, V. V. and Bruynseraede, Y. (1995b). Flux phases and quantized pinning force in superconductor with a periodic lattice of pinning centres, *Europhys. Lett.* **29**, 16, pp. 157–162.

[31] Baibich, M. N., Broto, J. M., Fert, A., Van Dau, F. N., Petroff, F., Etienne, P., Creuzet, G., Friederich, A. and Chazelas, J. (1988). Giant Magnetoresistance of (001)Fe/(001)Cr Magnetic Superlattices, *Phys. Rev. Lett.* **61**, 21, pp. 2472–2475.

[32] Bakhanova, E. S. and Genkin, V. M. (1993). *Applied Superconductivity* (DGM Intoranationgesellschaft, Göttingen).

[33] Balicas, L., Brooks, J. S., Storr, K., Uji, S., Tokumoto, M., Tanaka, H., Kobayashi, H., Kobayashi, A., Barzykin, V. and Gor'kov, L. P. (2001). Superconductivity in an Organic Insulator at Very High Magnetic Fields, *Phys. Rev. Lett.* **87**, 6, p. 067002.

[34] Barba, J. J., de Souza Silva, C. C., Cabral, L. R. E. and Aguiar, J. A. (2008). Flux trapping and paramagnetic effects in superconducting thin films: The role of de Gennes boundary conditions, *Physica C* **468**, 7–10, pp. 718–721.

[35] Bardeen, J. (1961). Quantization of Flux in a Superconducting Cylinder, *Phys. Rev. Lett.* **7**, 5, pp. 162–163.

[36] Bardeen, J., Cooper, L. N. and Schrieffer, J. R. (1957a). Microscopic Theory of Superconductivity, *Phys. Rev.* **106**, 1, pp. 162–164.

[37] Bardeen, J., Cooper, L. N. and Schrieffer, J. R. (1957b). Theory of superconductivity, *Phys. Rev.* **108**, 5, pp. 1175–1204.

[38] Bardeen, J. and Stephen, M. J. (1965). Theory of the Motion of Vortices in Superconductors, *Phys. Rev.* **140**, 4A, pp. A1197–A1207.

[39] Bartussek, R., Hanggi, P. and Kissner, J. G. (1994). Periodically Rocked Thermal Ratchets, *Europhys. Lett.* **28**, 7, pp. 459–464.

[40] Baturina, T. I., Islamov, D. R. and Kvon, Z. D. (2002). Subgap Anomaly and Above-Energy-Gap Structure in Chains of Diffusive SNS Junctions, *JETP* **75**, pp. 326–330.

[41] Bean, C. P. (1962). Magnetization of hard superconductors, *Phys. Rev. Lett.* **8**, 6, pp. 250–253.

[42] Bean, C. P. and Livingston, J. D. (1964). Surface Barrier in Type-II Superconductors, *Phys. Rev. Lett.* **12**, 1, pp. 14–16.

[43] Bekaert, J., Morelle, M., Pogosov, W. V., Borghs, G. and Moshchalkov, V. V. (2004). Hall magnetometry of superconducting microstructures, *Physica C* **404**, 1–4, pp. 44–49.

[44] Belkin, A., Novosad, V., Iavarone, M., Fedor, J., Pearson, J. E., Petrean-Troncalli, A. and Karapetrov, G. (2008). Tunable transport in magnetically coupled MoGe/Permalloy hybrids, *Appl. Phys. Lett.* **93**, 7, p. 072510.

[45] Benoist, R. and Zwerger, W. (1997). Critical fields of mesoscopic superconductors, *Z. Phys. B* **103**, pp. 377–381.

[46] Berdiyorov, G. R., Milošević, M. V. and Peeters, F. M. (2006). Novel Com-

mensurability Effects in Superconducting Films with Antidot Arrays, *Phys. Rev. Lett.* **96**, 20, p. 207001.

[47] Berger, J. (2004). Noise rectification by a superconducting loop with two weak links, *Phys. Rev. B* **70**, 2, p. 024524.

[48] Berger, J. and Rubinstein, J. (1995). Topology of the Order Parameter in the Little–Parks Experiment, *Phys. Rev. Lett.* **75**, pp. 320–322.

[49] Berger, J. and Rubinstein, J. (1997a). Design for the detection of the singly connected superconducting state, *Physica C* **288**, pp. 105–114.

[50] Berger, J. and Rubinstein, J. (1997b). Formation of topological defects in thin superconducting rings, *Phil. Trans. R. Soc. Lond. A.* **355**, pp. 1969–1978.

[51] Berger, J. and Rubinstein, J. (1997c). Signatures for the second critical point in the phase diagram of a superconducting ring, *Phys. Rev. B* **56**, pp. 5124–5127.

[52] Berger, J. and Rubinstein, J. (1999). Flux-induced vortex in mesoscopic superconducting loops, *Phys. Rev. B* **59**, 13, pp. 8896–8901.

[53] Bergeret, F. S., Volkov, A. F. and Efetov, K. B. (2005). Odd triplet superconductivity and related phenomena in superconductor-ferromagnet structures, *Rev. Mod. Phys.* **77**, 4, pp. 1321–1373.

[54] Berghuis, P. and Kes, P. H. (1993). Two-dimensional collective pinning and vortex-lattice melting in a-Nb$_{1-x}$Ge$_x$ films, *Phys. Rev. B* **47**, 1, pp. 262–272.

[55] Bezryadin, A., Buzdin, A. and Pannetier, B. (1995a). *Macroscopic Quantum Phenomena and Coherence in Superconducting Arrays* (World Scientific, Singapore).

[56] Bezryadin, A., Buzdin, A. and Pannetier, B. (1995b). Phase diagram of multiply connected superconductors: A thin-wire loop and a thin film with a circular hole, *Phys. Rev. B* **51**, 6, pp. 3718–3724.

[57] Bezryadin, A., Ovchinnikov, Y. N. and Pannetier, B. (1996). Nucleation of vortices inside open and blind microholes, *Phys. Rev. B* **53**, 13, pp. 8553–8560.

[58] Bezryadin, A. and Pannetier, B. (1996). Role of edge superconducting states in trapping of multi-quanta vortices by microholes. Application of the bitter decoration technique, *J. Low Temp. Phys.* **102**, 1–2, pp. 73–94.

[59] Blatter, G., Feigel'man, M. V., Geshkenbein, V. B., Larkin, A. I. and Vinokur, V. M. (1994). Vortices in high-temperature superconductors, *Rev. Mod. Phys.* **66**, 4, pp. 1125–1388.

[60] Blatter, G. and Geshkenbein, V. B. (1997). Van der Waals attraction of vortices in superconductors, *Physica C* **282–287**, Part 1, pp. 319–322.

[61] Blonder, G. E., Tinkham, M. and Klapwijk, T. M. (1982). Transition from metallic to tunneling regimes in superconducting microconstrictions: Excess current, charge imbalance, and supercurrent conversion, *Phys. Rev. B* **25**, 7, pp. 4515–4532.

[62] Blumberg, G., Mialitsin, A., Dennis, B. S., Klein, M. V., Zhigadlo, N. D. and Karpinski, J. (2007). Observation of Leggett's Collective Mode in a Multiband MgB$_2$ Superconductor, *Phys. Rev. Lett.* **99**, 22, p. 227002.

[63] Böbel, G. (1965). Behaviour of type-II superconducting cylinders of small radius near the Lower Critical Field, *Il Nuovo Cimento* **38**, pp. 1740–1746.

[64] Bonča, J. and Kabanov, V. V. (2001). Phase transitions in the mesoscopic superconducting square, *Phys. Rev. B* **65**, 1, p. 012509.

[65] Bouquet, F., Fisher, R. A., Phillips, N. E., Hinks, D. G. and Jorgensen, J. D. (2001). Specific Heat of $Mg^{11}B_2$: Evidence for a Second Energy Gap, *Phys. Rev. Lett.* **87**, 4, p. 047001.

[66] Brandt, E. H. (1986). Elastic and plastic properties of the flux-line lattice in type-II superconductors, *Phys. Rev. B* **34**, 9, pp. 6514–6517.

[67] Brandt, E. H. (1995). The flux-line lattice in superconductors, *Rep. Prog. Phys.* **58**, 11, pp. 1465–1594.

[68] Brandt, E. H. and Zhou, S. P. (2009). Attractive vortices, *Physics* **2**, 22.

[69] Braun, F. and von Delft, J. (1999). Superconductivity in ultrasmall metallic grains, *Phys. Rev. B* **59**, 14, pp. 9527–9544.

[70] Braunisch, W., Knauf, N., Kataev, V., Neuhausen, S., Grütz, A., Kock, A., Roden, B., Khomskii, D. and Wohlleben, D. (1992). Paramagnetic Meissner effect in Bi high-temperature superconductors, *Phys. Rev. Lett.* **68**, 12, pp. 1908–1911.

[71] Brinkman, A., Golubov, A. A., Rogalla, H., Dolgov, O. V., Kortus, J., Kong, Y., Jepsen, O. and Andersen, O. K. (2002). Multiband model for tunneling in MgB_2 junctions, *Phys. Rev. B* **65**, 18, p. 180517.

[72] Bruyndoncx, V., Strunk, C., Moshchalkov, V. V., Haesendonck, C. V. and Bruynseraede, Y. (1996). Fluxoid quantization effects in superconducting mesoscopic Al multiloop structures, *Europhys. Lett.* **36**, 6, pp. 449–454.

[73] Bruyndoncx, V., Van Look, L., Verschuere, M. and Moshchalkov, V. V. (1999). Dimensional crossover in a mesoscopic superconducting loop of finite width, *Phys. Rev. B* **60**, 14, pp. 10468–10476.

[74] Bruynseraede, Y., Temst, K., Osquiguil, E., van Haesendonck, C., Gilabert, A. and Schuller, I. K. (1992). Dimensional and flux lattice transitions in artificially layered superconductors, *Phys. Scr.* **T42**, pp. 37–45.

[75] Buisson, O., Gandit, P., Rammal, R., Wang, Y. Y. and Pannetier, B. (1990). Magnetization oscillations of a superconducting disk, *Phys. Lett. A* **150**, 1, pp. 36–42.

[76] Bulaevskii, L. N., Buzdin, A. I., Kulic, M. L. and Panjukov, S. V. (1985). Coexistence of superconductivity and magnetism: theoretical predictions and experimental results, *Adv. Phys.* **134**, 2, pp. 175–261.

[77] Bulaevskii, L. N., Chudnovsky, E. M. and Maley, M. P. (2000). Magnetic pinning in superconductor–ferromagnet multilayers, *Appl. Phys. Lett.* **76**, 18, pp. 2594–2596.

[78] Buzdin, A. and Feinberg, D. (1996). Electromagnetic pinning of vortices by non-superconducting defects and their influence on screening, *Physica C* **303**, 3–4, pp. 303–311.

[79] Buzdin, A. I. (1993). Multiple-quanta vortices at columnar defects, *Phys. Rev. B* **47**, 17, pp. 11416–11419.

[80] Buzdin, A. I. (2005). Proximity effects in superconductor-ferromagnet heterostructures, *Rev. Mod. Phys.* **77**, 3, pp. 935–976.

[81] Buzdin, A. I. and Brison, J. P. (1994). Vortex structures in small super-conducting disks, *Phys. Lett. A* **196**, 1–2, pp. 267–271.

[82] Buzdin, A. I. and Mel'nikov, A. S. (2003). Domain wall superconductivity in ferromagnetic superconductors, *Phys. Rev. B* **67**, 2, p. 020503.

[83] Buzdin, A. I. and Simonov, A. Y. (1990). Penetration of inclined vortices into layered superconductors, *JETP Letters* **51**, pp. 191–195.

[84] Carballeira, C., Moshchalkov, V. V., Chibotaru, L. F. and Ceulemans, A. (2005). Multiquanta Vortex Entry and Vortex-Antivortex Pattern Expansion in a Superconducting Microsquare with a Magnetic Dot, *Phys. Rev. Lett.* **95**, 23, p. 237003.

[85] Carneiro, G. (2004). Pinning and creation of vortices in superconducting films by a magnetic dipole, *Phys. Rev. B* **69**, 21, p. 214504.

[86] Carneiro, G. (2005). Tunable ratchet effects for vortices pinned by periodic magnetic dipole arrays, *Physica C* **432**, 3–4, pp. 206–214.

[87] Castro, J. I. and López, A. (1995). Variational approach to superconductive networks, *Phys. Rev. B* **52**, 10, pp. 7495–7503.

[88] Chesca, B., Schulz, R. R., Goetz, B., Schneider, C. W., Hilgenkamp, H. and Mannhart, J. (2002). *d*-Wave Induced Zero-Field Resonances in dc π-Superconducting Quantum Interference Devices, *Phys. Rev. Lett.* **88**, 17, p. 177003.

[89] Chi, C. C., Santhanam, P. and Blchl, P. E. (1992). Interacting loop-current model of superconducting networks, *J. Low Temp. Phys.* **88**, 1–2, pp. 163–195.

[90] Chibotaru, L. F., Ceulemans, A., Bruyndoncx, V. and Moshchalkov, V. V. (2000). Symmetry-induced formation of antivortices in mesoscopic super-conductors, *Nature* **408**, pp. 833–835.

[91] Chibotaru, L. F., Ceulemans, A., Bruyndoncx, V. and Moshchalkov, V. V. (2001). Vortex Entry and Nucleation of Antivortices in a Mesoscopic Superconducting Triangle, *Phys. Rev. Lett.* **86**, 7, pp. 1323–1326.

[92] Chibotaru, L. F., Ceulemans, A., Morelle, M., Teniers, G., Carballeira, C. and Moshchalkov, V. V. (2005). Ginzburg–Landau description of confinement and quantization effects in mesoscopic superconductors, *J. Math. Phys.* **46**, 9, p. 095108.

[93] Chibotaru, L. F., Ceulemans, A., Teniers, G., Bruyndoncx, V. and Moshchalkov, V. V. (2002a). Vector potential gauge for superconducting regular polygons, *Eur. Phys. J. B* **27**, 3, pp. 341–346.

[94] Chibotaru, L. F., Ceulemans, A., Teniers, G. and Moshchalkov, V. V. (2002b). Nucleation of superconductivity in regular polygons: superconducting vector potential gauge approach, *Physica C* **369**, 1–4, pp. 149–157.

[95] Chibotaru, L. F., Dao, V. H. and Ceulemans, A. (2007). Thermodynamically stable noncomposite vortices in mesoscopic two-gap superconductors, *Europhys. Lett.* **78**, 4, p. 47001.

[96] Chibotaru, L. F., Teniers, G., Ceulemans, A. and Moshchalkov, V. V. (2004). Pseudo Jahn–Teller mechanism for symmetry-breaking phase transitions in vortex molecules, *Phys. Rev. B* **70**, 9, p. 094505.

[97] Chiorescu, I., Nakamura, Y., Harmans, C. J. P. M. and Mooij, J. E. (2003).

Coherent Quantum Dynamics of a Superconducting Flux Qubit, *Science* **299**, pp. 1869–1871.

[98] Choi, H. J., Roundy, D., Sun, H., Cohen, M. L. and Louie, S. G. (2002). The origin of the anomalous superconducting properties of MgB_2, *Nature* **418**, pp. 758–760.

[99] Chou, S. Y., Wei, M. S., Krauss, P. R. and Fischer, P. B. (1994). Single-domain magnetic pillar array of 35 nm diameter and 65 Gbits/in.^{2} density for ultrahigh density quantum magnetic storage, *J. Appl. Phys.* **76**, 10, pp. 6673–6675.

[100] Chun, C. S. L., Zheng, G.-G., Vincent, J. L. and Schuller, I. K. (1984). Dimensional crossover in superlattice superconductors, *Phys. Rev. B* **29**, 9, pp. 4915–4920.

[101] Civale, L., Marwick, A. D., McElfresh, M. W., Worthington, T. K., Malozemoff, A. P., Holtzberg, F. H., Thompson, J. R. and Kirk, M. A. (1990). Defect independence of the irreversibility line in proton-irradiated Y-Ba-Cu-O crystals, *Phys. Rev. Lett.* **65**, 9, pp. 1164–1167.

[102] Civale, L., Marwick, A. D., Worthington, T. K., Kirk, M. A., Thompson, J. R., Krusin-Elbaum, L., Sun, Y., Clem, J. R. and Holtzberg, F. (1991). Vortex confinement by columnar defects in $YBa_2Cu_3O_7$ crystals: Enhanced pinning at high fields and temperatures, *Phys. Rev. Lett.* **67**, 5, pp. 648–651.

[103] Clark, K., Hassanien, A., Khan, S., Braun, K.-F., Tanaka, H. and Hla, S.-W. (2010). Superconductivity in just four pairs of $(BETS)_2GaCl_4$ molecules, *Nature Nanotech.* **5**, pp. 261–265.

[104] Clarke, J. (1639). *Paroemiologia Anglo-Latina* (Microfilm. Ann Arbor).

[105] Clem, J. R. (1974). *Low Temperature Physics*, Vol. 3 (Plenum, New York).

[106] Clem, J. R. (1975). Simple model for the vortex core in a type II superconductor, *J. Low Temp. Phys.* **18**, 5–6, pp. 427–434.

[107] Cody, G. D. and Miller, R. E. (1966). Parallel and Perpendicular Magnetic Transitions of Superconducting Films and Foils of Lead, *Phys. Rev. Lett.* **16**, 16, pp. 697–701.

[108] Cohen, D. (1972). Magnetoencephalography: Detection of the Brain's Electrical Activity with a Superconducting Magnetometer, *Science* **175**, pp. 664–666.

[109] Cole, D., Bending, S., Savel'ev, S., Grigorenko, A., Tamegai, T. and Nori, F. (2006). Ratchet without spatial asymmetry for controlling the motion of magnetic flux quanta using time-asymmetric drives, *Nature Mater.* **5**, pp. 305–311.

[110] Cooley, L. D. and Grishin, A. M. (1995). Pinch effect in commensurate vortex-pin lattices, *Phys. Rev. Lett.* **74**, 14, pp. 2788–2791.

[111] Cooley, L. D., Lee, P. J., Larbalestier, D. C. and O'Larey, P. M. (1994). Periodic pin array at the fluxon lattice scale in a high-field superconducting wire, *Appl. Phys. Lett.* **64**, 10, pp. 1298–1300.

[112] Cyrot, M. and Pavuna, D. (1992). *Introduction to Superconductivity* (World Scientific, Singapore).

[113] Daldini, O., Martinoli, P., Olsen, J. L. and Berner, G. (1974). Vortex-Line

Pinning by Thickness Modulation of Superconducting Films, *Phys. Rev. Lett.* **32**, 5, pp. 218–221.

[114] Dao, V. H., Chibotaru, L. F., Nishio, T. and Moshchalkov, V. V. (2010). Giant vortices, vortex rings and reentrant behavior in type-1.5 superconductors, *arXiv:1007.1849v1 [cond-mat.supr-con]* .

[115] Davidovic-acute, D., Kumar, S., Reich, D. H., Siegel, J., Field, S. B., Tiberio, R. C., Hey, R. and Ploog, K. (1997). Magnetic correlations, geometrical frustration, and tunable disorder in arrays of superconducting rings, *Phys. Rev. B* **55**, 10, pp. 6518–6540.

[116] Davidović, D., Kumar, S., Reich, D. H., Siegel, J., Field, S. B., Tiberio, R. C., Hey, R. and Ploog, K. (1996). Correlations and Disorder in Arrays of Magnetically Coupled Superconducting Rings, *Phys. Rev. Lett.* **76**, 5, pp. 815–818.

[117] De Blois, R. W. and De Sorbo, W. (1964). Surface Barrier in Type-II Superconductors, *Phys. Rev. Lett.* **12**, 18, pp. 499–501.

[118] de Gennes, P. G. (1966). *Superconductivity of Metals and Alloys* (Benjamin, New York).

[119] de Gennes, P. G. (1981). Champ critique dune boucle supraconductrice ramifieé. *C. R. Acad. Sci. Ser. II* **292**, p. 279.

[120] de Souza Silva, C. C., Aguiar, J. A. and Moshchalkov, V. V. (2003). Linear ac dynamics of vortices in a periodic pinning array, *Phys. Rev. B* **68**, 13, p. 134512.

[121] de Souza Silva, C. C., Silhanek, A. V., Van de Vondel, J., Gillijns, W., Metlushko, V., Ilic, B. and Moshchalkov, V. V. (2007). Dipole-Induced Vortex Ratchets in Superconducting Films with Arrays of Micromagnets, *Phys. Rev. Lett.* **98**, 11, p. 117005.

[122] de Souza Silva, C. C., Van de Vondel, J., Morelle, M. and Moshchalkov, V. V. (2006). Controlled multiple reversals of a ratchet effect, *Nature* **440**, pp. 651–654.

[123] Dediu, V. I., Kabanov, V. V. and Sidorenko, A. S. (1994). Dimensional effects in V/Cu superconducting superlattices, *Phys. Rev. B* **49**, 6, pp. 4027–4032.

[124] Deo, P. S., Schweigert, V. A., Peeters, F. M. and Geim, A. K. (1997). Magnetization of Mesoscopic Superconducting Disks, *Phys. Rev. Lett.* **79**, 23, pp. 4653–4656.

[125] Derényi, I. and Vicsek, T. (1995). Cooperative Transport of Brownian Particles, *Phys. Rev. Lett.* **75**, 3, pp. 374–377.

[126] Dingle, R. B. (1952). Some Magnetic Properties of Metals. I. General Introduction, and Properties of Large Systems of Electrons, *Proc. R. Soc. Lond. A* **211**, pp. 500–516.

[127] Dolan, G. J. and Dunsmuir, J. H. (1988). Very small ($\gtrsim$ 20 nm) lithographic wires, dots, rings, and tunnel junctions, *Physica B* **152**, 1–2, pp. 7–13.

[128] Doria, M. M. (2004). Vortex matter in presence of nano-scale magnetic defects, *Physica C* **408–410**, pp. 466–469.

[129] Doria, M. M., de C. Romaguera, A. R., Milošević, M. V. and Peeters, F. M. (2007). Threefold onset of vortex loops in superconductors with a magnetic

core, *Europhys. Lett.* **79**, p. 47006.

[130] Doria, M. M. and Zebende, G. F. (2002). Multiple trapping of vortex lines by a regular array of pinning centers, *Phys. Rev. B* **66**, 6, p. 064519.

[131] Dorsey, A. T. (2000). Superconductivity: Geometry spawns vortices, *Nature* **408**, pp. 783–785.

[132] Dubonos, S. V., Kuznetsov, V., Zhilyaev, I. N., Nikulov, A. V. and Firsov, A. A. (2003). Observation of the external-ac-current-induced dc voltage proportional to the steady current in superconducting loops, *JETP* **77**, 7, pp. 371–375.

[133] Eisterer, M. (2007). Magnetic properties and critical currents of MgB_2, *Supercond. Sci. Technol.* **20**, 5, p. R47.

[134] Elion, W. J., Van der Zant, H. S. J. and Mooij, J. E. (1995). *Macroscopic Quantum Phenomena and Coherence in Superconducting Networks* (World Scientific, Singapore).

[135] Ensslin, K. and Petroff, P. M. (1990). Magnetotransport through an antidot lattice in $GaAs-Al_xGa_{1-x}As$ heterostructures, *Phys. Rev. B* **41**, 17, pp. 12307–12310.

[136] Erdin, S., Kayali, A. F., Lyuksyutov, I. F. and Pokrovsky, V. L. (2002). Interaction of mesoscopic magnetic textures with superconductors, *Phys. Rev. B* **66**, 1, p. 014414.

[137] Erdin, S., Lyuksyutov, I. F., Pokrovsky, V. L. and Vinokur, V. M. (2001). Topological Textures in a Ferromagnet–Superconductor Bilayer, *Phys. Rev. Lett.* **88**, 1, p. 017001.

[138] Eskildsen, M. R., Kugler, M., Tanaka, S., Jun, J., Kazakov, S. M., Karpinski, J. and Fischer, O. (2002). Vortex Imaging in the π-Band of Magnesium Diboride, *Phys. Rev. Lett.* **89**, 18, p. 187003.

[139] Falo, F., Martinez, P. J., Mazo, J. J. and Cilla, S. (1999). Ratchet potential for fluxons in Josephson-junction arrays, *Europhys. Lett.* **45**, 6, pp. 700–706.

[140] Fang, H., Zeller, R. and Stiles, P. J. (1989). Fabrication of quasi-zero-dimensional submicron dot array and capacitance spectroscopy in a GaAs/AlGaAs heterostructure, *Appl. Phys. Lett.* **55**, 14, pp. 1433–1435.

[141] Farkas, Z., Tegzes, P., Vukics, A. and Vicsek, T. (1999). Transitions in the horizontal transport of vertically vibrated granular layers, *Phys. Rev. E* **60**, 6, pp. 7022–7031.

[142] Fasano, Y., Herbsommer, J. A., de la Cruz, F., Pardo, F., Gammel, P. L., Bucher, E. and Bishop, D. J. (1999). Observation of periodic vortex pinning induced by Bitter decoration, *Phys. Rev. B* **60**, 22, pp. R15047–R15050.

[143] Fay, D. and Appel, J. (1980). Coexistence of p-state superconductivity and itinerant ferromagnetism, *Phys. Rev. B* **22**, 7, pp. 3173–3182.

[144] Field, S. B., James, S. S., Barentine, J., Metlushko, V., Crabtree, G., Shtrikman, H., Ilic, B. and Brueck, S. R. J. (2002). Vortex Configurations, Matching, and Domain Structure in Large Arrays of Artificial Pinning Centers, *Phys. Rev. Lett.* **88**, 6, p. 067003.

[145] Fink, H. J. (1969). Vortex Nucleation in a Superconducting Slab near a Second-Order Phase Transition and Excited States of the Sheath near h_{c3},

Phys. Rev. **177**, 2, pp. 732–737.

[146] Fink, H. J. (1992). Superconducting vortex with extended core, *Phys. Rev. B* **45**, 9, pp. 4799–4802.

[147] Fink, H. J., Buisson, O. and Pannetier, B. (1991). Temperature dependence of the critical current of the superconducting microladder in zero magnetic field: Theory and experiment, *Phys. Rev. B* **43**, 13, pp. 10144–10150.

[148] Fink, H. J. and Grünfeld, V. (1985). Critical current of thin superconducting wire with side branches, *Phys. Rev. B* **31**, 1, pp. 600–602.

[149] Fink, H. J., Grünfeld, V. and López, A. (1987). Quantum-interference device without Josephson junctions, *Phys. Rev. B* **35**, pp. 35–37.

[150] Fink, H. J. and Haley, S. B. (1991a). Critical transport currents of the superconducting ladder, *Phys. Rev. Lett.* **66**, 2, pp. 216–219.

[151] Fink, H. J. and Haley, S. B. (1991b). Superconducting micronets: A state-variable approach, *Phys. Rev. B* **43**, 13, pp. 10151–10163.

[152] Fink, H. J., López, A. and Maynard, R. (1982). Magnetic phase boundary of simple superconductive micronetworks, *Phys. Rev. B* **26**, pp. 5237–5240.

[153] Flouquet, J. and Buzdin, A. I. (2002). Ferromagnetic superconductors, *Phys. World* **15**, pp. 41–46.

[154] Fomin, V. M., Devreese, J. T., Bruyndoncx, V. and Moshchalkov, V. V. (2000). Superconductivity in a mesoscopic double square loop: Effect of imperfections, *Phys. Rev. B* **62**, 13, pp. 9186–9190.

[155] Fomin, V. M., Devreese, J. T. and Moshchalkov, V. V. (1998a). Surface Superconductivity in a wedge, *Europhys. Lett.* **42**, 5, pp. 553–558.

[156] Fomin, V. M., Devreese, J. T. and Moshchalkov, V. V. (1999). Surface Superconductivity in a wedge, *Europhys. Lett.* **46**, 1, pp. 118–119.

[157] Fomin, V. M., Misko, V. R., Devreese, J. T. and Moshchalkov, V. V. (1997). On the superconducting phase boundary for a mesoscopic square loop, *Solid State Commun.* **101**, pp. 303–308.

[158] Fomin, V. M., Misko, V. R., Devreese, J. T. and Moshchalkov, V. V. (1998b). Superconducting mesoscopic square loop, *Phys. Rev. B* **58**, 17, pp. 11703–11715.

[159] Frenkel, A. (1993). Mechanism of nonequilibrium optical response of high-temperature superconductors, *Phys. Rev. B* **48**, 13, pp. 9717–9725.

[160] Friedman, J. R., Patel, V., Chen, W., Tolpygo, S. K. and Lukens, J. E. (2000). Quantum superposition of distinct macroscopic states, *Nature* **406**, pp. 43–46.

[161] Fritzsche, J., Kramer, R. B. G. and Moshchalkov, V. V. (2009a). Highly transparent superconducting-normal junctions induced by local fields of magnetic domains in a homogeneous superconductor, *Phys. Rev. B* **80**, 9, p. 094514.

[162] Fritzsche, J., Kramer, R. B. G. and Moshchalkov, V. V. (2009b). Visualization of the vortex-mediated pinning of ferromagnetic domains in superconductor-ferromagnet hybrids, *Phys. Rev. B* **79**, 13, p. 132501.

[163] Fritzsche, J., Moshchalkov, V. V., Eitel, H., Koelle, D., Kleiner, R. and Szymczak, R. (2006). Local observation of reverse-domain superconduc-

tivity in a superconductor-ferromagnet hybrid, *Phys. Rev. Lett.* **96**, 24, p. 247003.

[164] Fulde, P. and Ferrell, R. A. (1964). Superconductivity in a Strong Spin-Exchange Field, *Phys. Rev.* **135**, 3A, pp. A550–A563.

[165] Geim, A. K., Dubonos, S. V., Grigorieva, I. V., Novoselov, K. S., Peeters, F. M. and A., S. V. (2000). Non-quantized penetration of magnetic field in the vortex state of superconductors, *Nature* **407**, pp. 55–57.

[166] Geim, A. K., Dubonos, S. V., Lok, J. G. S., Grigorieva, I. V., Maan, J. C., Theil Hansen, L. and Lindelof, P. E. (1997a). Ballistic Hall micromagnetometry, *Appl. Phys. Lett.* **71**, 16, pp. 2379–2381.

[167] Geim, A. K., Dubonos, S. V., Lok, J. G. S., Henini, M. and Maan, J. C. (1993). Paramagnetic Meissner effect in small superconductors, *Nature* **396**, pp. 144–146.

[168] Geim, A. K., Grigorieva, I. V., Dubonos, S. V., Lok, J. G. S., Maan, J. C., Filippov, A. E. and Peeters, F. M. (1997b). Phase transitions in individual sub-micrometre superconductors, *Nature* **390**, pp. 259–262.

[169] Geim, A. K., Grigorieva, I. V., Lok, J. G. S., Maan, J. C., Dubonos, S. V., Li, X. Q., Peeters, F. M. and Nazarov, Y. V. (1998). Precision magnetometry on a submicron scale: magnetisation of superconducting quantum dots, *Superlattices Microstruct.* **23**, 1, pp. 151–160.

[170] Geoffroy, O., Givord, D., Otani, Y., Pannetier, B. and Ossart, F. (1993). Magnetic and transport properties of ferromagnetic particulate arrays fabricated on superconducting thin films, *J. Magn. Magn. Mater.* **121**, 1–3, pp. 223–226.

[171] Gerhäuser, W., Neumüller, H. W., Schmidt, W., Ries, G., Saemann-Ischenko, G., Gerstenberg, H. and Sauerzopf, F. M. (1991). Comparison of flux pinning enhancement in fast neutron irradiated Bi-2212 single crystals and polycrystalline melt samples, *Physica C* **185–189**, Part 4, pp. 2273–2274.

[172] Gerhäuser, W., Ries, G., Neumüller, H. W., Schmidt, W., Eibl, O., Saemann-Ischenko and Klaumünzer, S. (1992). Flux-line pinning in $Bi_2Sr_2Ca_1Cu_2O_x$ crystals: Interplay of intrinsic 2D behavior and irradiation-induced columnar defects, *Phys. Rev. Lett.* **68**, 6, pp. 879–882.

[173] Geurts, R., Milošević, M. V. and Peeters, F. M. (2006). Symmetric and Asymmetric Vortex-Antivortex Molecules in a Fourfold Superconducting Geometry, *Phys. Rev. Lett.* **97**, 13, p. 137002.

[174] Geurts, R., Milošević, M. V. and Peeters, F. M. (2008). Topologically trapped vortex molecules in Bose–Einstein condensates, *Phys. Rev. A* **78**, 5, p. 053610.

[175] Gheorghe, D. G., Menghini, M., Wijngaarden, R. J., Raedts, S., Silhanek, A. V. and Moshchalkov, V. V. (2006). Anisotropic avalanches and flux penetration in patterned superconductors, *Physica C* **437–438**, pp. 69–72.

[176] Gheorghe, D. G., Wijngaarden, R. J., Gillijns, W., Silhanek, A. V. and Moshchalkov, V. V. (2008). Magnetic flux patterns in superconductors deposited on a lattice of magnetic dots: A magneto-optical imaging study, *Phys. Rev. B* **77**, 5, p. 054502.

[177] Giaever, I. and Megerle, K. (1961). Study of Superconductors by Electron Tunneling, *Phys. Rev.* **122**, 4, pp. 1101–1111.

[178] Giaever, I. and Zeller, H. R. (1968). Superconductivity of Small Tin Particles Measured by Tunneling, *Phys. Rev. Lett.* **20**, 26, pp. 1504–1507.

[179] Gillijns, W., Aladyshkin, A. Y., Silhanek, A. V. and Moshchalkov, V. V. (2007a). Magnetic confinement of the superconducting condensate in superconductor-ferromagnet hybrid composites, *Phys. Rev. B* **76**, 6, p. 060503.

[180] Gillijns, W., Milošević, M. V., Silhanek, A. V., Moshchalkov, V. V. and Peeters, F. M. (2007b). Influence of magnet size on magnetically engineered field-induced superconductivity, *Phys. Rev. B* **76**, 18, p. 184516.

[181] Gillijns, W., Silhanek, A. V. and Moshchalkov, V. V. (2006). Tunable field-induced superconductivity, *Phys. Rev. B* **74**, 22, p. 220509.

[182] Gillijns, W., Silhanek, A. V., Moshchalkov, V. V., Olson Reichhardt, C. J. and Reichhardt, C. (2007c). Origin of Reversed Vortex Ratchet Motion, *Phys. Rev. Lett.* **99**, 24, p. 247002.

[183] Ginzburg, V. L. (1957). Ferromagnetic superconductors, *Sov. Phys. JETP* **4**, pp. 153–161.

[184] Giroud, M., Genicon, J. L., Tournier, R., Geantet, C., Peña, O., Horyn, R. and Sergent, M. (1987). Magnetic field-induced superconductivity in the ferromagnetic state of $HoMo_6S_8$, *J. Low Temp. Phys.* **69**, 5–6, pp. 419–450.

[185] Giubileo, F., Roditchev, D., Sacks, W., Lamy, R., Thanh, D. X., Klein, J., Miraglia, S., Fruchart, D., Marcus, J. and Monod, P. (2001). Two-Gap State Density in MgB_2: A True Bulk Property Or A Proximity Effect? *Phys. Rev. Lett.* **87**, 17, p. 177008.

[186] Gladilin, V. N., Fomin, V. M. and Devreese, J. T. (2002). Shape of nanosize superconducting grains: does it influence pairing characteristics? *Solid State Comm.* **121**, 9–10, pp. 519–523.

[187] Goldberg, S., Segev, Y., Myasoedov, Y., Gutman, I., Avraham, N., Rappaport, M., Zeldov, E., Tamegai, T., Hicks, C. W. and Moler, K. A. (2009). Mott insulator phases and first-order melting in $Bi_2Sr_2CaCu_2O_{8+\delta}$ crystals with periodic surface holes, *Phys. Rev. B* **79**, 6, p. 064523.

[188] Goldobin, E., Sterck, A., Gaber, T., Koelle, D. and Kleiner, R. (2004). Dynamics of Semifluxons in Nb Long Josephson $0 - \pi$ Junctions, *Phys. Rev. Lett.* **92**, 5, p. 057005.

[189] Golubov, A. A., Kortus, J., Dolgov, O. V., Jepsen, O., Kong, Y., Andersen, O. K., Gibson, B. J., Ahn, K. and Kremer, R. K. (2002). Specific heat of MgB_2 in a one- and a two-band model from first-principles calculations, *J. Phys.: Condens. Matter* **14**, pp. 1353–1360.

[190] Golubović, D. S., Milošević, M. V., Peeters, F. M. and Moshchalkov, V. V. (2005). Magnetically induced splitting of a giant vortex state in a mesoscopic superconducting disk, *Phys. Rev. B* **71**, 18, p. 180502.

[191] Golubović, D. S., Pogosov, W. V., Morelle, M. and Moshchalkov, V. V. (2003a). Little-parks effect in a superconducting loop with a magnetic dot, *Phys. Rev. B* **68**, 17, p. 172503.

[192] Golubović, D. S., Pogosov, W. V., Morelle, M. and Moshchalkov, V. V.

(2003b). Nucleation of superconductivity in an Al mesoscopic disk with magnetic dot, *Appl. Phys. Lett.* **83**, 8, p. 1593.

[193] Golubović, D. S., Pogosov, W. V., Morelle, M. and Moshchalkov, V. V. (2004). Magnetic Phase Shifter for Superconducting Qubits, *Phys. Rev. Lett.* **92**, 17, p. 177904.

[194] Gommers, R., Denisov, S. and Renzoni, F. (2006). Quasiperiodically Driven Ratchets for Cold Atoms, *Phys. Rev. Lett.* **96**, 24, p. 240604.

[195] Goodenough, J. B. (1956). Interpretation of Domain Patterns Recently Found in BiMn and SiFe Alloys, *Phys. Rev.* **102**, 2, pp. 356–365.

[196] Goyal, A., Kang, S., Leonard, K. J., Martin, P. M., Gapud, A. A., Varela, M., Paranthaman, M., Ijaduola, A. O., Specht, E. D., Thompson, J. R., Christen, D. K., Pennycook, S. J. and List, F. A. (2005). Irradiation-free, columnar defects comprised of self-assembled nanodots and nanorods resulting in strongly enhanced flux-pinning in $YBa_2Cu_3O_{7-\delta}$ films, *Supercond. Sci. Technol.* **18**, pp. 1533–1538.

[197] Grigorenko, A. N., Bending, S. J., Van Bael, M. J., Lange, M., Moshchalkov, V. V., Fangohr, H. and de Groot, P. A. J. (2003). Symmetry Locking and Commensurate Vortex Domain Formation in Periodic Pinning Arrays, *Phys. Rev. Lett.* **90**, 23, p. 237001.

[198] Grigorenko, A. N., Howells, G. D., Bending, S. J., Bekaert, J., Van Bael, M. J., Van Look, L., Moshchalkov, V. V., Bruynseraede, Y., Borghs, G., Kaya, I. I. and Stradling, R. A. (2001). Direct imaging of commensurate vortex structures in ordered antidot arrays, *Phys. Rev. B* **63**, 5, p. 052504.

[199] Grigorieva, I. V., Escoffier, W., Richardson, J., Vinnikov, L. Y., Dubonos, S. and Oboznov, V. (2006). Direct Observation of Vortex Shells and Magic Numbers in Mesoscopic Superconducting Disks, *Phys. Rev. Lett.* **96**, 7, p. 077005.

[200] Groff, R. P. and Parks, R. D. (1968). Fluxoid Quantization and Field-Induced Depairing in a Hollow Superconducting Microcylinder, *Phys. Rev.* **176**, pp. 567–580.

[201] Gross, R. (2005). Grain boundaries in high temperature superconductors: A retrospective view, *Physica C* **432**, 3–4, pp. 105–115.

[202] Grundy, P. J. (1965). Observation of magnetic domains in magnetoplumbite by Lorentz electron microscopy, *Br. J. Appl. Phys.* **16**, pp. 409–410.

[203] Gschneidner, K. A. (1964). Physical properties and interrelations of metallic and semimetallic elements, *Solid State Phys.* **16**, pp. 275–426.

[204] Gu, J. Y., You, C.-Y., Jiang, J. S., Pearson, J., Bazaliy, Y. B. and Bader, S. D. (2002). Magnetization-Orientation Dependence of the Superconducting Transition Temperature in the Ferromagnet-Superconductor-Ferromagnet System: CuNi/Nb/CuNi, *Phys. Rev. Lett.* **89**, 26, p. 267001.

[205] Gutierrez, J., Llordes, A., Gazquez, J., Gibert, M., Roma, N., Ricart, S., Pomar, A., Sandiumenge, F., Mestres, N. and Puig, X., T.and Obradors (2007). Strong isotropic flux pinning in solution-derived $YBa_2Cu_3O_{7-x}$ nanocomposite superconductor films, *Nat. Mater.* **6**, pp. 367–373.

[206] H, Y., Sata, K., Nakata, S., Sato, O., Kato, M., Kasai, J., Hasegawa, T., Satoh, K., Yotsuya, T. and Ishida, T. (2004). Vortex configurations

in the nanofabricated network of Nb: direct observations and calculations, *Physica C* **412–414**, 1, pp. 552–556.

[207] Hakonen, P. J., Ikkala, O. T., Islander, S. T., Lounasmaa, O. V., Markkula, T. K., Roubeau, P., Saloheimo, K. M., Volovik, G. E., Andronikashvili, E. L., Garibashvili, D. I. and Tsakadze, J. S. (1982). NMR Experiments on Rotating Superfluid ^{3}He-a: Evidence for Vorticity, *Phys. Rev. Lett.* **48**, 26, pp. 1838–1841.

[208] Haley, S. B. and Fink, H. J. (1984). Nonuniform superconducting ring with dangling side branches, *Phys. Lett. A* **102**, 9, pp. 431–433.

[209] Hallet, X., Mátéfi-Tempfli, M., Michotte, S., Piraux, L., Vanacken, J., Moshchalkov, V. V. and Mátéfi-Tempfli, S. (2009). High magnetic field matching effects in NbN films induced by template grown dense ferromagnetic nanowires arrays, *Appl. Phys. Lett.* **95**, 25, p. 252503.

[210] Harada, K., Kamimura, O., Kasai, H., Matsuda, T., Tonomura, A. and Moshchalkov, V. V. (1996). Direct Observation of Vortex Dynamics in Superconducting Films with Regular Arrays of Defects, *Science* **274**, 5290, pp. 1167–1170.

[211] Harada, K., Matsuda, T., Bonevich, J., Igarashi, M., Kondo, S., Pozzi, G., Kawabe, U. and Tonomura, A. (1992). Real-time observation of vortex lattices in a superconductor by electron microscopy, *Nature* **360**, pp. 51–53.

[212] Haviland, D. B., Liu, Y. and Goldman, A. M. (1989). Onset of superconductivity in the two-dimensional limit, *Phys. Rev. Lett.* **62**, 18, pp. 2180–2183.

[213] Hebard, A., Fiory, A. and Somekh, S. (1977). Critical currents in Al films with a triangular lattice of 1 µm holes, *IEEE Trans. Magn.* **13**, 1, pp. 589–592.

[214] Heinzel, C., Theilig, T. and Ziemann, P. (1993). Paramagnetic Meissner effect analyzed by second harmonics of the magnetic susceptibility: Consistency with a ground state carrying spontaneous currents, *Phys. Rev. B* **48**, 5, pp. 3445–3454.

[215] Helseth, L. E. (2002). Interaction between superconducting films and magnetic nanostructures, *Phys. Rev. B* **66**, 10, p. 104508.

[216] Higgins, M. J. and Bhattacharya, S. (1996). Varieties of dynamics in a disordered flux-line lattice, *Physica C* **257**, 3, pp. 232–254.

[217] Hilgenkamp, H., Ariando, Smilde, H.-J. H., Blank, D. H. A., Rijnders, G., Rogalla, H., Kirtley, J. R. and Tsuei, C. C. (2003). Ordering and manipulation of the magnetic moments in large-scale superconducting π-loop arrays, *Nature* **422**, pp. 50–53.

[218] Himpsel, F. J., Ortega, J. E., Mankey, G. J. and Willis, R. F. (1998). Magnetic nanostructures, *Adv. Phys.* **47**, 4, pp. 511–597.

[219] Hoffmann, A., Prieto, P. and Schuller, I. K. (2000). Periodic vortex pinning with magnetic and nonmagnetic dots: The influence of size, *Phys. Rev. B* **61**, 10, pp. 6958–6965.

[220] Hofstadter, D. R. (1976). Energy levels and wave functions of bloch electrons in rational and irrational magnetic fields, *Phys. Rev. B* **14**, 6, pp. 2239–2249.

[221] Horane, E. M., Castro, J. I., Buscaglia, G. C. and López, A. (1996). Transition between different quantum states in a mesoscopic system: The superconducting ring, *Phys. Rev. B* **53**, pp. 9296–9300.

[222] Houghton, A. and McLean, F. B. (1965). Nucleation of superconductivity in wedge geometry, *Phys. Lett.* **19**, 3, pp. 172–174.

[223] Houzet, M., Buzdin, A., Bulaevskii, L. and Maley, M. (2002). New Superconducting Phases in Field-Induced Organic Superconductor $\lambda -$ $(BETS)_2FeCl_4$, *Phys. Rev. Lett.* **88**, 22, p. 227001.

[224] Hubert, A. and Schfer, R. (1998). *Magnetic Domains* (Springer, Berlin/Heidelberg).

[225] Huebener, R. P. (1979). *Magnetic flux structures in superconductors* (Springer-Verlag, Berlin – New York).

[226] Huebener, R. P. (1990). *Magnetic Flux Structures of Superconductors* (Springer–Verlag, New York).

[227] Hunte, F., Jaroszynski, J., Gurevich, A., Larbalestier, D. C., Jin, R., Sefat, A. S., McGuire, M. A., Sales, B. C., Christen, D. K. and Mandrus, D. (2008). Two-band superconductivity in $LaFeAsO_{0.89}F_{0.11}$ at very high magnetic fields, *Nature* **453**, pp. 903–905.

[228] Huy, N. T., Gasparini, A., de Nijs, D. E., Huang, Y., Klaasse, J. C. P., Gortenmulder, T., de Visser, A., Hamann, A., Görlach, T. and Löhneysen, H. v. (2007). Superconductivity on the Border of Weak Itinerant Ferromagnetism in UCoGe, *Phys. Rev. Lett.* **99**, 6, p. 067006.

[229] Hyndman, R., Mougin, A., Sampaio, L. C., Ferré, J., Jamet, J. P., Meyer, P., Mathet, V., Chappert, C., Mailly, D. and Gierak, J. (2002). Magnetization reversal in weakly coupled magnetic patterns, *J. Magn. Magn. Mater.* **240**, 1–3, pp. 34–36.

[230] Iavarone, M., Karapetrov, G., Koshelev, A. E., Kwok, W. K., Crabtree, G. W., Hinks, D. G., Kang, W. N., Choi, E.-M., Kim, H. J., Kim, H.-J. and Lee, S. I. (2002). Two-band superconductivity in MgB_2, *Phys. Rev. Lett.* **89**, 18, p. 187002.

[231] Ibrahim, I. S., Schweigert, V. A. and Peeters, F. M. (1998). Diffusive transport in a Hall junction with a microinhomogeneous magnetic field, *Phys. Rev. B* **57**, 24, pp. 15416–15427.

[232] Imre, A., Varga, E., Lili, J., Ilic, B., Metlushko, V., Csaba, G., Orlov, A., Bernstein, G. H. and Porod, W. (2006). Flux-Closure Magnetic States in Triangular Cobalt Ring Elements, *IEEE Trans. Magn.* **42**, 11, pp. 3641–3644.

[233] Ioffe, L. B., Geshkenbein, V. B., Feigel'man, M. V., Fauchére, A. L. and Blatter, G. (1999). Environmentally decoupled sds-wave Josephson junctions for quantum computing, *Nature* **398**, pp. 679–681.

[234] Izyumov, Y. N. P., Yu. A. and Khusainov, M. G. (2002). Competition between superconductivity and magnetism in ferromagnet/superconductor heterostructures, *Usp. Fiz. Nauk* **172**, pp. 113–154.

[235] Jaccard, Y., Martín, J. I., Cyrille, M.-C., Vélez, M., Vicent, J. L. and Schuller, I. K. (1998). Magnetic pinning of the vortex lattice by arrays of submicrometric dots, *Phys. Rev. B* **58**, 13, pp. 8232–8235.

[236] Jaccarino, V. and Peter, M. (1962). Ultra-High-Field Superconductivity, *Phys. Rev. Lett.* **9**, 7, pp. 290–292.

[237] Jacobs, L. and Rebbi, C. (1979). Interaction energy of superconducting vortices, *Phys. Rev. B* **19**, 9, pp. 4486–4494.

[238] Jahn, H. A. and Teller, E. (1937). Stability of polyatomic molecules in degenerate electronic states. I. Orbital degeneracy, *Proc. R. Soc. (London) A* **161**, pp. 220–235.

[239] Jiang, J. S., Davidović, D., Reich, D. H. and Chien, C. L. (1995). Oscillatory Superconducting Transition Temperature in Nb/Gd Multilayers, *Phys. Rev. Lett.* **74**, 2, pp. 314–317.

[240] Joseph, A. S. and Tomasch, W. J. (1964). Experimental Evidence for Delayed Entry of Flux Into a Type-II Superconductor, *Phys. Rev. Lett.* **12**, 9, pp. 219–222.

[241] Kanoda, K., Mazaki, H., Hosoito, N. and Shinjo, T. (1987). Upper critical field of V-Ag multilayered superconductors, *Phys. Rev. B* **35**, 13, pp. 6736–6748.

[242] Karapetrov, G., Fedor, J., Iavarone, M., Marshall, M. T. and Divan, R. (2005). Imaging of vortex states in mesoscopic superconductors, *Appl. Phys. Lett.* **87**, 16, p. 162515.

[243] Karpinski, J., Angst, M., Jun, J., Kazakov, S. M., Puzniak, R., Wisniewski, A., Roos, J., Keller, H., Perucchi, A., Degiorgi, L., Eskildsen, M. R., Bordet, P., Vinnikov, L. and Mironov, A. (2003). MgB$_2$ single crystals: high pressure growth and physical properties, *Supercond. Sci. Technol.* **16**, pp. 221–230.

[244] Kastalsky, A., Kleinsasser, A. W., Greene, L. H., Bhat, R., Milliken, F. P. and Harbison, J. P. (1991). Observation of pair currents in superconductor-semiconductor contacts, *Phys. Rev. Lett.* **67**, 21, pp. 3026–3029.

[245] Kato, M. (2009). Private communication.

[246] Kato, M., Koyama, T., Machida, M. and Ishida, T. (2009). Magnetic flux structures of composite superconducting structures with d- and s-waves superconductors (d-dots), *Physica C* **469**, 15–20, pp. 1067–1070.

[247] Kawasaki, T., Ru, Q. X., Matsuda, T., Bando, Y. and Tonomura, A. (1991). High-Resolution Holography Observation of H-Nb$_2$O$_5$, *Jpn. J. Appl. Phys.* **30**, Part 2, No. 10B, pp. L1830–L1832.

[248] Kemmler, M., Gürlich, C., Sterck, A., Pöhler, H., Neuhaus, M., Siegel, M., Kleiner, R. and Koelle, D. (2006). Commensurability Effects in Superconducting Nb Films with Quasiperiodic Pinning Arrays, *Phys. Rev. Lett.* **97**, 14, p. 147003.

[249] Khalfin, I. B. and Shapiro, B. Y. (1993). Relaxation of magnetic flux in a superconductor with a system of columnar defects, *Physica C* **207**, 3–4, pp. 359–365.

[250] Kim, Y. B. and Stephen, M. J. (1969). *Superconductivity* (Marcel Dekker, New York).

[251] Kittel, C. (1976). *Introduction to Solid State Physics* (Wiley & Sons Inc).

[252] Kläui, M., Jubert, P.-O., Allenspach, R., Bischof, A., Bland, J. A. C., Faini, G., Rüdiger, U., Vaz, C. A. F., Vila, L. and Vouille, C. (2005).

Direct Observation of Domain-Wall Configurations Transformed by Spin Currents, *Phys. Rev. Lett.* **95**, 2, p. 026601.

[253] Kleinsasser, A. W., Jackson, T. N., McInturff, D., Rammo, F., Pettit, G. D. and Woodall, J. M. (1990). Crossover from tunneling to metallic behavior in superconductor-semiconductor contacts, *Appl. Phys. Lett.* **57**, 17, pp. 1811–1813.

[254] Klimin, S. N., Fomin, V. M., Devreese, J. T. and Moshchalkov, V. V. (1999). Superconductivity in a wedge with a small angle: an analytical treatment, *Solid State Commun.* **111**, 10, pp. 589–594.

[255] Koike, K., Matsuyama, H., Hirayama, Y., Tanahashi, K., Kanemura, T., Kitakami, O. and Shimada, Y. (2001). Magnetic block array for patterned magnetic media, *Appl. Phys. Lett.* **78**, 6, pp. 784–786.

[256] König, C., Sperlich, M., Heinesch, R., Calarco, R., Hauch, J. O., Rüdiger, U., Güntherodt, G., Kirsch, S., Özyilmaz, B. and Kent, A. D. (2001). Shape-dependent magnetization reversal processes and flux-closure configurations of microstructured epitaxial Fe(110) elements, *Appl. Phys. Lett.* **79**, 22, pp. 3648–3650.

[257] Koorevaar, P., Suzuki, Y., Coehoorn, R. and Aarts, J. (1994). Decoupling of superconducting V by ultrathin Fe layers in V/Fe multilayers, *Phys. Rev. B* **49**, 1, pp. 441–449.

[258] Koshelev, A. E. and Golubov, A. A. (2003). Mixed State of a Dirty Two-Band Superconductor: Application to MgB_2, *Phys. Rev. Lett.* **90**, 17, p. 177002.

[259] Kostić, P., Veal, B., Paulikas, A. P., Welp, U., Todt, V. R., Gu, C., Geiser, U., Williams, J. M., Carlson, K. D. and Klemm, R. A. (1996). Paramagnetic Meissner effect in Nb, *Phys. Rev. B* **53**, 2, pp. 791–801.

[260] Koyama, T., Machida, M., Kato, M. and Ishida, T. (2004). Phase dynamics in a *d*-dot embedded in a *s*-wave superconductor, *Physica C* **412–414**, Part 1, pp. 358–361.

[261] Krägeloh, U. (1970). Der Zwischenzustand bei Supraleitern zweiter Art, *Phys. Stat. Sol (b)* **42**, 2, pp. 559–576.

[262] Kramer, L. (1971). Thermodynamic Behavior of Type-II Superconductors with Small κ near the Lower Critical Field, *Phys. Rev. B* **3**, 11, pp. 3821–3825.

[263] Kramer, R. B. G., Silhanek, A. V., Van de Vondel, J., Raes, B. and Moshchalkov, V. V. (2009). Symmetry-Induced Giant Vortex State in a Superconducting Pb Film with a Fivefold Penrose Array of Magnetic Pinning Centers, *Phys. Rev. Lett.* **103**, 6, p. 067007.

[264] Krasil'nikov, A. S., Mamsurova, L. G., Trusevich, N. G., Shcherbakov, L. G. and Pukhov, K. K. (1995). Magnetization of fine-grained YBaCuO near the lower critical field, *Low. Temp. Phys.* **21**, 1, pp. 38–44.

[265] Krauth, W., Trivedi, N. and Ceperley, D. (1991). Superfluid-insulator transition in disordered boson systems, *Phys. Rev. Lett.* **67**, 17, pp. 2307–2310.

[266] Krutzler, C., Zehetmayer, M., Eisterer, M., Weber, H. W., Zhigadlo, N. D., Karpinski, J. and Wisniewski, A. (2006). Anisotropic reversible mixed-state

properties of superconducting carbon-doped $Mg(B_{1-x}C_x)_2$ single crystals, *Phys. Rev. B* **74**, 14, p. 144511.

[267] Kuzmin, L. S., Nazarov, Y. V., Haviland, D. B., Delsing, P. and Claeson, T. (1991). Coulomb blockade and incoherent tunneling of Cooper pairs in ultrasmall junctions affected by strong quantum fluctuations, *Phys. Rev. Lett.* **67**, 9, pp. 1161–1164.

[268] Kvon, Z. D., Baturina, T. I., Donaton, R. A., Baklanov, M. R., Maex, K., Olshanetsky, E. B., Plotnikov, A. E. and Portal, J. C. (2000). Proximity effects and Andreev reflection in a mesoscopic SNS junction with perfect NS interfaces, *Phys. Rev. B* **61**, 17, pp. 11340–11343.

[269] Kwok, W. K., Fendrich, J., Fleshler, S., Welp, U., Downey, J. and Crabtree, G. W. (1994). Vortex liquid disorder and the first order melting transition in $YBa_2Cu_3O_{7-\delta}$, *Phys. Rev. Lett.* **72**, 7, pp. 1092–1095.

[270] Laiho, R., Lähderanta, E., Sonin, E. B. and Traito, K. B. (2003). Penetration of vortices into the ferromagnet/type-II superconductor bilayer, *Phys. Rev. B* **67**, 14, p. 144522.

[271] Landau, L. D. and Lifshitz, E. M. (1975). *Quantum Mechanics*, 2nd edn. (Pergamon, Oxford).

[272] Landau, L. D. and Lifshitz, E. M. (1984). *Electrodynamics of Continuum Media*, 2nd edn. (Pergamon, Oxford).

[273] Lange, M. (2003). *Vortex Matter in Hybrid Superconductor/Ferromagnet Nanosystems*, Ph.D. thesis, Katholieke Universiteit Leuven.

[274] Lange, M., Bael, M. J. V., Bruynseraede, Y. and Moshchalkov, V. V. (2003a). Nanoengineered Magnetic-Field-Induced Superconductivity, *Phys. Rev. Lett.* **90**, 19, p. 197006.

[275] Lange, M., Van Bael, M. J. and Moshchalkov, V. V. (2003b). Phase diagram of a superconductor/ferromagnet bilayer, *Phys. Rev. B* **68**, 17, p. 174522.

[276] Larkin, A. I. and Ovchinnikov, Y. N. (1965). Inhomogeneous state of superconductors, *Sov. Phys. JETP* **47**, pp. 1136–1146.

[277] Larson, T. and Haley, S. B. (1995). Mesoscopic superconducting devices and circuits: Variable squid and superconducting micronet current amplifier, *Appl. Supercond.* **3**, 11–12, pp. 573–583.

[278] Larson, T. and Haley, S. B. (1997). A superconducting micronet current amplifier, *J. Low Temp. Phys.* **107**, 1–2, pp. 3–19.

[279] Larson, T., Haley, S. B. and Erdös, P. (1994). Current vortex patterns in the superconducting microladder, *Physica B* **194–196**, Part 2, pp. 1425–1426.

[280] Lee, C. S., Jankó, B., Derényi, I. and Barabási, A. L. (1999). Reducing vortex density in superconductors using the "ratchet effect", *Nature* **400**, pp. 337–340.

[281] Leggett, A. J. (2002). Superconducting Qubits – a Major Roadblock Dissolved? *Science* **296**, pp. 861–862.

[282] Levitov, L. S. (1991). Phyllotaxis of flux lattices in layered superconductors, *Phys. Rev. Lett.* **66**, 2, pp. 224–227.

[283] Lévy, F., Sheikin, I., Grenier, B. and Huxle, A. D. (2005). Magnetic Field-Induced Superconductivity in the Ferromagnet URhGe, *Scinece* **309**, pp. 1343–1346.

[284] Lévy, F., Sheikin, I., Grenier, B., Marcenat, C. and Huxle, A. D. (2009). Coexistence and interplay of superconductivity and ferromagnetism in URhGe, *J. Phys.: Condens. Matter* **21**, p. 164211.

[285] Li, M. S. (2003). Paramagnetic meissner effect and related dynamical phenomena, *Phys. Reports* **376**, 3, pp. 133–223.

[286] Li, W.-K. and Blinder, S. M. (1985). Solution of the Schrödinger equation for a particle in an equilateral triangle, *J. Math. Phys.* **26**, 11, pp. 2784–2786.

[287] Libál, A., Reichhardt, C., Jankó, B. and Olson Reichhardt, C. J. (2006). Dynamics, Rectification, and Fractionation for Colloids on Flashing Substrates, *Phys. Rev. Lett.* **96**, 18, p. 188301.

[288] Linke, H., Humphrey, T. E., Lindelof, P. E., Löfgren, A., Newbury, R., Omling, P., Sushkov, A. O., Taylor, P. R. and Xu, H. (2002). Quantum ratchets and quantum heat pumps, *Appl. Phys. A* **75**, 2, pp. 237–246.

[289] Linke, H., Humphrey, T. E., Löfgren, A., Sushkov, A. O., Newbury, R., Taylor, R. P. and Omling, P. (1999). Experimental Tunneling Ratchets, *Science* **286**, 5448, pp. 2314–2317.

[290] Little, W. A. and Parks, R. D. (1962). Observation of Quantum Periodicity in the Transition Temperature of a Superconducting Cylinder, *Phys. Rev. Lett.* **9**, pp. 9–12.

[291] Liu, Y., Zadorozhny, Y., Rosario, M. M., Rock, B. Y., Carrigan, P. T. and Wang, H. (2001). Destruction of the Global Phase Coherence in Ultrathin, Doubly Connected Superconducting Cylinders, *Science* **294**, pp. 2332–2334.

[292] London, F. and London, H. (1935). The electromagnetic equations of the supraconductor, *Proc. Roy. Soc. A* **149**, 8, pp. 71–88.

[293] Lozovik, Y. E. and Rakoch, E. A. (1998). Energy barriers, structure, and two-stage melting of microclusters of vortices, *Phys. Rev. B* **57**, 2, pp. 1214–1225.

[294] Lu, Q., Olson Reichhardt, C. J. and Reichhardt, C. (2007). Reversible vortex ratchet effects and ordering in superconductors with simple asymmetric potential arrays, *Phys. Rev. B* **75**, 5, p. 054502.

[295] Lukashenko, A., Wördenweber, R. and Ustinov, A. V. (2008). Imaging of vortex flow in microstructured high-T_c films by laser scanning microscope, *Physica C* **468**, 7–10, pp. 552–556.

[296] Lykov, A. N. (1993). Pinning in superconducting films with triangular lattice of holes, *Solid State Commun.* **86**, 8, pp. 531–533.

[297] Lyuksyutov, I. F. and Pokrovsky, V. (1998). Magnetization Controlled Superconductivity in a Film with Magnetic Dots, *Phys. Rev. Lett.* **81**, 11, pp. 2344–2347.

[298] Lyuksyutov, I. F. and Pokrovsky, V. L. (2005). Ferromagnetsuperconductor hybrids, *Adv. Phys.* **54**, 1, pp. 67–136.

[299] Marconi, V. I. (2007). Rocking Ratchets in Two-Dimensional Josephson Networks: Collective Effects and Current Reversal, *Phys. Rev. Lett.* **98**, 4, p. 047006.

[300] Marmorkos, I. K., Matulis, A. and Peeters, F. M. (1996). Vortex structure

around a magnetic dot in planar superconductors, *Phys. Rev. B* **53**, 5, pp. 2677–2685.

[301] Martín, J. I., Vélez, M., Hoffmann, A., Schuller, I. K. and Vicent, J. L. (1999). Artificially Induced Reconfiguration of the Vortex Lattice by Arrays of Magnetic Dots, *Phys. Rev. Lett.* **83**, 5, pp. 1022–1025.

[302] Martín, J. I., Vélez, M., Hoffmann, A., Schuller, I. K. and Vicent, J. L. (2000). Temperature dependence and mechanisms of vortex pinning by periodic arrays of Ni dots in Nb films, *Phys. Rev. B* **62**, 13, pp. 9110–9116.

[303] Martín, J. I., Vélez, M., Nogués, J. and Schuller, I. K. (1997). Flux pinning in a superconductor by an array of submicrometer magnetic dots, *Phys. Rev. Lett.* **79**, 10, pp. 1929–1932.

[304] Mathai, A., Gim, Y., Black, R. C., Amar, A. and Wellstood, F. C. (1995). Experimental Proof of a Time-Reversal-Invariant Order Parameter with a π shift in $YBa_2Cu_3O_{7-\delta}$, *Phys. Rev. Lett.* **74**, 22, pp. 4523–4526.

[305] Matsuura, T., Tsuneta, T., Inagaki, K. and Tanda, S. (2005). Dynamics of charge density wave ring, *Physica C* **426–431**, 1, pp. 431–435.

[306] Matthias, B. T. and Suhl, H. (1960). Possible Explanation of the "coexistence" of Ferromagnetism and Superconductivity, *Phys. Rev. Lett.* **4**, 2, pp. 51–52.

[307] Matthias, B. T., Suhl, H. and Corenzwit, E. (1958a). Spin Exchange in Superconductors, *Phys. Rev. Lett.* **1**, 3, pp. 92–94.

[308] Matthias, B. T., Suhl, H. and Corenzwit, E. (1958b). Spin Exchange in Superconductors, *Phys. Rev. Lett.* **1**, 4, p. 152.

[309] Matthias, S. and Müller, F. (2003). Asymmetric pores in a silicon membrane acting as massively parallel brownian ratchets, *Nature* **424**, pp. 53–57.

[310] Mattson, J. E., Sowers, C. H., Berger, A. and Bader, S. D. (1992). Magnetoresistivity and oscillatory interlayer magnetic coupling of sputtered Fe/Nb superlattices, *Phys. Rev. Lett.* **68**, 21, pp. 3252–3255.

[311] Mazin, I. I., Andersen, O. K., Jepsen, O., Dolgov, O. V., Kortus, J., Golubov, A. A., Kuz'menko, A. B. and van der Marel, D. (2002). Superconductivity in MgB_2: Clean or dirty? *Phys. Rev. Lett.* **89**, 10, p. 107002.

[312] Mel'nikov, A. S., Nefedov, I. M., Ryzhov, D. A., Shereshevskii, I. A., Vinokur, V. M. and Vysheslavtsev, P. P. (2002). Vortex states and magnetization curve of square mesoscopic superconductors, *Phys. Rev. B* **65**, 14, p. 140503.

[313] Mel'nikov, A. S. and Vinokur, V. M. (2002). Mesoscopic superconductor as a ballistic quantum switch, *Nature* **415**, pp. 60–62.

[314] Menghini, M., Kramer, R. B. G., Silhanek, A. V., Sautner, J., Metlushko, V., De Keyser, K., Fritzsche, J., Verellen, N. and Moshchalkov, V. V. (2009). Direct visualization of magnetic vortex pinning in superconductors, *Phys. Rev. B* **79**, 14, p. 144501.

[315] Menghini, M., Van de Vondel, J., Gheorghe, D. G., Wijngaarden, R. J. and Moshchalkov, V. V. (2007). Asymmetry reversal of thermomagnetic avalanches in Pb films with a ratchet pinning potential, *Phys. Rev. B* **76**, 18, p. 184515.

[316] Menghini, M., Wijngaarden, R. J., Silhanek, A. V., Raedts, S. and

Moshchalkov, V. V. (2005). Dendritic flux penetration in Pb films with a periodic array of antidots, *Phys. Rev. B* **71**, 10, p. 104506.

[317] Mertelj, T. and Kabanov, V. V. (2003). Vortex-antivortex configurations and its stability in a mesoscopic superconducting square, *Phys. Rev. B* **67**, 13, p. 134527.

[318] Metlushko, V., Welp, U., Crabtree, G. W., Osgood, R., Bader, S. D., De-Long, L. E., Zhang, Z., Brueck, S. R. J., Ilic, B., Chung, K. and Hesketh, P. J. (1999). Interstitial flux phases in a superconducting niobium film with a square lattice of artificial pinning centers, *Phys. Rev. B* **60**, 18, pp. R12585–R12588.

[319] Metlushko, V. V., Baert, M., Jonckheere, R., Moshchalkov, V. V. and Bruynseraede, Y. (1994). Matching effects in Pb/Ge multilayers with the lattice of submicron holes, *Solid State Commun.* **91**, 5, pp. 331–335.

[320] Metlushko, V. V., DeLong, L. E., Baert, M., Rosseel, E., Van Bael, M. J., Temst, K., Moshchalkov, V. V. and Bruynseraede, Y. (1998). Supermatching vortex phases in superconducting thin films with antidot lattices, *Europhys. Lett.* **41**, 3, pp. 333–338.

[321] Metlushko, V. V., DeLong, L. E., Moshchalkov, V. V. and Bruynseraede, Y. (2003). Proximity effect in the superconducting magnetization of a Pb/Cu bilayer film with an antidot lattice, *Physica C* **391**, 2, pp. 196–202.

[322] Meul, H. W., Rossel, C., Decroux, M., Fischer, O., Remenyi, G. and Briggs, A. (1984). Observation of Magnetic-Field-Induced Superconductivity, *Phys. Rev. Lett.* **53**, 5, pp. 497–500.

[323] Milošević, M. V. and Peeters, F. M. (2003). Interaction between a superconducting vortex and an out-of-plane magnetized ferromagnetic disk: Influence of the magnet geometry, *Phys. Rev. B* **68**, 9, p. 094510.

[324] Milošević, M. V. and Peeters, F. M. (2005). Field-enhanced critical parameters in magnetically nanostructured superconductors, *Europhys. Lett.* **70**, 5, pp. 670–676.

[325] Milošević, M. V., Yampolskii, S. V. and Peeters, F. M. (2002a). Magnetic pinning of vortices in a superconducting film: The (anti)vortex-magnetic dipole interaction energy in the London approximation, *Phys. Rev. B* **66**, 17, p. 174519.

[326] Milošević, M. V., Yampolskii, S. V. and Peeters, F. M. (2002b). Vortex structure of thin mesoscopic disks in the presence of an inhomogeneous magnetic field, *Phys. Rev. B* **66**, 2, p. 024515.

[327] Milošević, M. V., Yampolskii, S. V. and Peeters, F. M. (2003). The vortex-magnetic dipole interaction in the London approximation, *J. Low Temp. Phys.* **130**, pp. 321–31.

[328] Minnhagen, P. (1987). The two-dimensional Coulomb gas, vortex unbinding, and superfluid-superconducting films, *Rev. Mod. Phys.* **59**, 4, pp. 1001–1066.

[329] Mirkovic, J., Savel'ev, H., S. Sato, Nori, F. and Kadowaki, K. (2006). Melting of the vortex-solid in irradiated $Bi_2Sr_2CaCu_2O_{8+\delta}$ single crystals in tilted magnetic fields, *New J. Phys.* **8**, pp. 226–239.

[330] Misko, V., Savel'ev, S. and Nori, F. (2005). Critical Currents in Quasiperi-

odic Pinning Arrays: Chains and Penrose Lattices, *Phys. Rev. Lett.* **95**, 17, p. 177007.

[331] Misko, V. R., Savel'ev, S. and Nori, F. (2006). Critical currents in superconductors with quasiperiodic pinning arrays: One-dimensional chains and two-dimensional Penrose lattices, *Phys. Rev. B* **74**, 2, p. 024522.

[332] Mkrtchyan, G. S. and Shmidt, V. V. (1972). Interaction between a cavity and a vortex in a superconductor of the second kind, *Sov. Phys. JETP* **34**, p. 195.

[333] Mooij, J. E., Orlando, T. P., Levitov, L., Tian, L., van der Wal, C. H. and Lloyd, S. (1999). Josephson Persistent-Current Qubit, *Science* **285**, pp. 1036–1039.

[334] Morelle, M. (2003). *Confinement and quantization effects in superconducting nanostructures*, Ph.D. thesis, Katholieke Universiteit Leuven.

[335] Morelle, M., Bekaert, J. and Moshchalkov, V. V. (2004). Influence of sample geometry on vortex matter in superconducting microstructures, *Phys. Rev. B* **70**, 9, p. 094503.

[336] Morelle, M., Bruyndoncx, V., Jonckheere, R. and Moshchalkov, V. V. (2001). Critical temperature oscillations in magnetically coupled superconducting mesoscopic loops, *Phys. Rev. B* **64**, 6, p. 064516.

[337] Morelle, M., Bruynseraede, Y. and Moshchalkov, V. V. (2003). Effect of current and voltage leads on the superconducting properties of mesoscopic triangles, *Phys. Stat. Sol. B* **237**, 1, pp. 365–373.

[338] Morelle, M., Schildermans, N. and Moshchalkov, V. V. (2006). Rectification effects in superconducting triangles, *Applied Physics Letters* **89**, 11, p. 112512.

[339] Morgan, D. J. and Ketterson, J. B. (1998). Asymmetric flux pinning in a regular array of magnetic dipoles, *Phys. Rev. Lett.* **80**, 16, pp. 3614–3617.

[340] Moshchalkov, V., Bruyndoncx, V. and Van Look, L. (2000). Connectivity and Flux Confinement Phenomena in Nanostructured Superconductors, in J. Berger and J. Rubinstein (eds.), *Connectivity and Superconductivity, Lecture Notes in Physics monographs*, Vol. 62 (Springer Berlin / Heidelberg), pp. 87–137.

[341] Moshchalkov, V. V. (1991). Multiple-ϕ_0 concentric vortex state in superconductors, *Solid State Commun.* **78**, 8, pp. 711–715.

[342] Moshchalkov, V. V., Baert, M., Metlushko, V. V., Rosseel, E., Van Bael, M. J., Temst, K., Bruynseraede, Y. and Jonckheere, R. (1998a). Pinning by an antidot lattice: The problem of the optimum antidot size, *Phys. Rev. B* **57**, 6, pp. 3615–3622.

[343] Moshchalkov, V. V., Baert, M., Metlushko, V. V., Rosseel, E., Van Bael, M. J., Temst, K., Jonckheere, R. and Bruynseraede, Y. (1996). Magnetization of multiple-quanta vortex lattices, *Phys. Rev. B* **54**, 10, pp. 7385–7393.

[344] Moshchalkov, V. V., Baert, M., Metlushko, V. V., Rosseel, E., Van Bael, M. J., Temst, K., Qiu, X., Jonckheere, R. and Bruynseraede, Y. (1995a). Quantization and Confinement Effects in Superconducting Films with an Antidot Lattice, *Jpn. J. Appl. Phys.* **34**, Part 1, No. 8B, pp. 4559–4561.

[345] Moshchalkov, V. V., Bruyndoncx, V., Rosseel, E., Van Look, L., Baert, M., Van Bael, M. J., Puig, T., Strunk, C. and Bruynseraede, Y. (1998b). Confinement and Quantization Effects in Mesoscopic Superconducting Structures, in *Lectures on Superconductivity in Networks and Mesoscopic Systems* (New York), pp. 171–199.

[346] Moshchalkov, V. V., Dhallé, M. and Bruynseraede, Y. (1993a). Numerical simulation of aperiodic vortex states in superconductors, *Physica C* **207**, pp. 307–317.

[347] Moshchalkov, V. V., Gielen, L., Baert, M., Metlushko, V., Neuttiens, G., Strunk, C., Bruyndoncx, V., Qiu, X., Dhalle, M., Temst, K., Potter, C., Jonckheere, R., Stockman, L., Van Bael, M., van Haesendonck, C. and Bruynseraede, Y. (1994). Quantum interference and confinement phenomena in mesoscopic superconducting systems, *Phys. Scr.* **T55**, pp. 168–176.

[348] Moshchalkov, V. V., Gielen, L., Dhallé, M., Van Haesendonck, C. and Bruynseraede, Y. (1993b). Quantum interference in a mesoscopic superconducting loop, *Nature* **361**, pp. 617–620.

[349] Moshchalkov, V. V., Gielen, L., Strunk, C., Jonckheere, R., Qiu, X., Van Haesendonck, C. and Bruynseraede, Y. (1995b). Effect of sample topology on the critical fields of mesoscopic superconductors, *Nature* **373**, pp. 319–322.

[350] Moshchalkov, V. V., Henry, J. Y., Marin, C., Rossat-Mignod, J. and Jacquot, J. F. (1991). Anisotropy of the first critical field and critical current in $YBa_2Cu_3O_{6.9}$ single crystals, *Physica C* **175**, pp. 407–418.

[351] Moshchalkov, V. V., Menghini, M., Nishio, T., Chen, Q. H., Silhanek, A. V., Dao, V. H., Chibotaru, L. F., Zhigadlo, N. D. and Karpinski, J. (2009). Type-1.5 Superconductivity, *Phys. Rev. Lett.* **102**, 11, p. 117001.

[352] Moshchalkov, V. V., Qiu, X. G. and Bruyndoncx, V. (1997). Paramagnetic Meissner effect from the self-consistent solution of the Ginzburg–Landau equations, *Phys. Rev. B* **55**, 17, pp. 11793–11801.

[353] Mott, N. E. and Davis, E. A. (1979). *Electronic Processes in Non-Crystalline Materials* (Clarendon Press, Oxford, U.K.).

[354] Mühge, T., Garif'yanov, N. N., Goryunov, Y. V., Khaliullin, G. G., Tagirov, L. R., Westerholt, K., Garifullin, I. A. and Zabel, H. (1996). Possible origin for oscillatory superconducting transition temperature in superconductor/ferromagnet multilayers, *Phys. Rev. Lett.* **77**, 9, pp. 1857–1860.

[355] Muller, M. W. (1967). Nucleation of the Honeycomb Domain Structure, *Phys. Rev.* **162**, 2, pp. 423–430.

[356] Murakami, M., Morita, M., Doi, K. and Miyamoto, K. (1989). A new process with the promise of high j_c in oxide superconductors, *Jpn. J. of Appl. Phys.* **28**, Part 1, No. 7, pp. 1189–1194.

[357] Nagamatsu, J., Nakagawa, N., Muranaka, T., Zenitani, Y. and Akimitsu, J. (2001). Superconductivity at 39 K in magnesium diboride, *Nature* **410**, pp. 63–64.

[358] Nakai, N., Ichioka, M. and Machida, K. (2002). Field Dependence of Electronic Specific Heat in Two-Band Superconductors, *J. Phys. Soc. Jpn.* **71**, pp. 23–26.

[359] Nakamura, K. and Thomas, H. (1988). Quantum Billiard in a Magnetic Field: Chaos and Diamagnetism, *Phys. Rev. Lett.* **61**, 3, pp. 247–250.

[360] Nazarov, Y. V. and Stoof, T. H. (1996). Diffusive Conductors as Andreev Interferometers, *Phys. Rev. Lett.* **76**, 5, pp. 823–826.

[361] Nelson, D. R. and Vinokur, V. M. (1993). Boson localization and correlated pinning of superconducting vortex arrays, *Phys. Rev. B* **48**, 17, pp. 13060–13097.

[362] Nguyen, C., Kroemer, H. and Hu, E. L. (1992). Anomalous Andreev conductance in InAs-AlSb quantum well structures with Nb electrodes, *Phys. Rev. Lett.* **69**, 19, pp. 2847–2850.

[363] Nielsen, M. A. and Chuang, I. L. (2000). *Quantum Computation and Quantum Information* (Cambridge Univeristy, Cambridge).

[364] Nishio, T., Dao, V. H., Chen, Q., Chibotaru, L. F., Kadowaki, K. and Moshchalkov, V. V. (2010). Scanning SQUID microscopy of vortex clusters in multiband superconductors, *Phys. Rev. B* **81**, 2, p. 020506.

[365] Nordborg, H. and Vinokur, V. M. (2000). Interaction between a vortex and a columnar defect in the London approximation, *Phys. Rev. B* **62**, 18, pp. 12408–12412.

[366] Octavio, M., Tinkham, M., Blonder, G. E. and Klapwijk, T. M. (1983). Subharmonic energy-gap structure in superconducting constrictions, *Phys. Rev. B* **27**, 11, pp. 6739–6746.

[367] O'Hare, A., Kusmartsev, F. V., Kugel, K. I. and Laad, M. S. (2007). Two-dimensional Ising model with competing interactions and its application to clusters and arrays of π-rings and adiabatic quantum computing, *Phys. Rev. B* **76**, 6, p. 064528.

[368] Olson, C. J., Reichhardt, C., Jankó, B. and Nori, F. (2001). Collective Interaction-Driven Ratchet for Transporting Flux Quanta, *Phys. Rev. Lett.* **87**, 17, p. 177002.

[369] Orlande, T. P. and Delin, K. A. (1991). *Foundations of Applied Superconductivity* (Prentice Hall, New Jersey).

[370] Orlando, T. P., Lloyd, S., Levitov, L. S., Berggren, K. K., Feldman, M. J., Bocko, M. F., Mooij, J. E., Harmans, C. J. P. and van der Wal, C. H. (2002). Flux-based superconducting qubits for quantum computation, *Physica C* **372–376**, Part 1, pp. 194–200.

[371] Orlando, T. P., Mooij, J. E., Tian, L., van der Wal, C. H., Levitov, L. S., Lloyd, S. and Mazo, J. J. (1999). Superconducting persistent-current qubit, *Phys. Rev. B* **60**, 22, pp. 15398–15413.

[372] Otani, Y., Pannetier, B., Nozières, J. P. and Givord, D. (1993). Magnetostatic interactions between magnetic arrays and superconducting thin films, *J. Magn. Magn. Mater.* **126**, 1–3, pp. 662–625.

[373] Palacios, J. J. (1998). Vortex matter in superconducting mesoscopic disks: Structure, magnetization, and phase transitions, *Phys. Rev. B* **58**, 10, pp. R5948–R5951.

[374] Palacios, J. J. (2000). Metastability and Paramagnetism in Superconducting Mesoscopic Disks, *Phys. Rev. Lett.* **84**, 8, pp. 1796–1799.

[375] Palacios, J. J., Peeters, F. M. and Baelus, B. J. (2001). Effective lowest

Landau level treatment of demagnetization in superconducting mesoscopic disks, *Phys. Rev. B* **64**, 13, p. 134514.

[376] Pannetier, B. (1991). *Quantum Coherence in Mesoscopic Systems* (Plenum Press, New York).

[377] Pannetier, B., Chaussy, J., Rammal, R. and Villegier, J. C. (1984). Experimental fine tuning of frustration: Two-dimensional superconducting network in a magnetic field, *Phys. Rev. Lett.* **53**, 19, pp. 1845–1848.

[378] Pannetier, M., Wijngaarden, R. J., Fløan, I., Rector, J., Dam, B., Griessen, R., Lahl, P. and Wördenweber, R. (2003). Unexpected fourfold symmetry in the resistivity of patterned superconductors, *Phys. Rev. B* **67**, 21, p. 212501.

[379] Parks, R. D. (1964). Quantized Magnetic Flux in Superconductors: Experiments confirm Fritz London's early concept that superconductivity is a macroscopic quantum phenomenon, *Science* **146**, pp. 1429–1435.

[380] Parks, R. D. and Little, W. A. (1964). Fluxoid Quantization in a Multiply-Connected Superconductor, *Phys. Rev.* **133**, pp. A97–A103.

[381] Pastoriza, H., Goffman, M. F., Arribére, A. and de la Cruz, F. (1994). First order phase transition at the irreversibility line of $Bi_2Sr_2CaCu_2O_8$, *Phys. Rev. Lett.* **72**, 18, pp. 2951–2954.

[382] Pearl, J. (1964). Current distribution in superconducting films carrying quantized fluxoids, *Appl. Phys. Lett.* **5**, 4, pp. 65–66.

[383] Peeters, F. M. and Li, X. Q. (1998). Hall magnetometer in the ballistic regime, *Appl. Phys. Lett.* **72**, 5, pp. 572–574.

[384] Pekola, J. P., Simola, J. T., Hakonen, P. J., Krusius, M., Lounasmaa, O. V., Nummila, K. K., Mamniashvili, G., Packard, R. E. and Volovik, G. E. (1984). Phase Diagram of the First-Order Vortex-Core Transition in Superfluid ^{3}He-B, *Phys. Rev. Lett.* **53**, 6, pp. 584–587.

[385] Perkins, G. K., Moore, J., Bugoslavsky, Y., Cohen, L. F., Jun, J., Kazakov, S. M., Karpinski, J. and Caplin, A. D. (2002). Superconducting critical fields and anisotropy of a MgB_2 single crystal, *Supercond. Sci. Technol.* **15**, pp. 1156–1159.

[386] Pogosov, W. V., Kugel, K. I., Rakhmanov, A. L. and Brandt, E. H. (2001). Approximate Ginzburg–Landau solution for the regular flux-line lattice: Circular cell method, *Phys. Rev. B* **64**, 6, p. 064517.

[387] Poole, C. P., Farach Jr., H. A. and Creswick, R. J. (1995). *Superconductivity* (Academic Press, San Diego).

[388] Priour, D. J. and Fertig, H. A. (2003). Deformation and depinning of superconducting vortices from artificial defects: A Ginzburg–Landau study, *Phys. Rev. B* **67**, 5, p. 054504.

[389] Pryadun, V. V., Sierra, J., Aliev, F. G., Golubovic, D. S. and Moshchalkov, V. V. (2006). Plain superconducting films as magnetic field tunable two-dimensional rectifiers, *Appl. Phys. Lett.* **88**, 6, p. 062517.

[390] Puig, T., Rosseel, E., Baert, M., Van Bael, M. J., Moshchalkov, V. V. and Bruynseraede, Y. (1997). Stable vortex configurations in superconducting 2×2 antidot clusters, *Appl. Phys. Lett.* **70**, 23, pp. 3155–3157.

[391] Puig, T., Rosseel, E., Van Look, L., Van Bael, M. J., Moshchalkov, V. V.,

Bruynseraede, Y. and Jonckheere, R. (1998). Vortex configurations in a Pb/Cu microdot with a 2 × 2 antidot cluster, *Phys. Rev. B* **58**, 9, pp. 5744–5756.

[392] Radović, Z., Ledvij, M., Dobrosavljević-Grujić, L., Buzdin, A. I. and Clem, J. R. (1991). Transition temperatures of superconductor-ferromagnet superlattices, *Phys. Rev. B* **44**, 2, pp. 759–764.

[393] Raedts, S., Silhanek, A. V., Moshchalkov, V. V., Moonens, J. and Leunissen, L. H. A. (2006). Crossover from intravalley to intervalley vortex motion in type-II superconductors with a periodic pinning array, *Phys. Rev. B* **73**, 17, p. 174514.

[394] Raedts, S., Silhanek, A. V., Van Bael, M. J. and Moshchalkov, V. V. (2004). Flux-pinning properties of superconducting films with arrays of blind holes, *Phys. Rev. B* **70**, 2, p. 024509.

[395] Ralph, D. C., Black, C. T. and Tinkham, M. (1997). Gate-Voltage Studies of Discrete Electronic States in Aluminum Nanoparticles, *Phys. Rev. Lett.* **78**, 21, pp. 4087–4090.

[396] Range, S. K. (2004). Gravity Probe B: Exploring Einstein's Universe with Gyroscopes, *NASA* **26**.

[397] Reichhardt, C. and Grønbech-Jensen, N. (2001). Critical currents and vortex states at fractional matching fields in superconductors with periodic pinning, *Phys. Rev. B* **63**, 5, p. 054510.

[398] Reichhardt, C., Groth, J., Olson, C. J., Field, S. B. and Nori, F. (1996). Spatiotemporal dynamics and plastic flow of vortices in superconductors with periodic arrays of pinning sites, *Phys. Rev. B* **54**, 22, pp. 16108–16115.

[399] Reichhardt, C., Olson, C. J. and Nori, F. (1998a). Commensurate and incommensurate vortex states in superconductors with periodic pinning arrays, *Phys. Rev. B* **57**, 13, pp. 7937–7943.

[400] Reichhardt, C., Olson, C. J. and Nori, F. (1998b). Nonequilibrium dynamic phases and plastic flow of driven vortex lattices in superconductors with periodic arrays of pinning sites, *Phys. Rev. B* **58**, 10, pp. 6534–6564.

[401] Reichhardt, C. and Olson Reichhardt, C. J. (2009). Transport anisotropy as a probe of the interstitial vortex state in superconductors with artificial pinning arrays, *Phys. Rev. B* **79**, 13, p. 134501.

[402] Reichhardt, C. and Reichhardt, C. J. O. (2007). Commensurability effects at nonmatching fields for vortices in diluted periodic pinning arrays, *Phys. Rev. B* **76**, 9, p. 094512.

[403] Reimann, P. (2002). Brownian Motors: Noisy Transport far from Equilibrium, *Phys. Rep.* **316**, p. 57.

[404] Richardson, R. W. (1963). A restricted class of exact eigenstates of the pairing-force Hamiltonian, *Phys. Lett.* **3**, 6, pp. 277–279.

[405] Richardson, R. W. (1966). Numerical Study of the 8-32-Particle Eigenstates of the Pairing Hamiltonian, *Phys. Rev.* **141**, 3, pp. 949–956.

[406] Rodrigo, J. G., Suderow, H. and Vieira, S. (2003). Superconducting nanobridges under magnetic fields, *Phys. Stat. Sol (b)* **237**, 1, pp. 386–393.

[407] Romijn, J., Klapwijk, T. M., Renne, M. J. and Mooij, J. E. (1982). Criti-

cal pair-breaking current in superconducting aluminum strips far below T_c, *Phys. Rev. B* **26**, pp. 3648–3655.

[408] Rose-Innes, A. C. and Rhoderick, E. H. (1978). *Introduction to Superconductivity* (Pergamon Press, Oxford, U.K.).

[409] Rosenstein, B., Shapiro, I. and Shapiro, B. Y. (2010). Maximal persistent current in a type-II superconductor with an artificial pinning array at the matching magnetic field, *Phys. Rev. B* **81**, 6, p. 064507.

[410] Ross, C. A., Haratani, S., Castaño, F. J., Hao, Y., Hwang, M., Shima, M., Cheng, J. Y., Vögeli, B., Farhoud, M., Walsh, M. and Smith, H. I. (2002). Magnetic behavior of lithographically patterned particle arrays, *J. Appl. Phys.* **91**, 10, pp. 6848–6853.

[411] Rosseel, E. (1998). *Critical parameters of superconductors with an antidot lattice*, Ph.D. thesis, Katholieke Universiteit Leuven.

[412] Rosseel, E., Puig, T., Baert, M., Van Bael, M. J., Moshchalkov, V. V. and Bruynseraede, Y. (1997). Upper critical field of Pb films with an antidot lattice, *Physica C* **282–287**, 3, pp. 1567–1568.

[413] Rosseel, E., Van Bael, M., Baert, M., Jonckheere, R., Moshchalkov, V. V. and Bruynseraede, Y. (1996). Depinning of caged interstitial vortices in superconducting a-WGe films with an antidot lattice, *Phys. Rev. B* **53**, 6, pp. R2983–R2986.

[414] Rubio-Bollinger, G., Suderow, H. and Vieira, S. (2001). Tunneling Spectroscopy in Small Grains of Superconducting MgB_2, *Phys. Rev. Lett.* **86**, 24, pp. 5582–5584.

[415] Saint-James, D. (1965). Etude du champ critique H_{c3} dans une geometrie cylindrique, *Phys. Lett.* **15**, 1, pp. 13–15.

[416] Sardella, E., Lisboa Filho, P. N., de Souza Silva, C. C., Eulálio Cabral, L. R. and Aires Ortiz, W. (2009). Vortex-antivortex annihilation dynamics in a square mesoscopic superconducting cylinder, *Phys. Rev. B* **80**, 1, p. 012506.

[417] Sardella, E. and Oliveira, T. M. (2004). Depairing Critical Current Density of a Mesoscopic Square Superconductor, *Braz. J. Phys.* **34**, 3B, pp. 1265–1269.

[418] Sarma, N. V. (1968). Transition from the flux lattice to the intermediate state structures in a lead-indium alloy, *Philos. Mag.* **18**, 151, pp. 171–176.

[419] Saxena, S. S., Agarwal, P., Ahilan, K., Grosche, F. M., Haselwimmer, R. K. W., Steiner, M. J., Pugh, E., Walker, I. R., Julian, S. R., Monthoux, P., Lonzarich, G. G., Huxley, A., Sheikin, I., Braithwaite, D. and Flouquet, J. (2000). Superconductivity on the border of itinerant-electron ferromagnetism in UGe_2, *Nature* **406**, pp. 587–592.

[420] Schechter, M., Imry, Y., Levinson, Y. and Delft, J. v. (2001). Thermodynamic properties of a small superconducting grain, *Phys. Rev. B* **63**, 21, p. 214518.

[421] Schildermans, N. (2008). *Rectification effects and nucleation of superconductivity in superconducting nanostructures*, Ph.D. thesis, Katholieke Universiteit Leuven.

[422] Schildermans, N., Kolton, A. B., Salenbien, R., Marconi, V. I., Silhanek, A. V. and Moshchalkov, V. V. (2007). Voltage rectification effects in meso-

scopic superconducting triangles: Experiment and modeling, *Phys. Rev. B* **76**, 22, p. 224501.

[423] Schilling, A., Fisher, R. A., Phillips, N. E., Welp, U., Dasgupta, D., Kwok, W. K. and Crabtree, G. W. (1996). Calorimetric measurement of the latent heat of vortex-lattice melting in untwinned $YBa_2Cu_3O_{7d}$, *Nature* **382**, pp. 791–793.

[424] Schliepe, B., Stindtmann, M., Nikolic, I. and Baberschke, K. (1993). Positive field-cooled susceptibility in high-T_c superconductors, *Phys. Rev. B* **47**, 13, pp. 8331–8334.

[425] Schultens, H. A. (1970). Current vortices in superconducting films at the surface parallel critical field, *Z. Physik* **232**, 5, pp. 430–438.

[426] Schulz, R. R., Chesca, B., Goetz, B., Schneider, C. W., Schmehl, A., Bielefeldt, H., Hilgenkamp, H., Mannhart, J. and Tsuei, C. C. (2000). Design and realization of an all d-wave dc π-superconducting quantum interference device, *Appl. Phys. Lett.* **76**, 7, pp. 912–914.

[427] Schuster, R., Ensslin, K., Wharam, D., Kühn, S., Kotthaus, J. P., Böhm, G., Klein, W., Tränkle, G. and Weimann, G. (1994). Phase-coherent electrons in a finite antidot lattice, *Phys. Rev. B* **49**, 12, pp. 8510–8513.

[428] Schweigert, V. A. and Peeters, F. M. (1998). Phase transitions in thin mesoscopic superconducting disks, *Phys. Rev. B* **57**, 21, pp. 13817–13832.

[429] Schweigert, V. A. and Peeters, F. M. (1999). Influence of the confinement geometry on surface superconductivity, *Phys. Rev. B* **60**, 5, pp. 3084–3087.

[430] Selders, P., Castellanos, A., Vaupel, M. and Wördenweber, R. (1997). Observation of and noise reduction by vortex lattice matching in $YBa_2Cu_3O_7$ thin films and rf-SQUIDs with regular arrays of antidots, *Appl. Supercond.* **5**, 7–12, pp. 269–276.

[431] Seynaeve, E., Rens, G., Volodin, A. V., Temst, K., Van Haesendonck, C. and Bruynseraede, Y. (2001). Transition from a single-domain to a multidomain state in mesoscopic ferromagnetic Co structures, *J. Appl. Phys.* **89**, 1, pp. 531–534.

[432] Shalóm, D. E. and Pastoriza, H. (2005). Vortex Motion Rectification in Josephson Junction Arrays with a Ratchet Potential, *Phys. Rev. Lett.* **94**, 17, p. 177001.

[433] Shan, L., Tao, H. J., Gao, H., Li, Z. Z., Ren, Z. A., Che, G. C. and Wen, H. H. (2003). s-wave pairing in $MgCNi_3$ revealed by point contact tunneling, *Phys. Rev. B* **68**, 14, p. 144510.

[434] Shoenberg, D. (1965). *Superconductivity* (Cambridge University Press, Cambridge, MA).

[435] Sierra, G., Dukelsky, J., Dussel, G. G., von Delft, J. and Braun, F. (2000). Exact study of the effect of level statistics in ultrasmall superconducting grains, *Phys. Rev. B* **61**, 18, pp. R11890–R11893.

[436] Silhanek, A. V., Gillijns, W., Milošević, M. V., Volodin, A., Moshchalkov, V. V. and Peeters, F. M. (2007a). Optimization of superconducting critical parameters by tuning the size and magnetization of arrays of magnetic dots, *Phys. Rev. B* **76**, 10, p. 100502.

[437] Silhanek, A. V., Gillijns, W., Moshchalkov, V. V., Metlushko, V., Gozzini,

F., Ilic, B., Uhlig, W. C. and Unguris, J. (2007b). Manipulation of the vortex motion in nanostructured ferromagnetic/superconductor hybrids, *Appl. Phys. Lett.* **90**, 18, p. 182501.

[438] Silhanek, A. V., Gillijns, W., Moshchalkov, V. V., Zhu, B. Y., Moonens, J. and Leunissen, L. H. A. (2006). Enhanced pinning and proliferation of matching effects in a superconducting film with a Penrose array of magnetic dots, *Appl. Phys. Lett.* **89**, 15, p. 152507.

[439] Silhanek, A. V., Van Look, L., Jonckheere, R., Zhu, B. Y., Raedts, S. and Moshchalkov, V. V. (2005). Enhanced vortex pinning by a composite antidot lattice in a superconducting Pb film, *Phys. Rev. B* **72**, 1, p. 014507.

[440] Silhanek, A. V., Verellen, N., Metlushko, V., Gillijns, W., Gozzini, F., Ilic, B. and Moshchalkov, V. V. (2008). Rectification effects in superconductors with magnetic pinning centers, *Physica C* **468**, 7–10, pp. 563–567.

[441] Simonin, J., Rodrigues, D. and López, A. (1982). Upper Critical Field of Regular Superconductive Networks, *Phys. Rev. Lett.* **49**, 13, pp. 944–947.

[442] Simonin, J. M., Wiecko, C. and López, A. (1983). Upper critical fields of regular superconductive networks. surfaces and impurities, *Phys. Rev. B* **28**, 5, pp. 2497–2504.

[443] Singha Deo, P., Schweigert, V. A. and Peeters, F. M. (1999). Hysteresis in mesoscopic superconducting disks: The Bean–Livingston barrier, *Phys. Rev. B* **59**, 9, pp. 6039–6042.

[444] Sivakov, A. G., Zhuravel', A. P., Turutanov, O. G. and Dmitrenko, I. M. (1996). Spatially resolved characterization of superconducting films and cryoelectronic devices by means of low temperature scanning laser microscope, *Appl. Surf. Sci.* **106**, pp. 390–395.

[445] Smoluchowski, M. (1912). Experimentall nachweisbare, der üblichen Thermodynamic widersprechende Molekular phänomene. *Phys. Z.* **13**, pp. 1069–1080.

[446] Soulen, R. J., Byers, J. M., Osofsky, M. S., Nadgorny, B., Ambrose, T., Cheng, S. F., Broussard, P. R., Tanaka, C. T., Nowak, J., Moodera, J. S., Barry, A. and Coey, J. M. D. (1998). Measuring the Spin Polarization of a Metal with a Superconducting Point Contact, *Science* **282**, pp. 85–88.

[447] Souma, S., Machida, Y., Sato, T., Takahashi, T., Matsui, H., Wang, S.-C., Ding, H., Kaminski, A., Campuzano, J. C., Sasaki, S. and Kadowaki, K. (2003). The origin of multiple superconducting gaps in MgB$_2$, *Nature* **423**, pp. 65–67.

[448] Stoll, O. M., Montero, M. I., Guimpel, J., Åkerman, J. J. and Schuller, I. K. (2002). Hysteresis and fractional matching in thin Nb films with rectangular arrays of nanoscaled magnetic dots, *Phys. Rev. B* **65**, 10, p. 104518.

[449] Straley, J. P. and Barnett, G. M. (1993). Phase diagram for a Josephson network in a magnetic field, *Phys. Rev. B* **48**, 5, pp. 3309–3315.

[450] Strunk, C., Bruyndoncx, V., Moshchalkov, V. V., Van Haesendonck, C., Bruynseraede, Y. and Jonckheere, R. (1996). Nonlocal effects in mesoscopic superconducting aluminum structures, *Phys. Rev. B* **54**, pp. R12701–R12704.

[451] Strunk, C., Sürgers, C., Paschen, U. and Löhneysen, H. v. (1994). Super-conductivity in layered Nb/Gd films, *Phys. Rev. B* **49**, 6, pp. 4053–4063.

[452] Suhl, H. (1965). Inertial Mass of a Moving Fluxoid, *Phys. Rev. Lett.* **14**, 7, pp. 226–229.

[453] Szabó, P., Samuely, P., Kačmarčík, J., Klein, T., Marcus, J., Fruchart, D., Miraglia, S., Marcenat, C. and Jansen, A. G. M. (2001). Evidence for Two superconducting Energy Gaps in MgB$_2$ by Point-Contact Spectroscopy, *Phys. Rev. Lett.* **87**, 13, p. 137005.

[454] Tachiki, M., Kotani, A., Matsumoto, H. and Umezawa, H. (1979). Super-conducting bloch-wall in ferromagnetic superconductors, *Solid State Commun.* **32**, 8, pp. 599–602.

[455] Tafuri, F. and Kirtley, J. R. (2005). Weak links in high critical temperature superconductors, *Rep. Prog. Phys.* **68**, pp. 2573–2663.

[456] Takezawa, N. and Fukushima, K. (1994). Optimal size of a cylindrical insulating inclusion acting as a pinning center for magnetic flux in superconductors, *Physica C* **228**, 1–2, pp. 149–159.

[457] Tanaka, K. and Eschrig, M. (2009). Abrikosov flux-lines in two-band superconductors with mixed dimensionality, *Supercond. Sci. Technol.* **22**, 1, p. 014001.

[458] Tanaka, K., Robel, I. and Jankó, B. (2002). Electronic structure of multiquantum giant vortex states in mesoscopic superconducting disks, *Phys. Rev. B* **99**, 8, pp. 5233–5236.

[459] Tang, H. X., Kawakami, R. K., Awschalom, D. D. and Roukes, M. L. (2006). Propagation dynamics of individual domain walls in Ga$_{1-x}$Mn$_x$As microdevices, *Phys. Rev. B* **74**, 4, p. 041310.

[460] Tempere, J., Gladilin, V. N., Silvera, I. F., Devreese, J. T. and Moshchalkov, V. V. (2009). Coexistence of the Meissner and vortex states on a nanoscale superconducting spherical shell, *Phys. Rev. B* **79**, 13, p. 134516.

[461] Teniers, G., Chibotaru, L. F., Ceulemans, A. and Moshchalkov, V. V. (2003). Nucleation of superconductivity in a mesoscopic rectangle, *Europhys. Lett.* **63**, 2, pp. 296–302.

[462] Théron, R., Simond, J.-B., Leemann, C., Beck, H., Martinoli, P. and Minnhagen, P. (1993). Evidence for nonconventional vortex dynamics in an ideal two-dimensional superconductor, *Phys. Rev. Lett.* **71**, 8, pp. 1246–1249.

[463] Thompson, D. J., Minhaj, M. S. M., Wenger, L. E. and Chen, J. T. (1995). Observation of Paramagnetic Meissner Effect in Niobium Disks, *Phys. Rev. Lett.* **75**, 3, pp. 529–532.

[464] Tilley, D. R. and Tilley, J. (1990). *Superfluidity and Superconductivity* (Hilger, Bristol, U.K.).

[465] Tinkham, M. (1975). *Introduction to Superconductivity* (McGraw Hill, New York).

[466] Tkaczyk, J. E., DeLuca, J. A., Karas, P. L., Bednarczyk, P. J., Christen, D. K., Klabunde, C. E. and Kerchner, H. R. (1993). Enhanced transport critical current at high fields after heavy ion irradiation of textured

TlBa$_2$Ca$_2$Cu$_3$O$_z$ thick films, *Appl. Phys. Lett.* **62**, 23, pp. 3031–3033.

[467] Träuble, H. and Essmann, U. (1967). Die Beobachtung magnetischer Strukturen von Supraleitern zweiter Art, *Phys. Stat. Sol. (b)* **20**, 1, pp. 95–111.

[468] Trías, E., Mazo, J. J., Falo, F. and Orlando, T. P. (2000). Depinning of kinks in a Josephson-junction ratchet array, *Phys. Rev. E* **61**, 3, pp. 2257–2266.

[469] Tsuei, C. C., Kirtley, J. R., Chi, C. C., Yu-Jahnes, L. S., Gupta, A., Shaw, T., Sun, J. Z. and Ketchen, M. B. (1994). Pairing Symmetry and Flux Quantization in a Tricrystal Superconducting Ring of YBa$_2$Cu$_3$O$_{7-\delta}$, *Phys. Rev. Lett.* **73**, 4, pp. 593–596.

[470] Uji, S., Shinagawa, H., Terashima, T., Yakabe, T., Terai, Y., Tokumoto, M., Kobayashi, A., Tanaka, H. and Kobayashi, H. (2001). Magnetic-field-induced superconductivity in a two-dimensional organic conductor, *Nature* **410**, pp. 908–910.

[471] Van Bael, M. J. (1998). *Regular arrays of magnetic dots and their flux pinning properties*, Ph.D. thesis, Katholieke Universiteit Leuven.

[472] Van Bael, M. J., Bekaert, J., Temst, K., Van Look, L., Moshchalkov, V. V., Bruynseraede, Y., Howells, G. D., Grigorenko, A. N., Bending, S. J. and Borghs, G. (2001). Local Observation of Field Polarity Dependent Flux Pinning by Magnetic Dipoles, *Phys. Rev. Lett.* **86**, 1, pp. 155–158.

[473] Van Bael, M. J., Lange, M., Raedts, S., Moshchalkov, V. V., Grigorenko, A. N. and Bending, S. J. (2003). Local visualization of asymmetric flux pinning by magnetic dots with perpendicular magnetization, *Phys. Rev. B* **68**, 1, p. 014509.

[474] Van Bael, M. J., Temst, K., Moshchalkov, V. V. and Bruynseraede, Y. (1999). Magnetic properties of submicron Co islands and their use as artificial pinning centers, *Phys. Rev. B* **59**, 22, pp. 14674–14679.

[475] Van de Vondel, J. (2007). *Vortex dynamics and rectification effects in superconducting films with periodic asymmetric pinning*, Ph.D. thesis, Katholieke Universiteit Leuven.

[476] Van de Vondel, J., de Souza Silva, C. C. and Moshchalkov, V. V. (2007). Diode effects in the surface superconductivity regime, *Europhys. Lett.* **80**, 1, p. 17006.

[477] Van de Vondel, J., de Souza Silva, C. C., Zhu, B. Y., Morelle, M. and Moshchalkov, V. V. (2005). Vortex-rectification effects in films with periodic asymmetric pinning, *Phys. Rev. Lett.* **94**, 5, p. 057003.

[478] van der Wal, C. H., ter Haar, A. C. J., Wilhelm, F. K., Schouten, R. N., Harmans, C. J. P. M., Orlando, T. P., Lloyd, S. and Mooij, J. E. (2000). Quantum Superposition of Macroscopic Persistent-Current States, *Science* **290**, pp. 773–777.

[479] van der Zant, H. S. J., Rijken, H. A. and Mooij, J. E. (1990). The superconducting transition of 2-D Josephson-junction arrays in a small perpendicular magnetic field, *J. Low Temp. Phys.* **79**, 5–6, pp. 289–310.

[480] van Dover, R. B., Gyorgy, E. M., White, A. E., Schneemeyer, L. F., Felder, R. J. and Waszczak, J. V. (1990). Critical currents in proton-irradiated single-crystal Ba$_2$YCu$_3$O$_{7-\delta}$, *Appl. Phys. Lett.* **56**, 26, pp. 2681–2683.

[481] van Gelder, A. P. (1968). Nucleation of Superconductivity above H_{c3}, *Phys. Rev. Lett.* **20**, 25, pp. 1435–1436.

[482] Van Look, L., Zhu, B. Y., Jonckheere, R., Zhao, B. R., Zhao, Z. X. and Moshchalkov, V. V. (2002). Anisotropic vortex pinning in superconductors with a square array of rectangular submicron holes, *Phys. Rev. B* **66**, 21, p. 214511.

[483] van Wees, B. J., de Vries, P., Magnée, P. and Klapwijk, T. M. (1992). Excess conductance of superconductor-semiconductor interfaces due to phase conjugation between electrons and holes, *Phys. Rev. Lett.* **69**, 3, pp. 510–513.

[484] Vavassori, P., Grimsditch, M., Novosad, V., Metlushko, V. and Ilic, B. (2003). Metastable states during magnetization reversal in square permalloy rings, *Phys. Rev. B* **67**, 13, p. 134429.

[485] Velez, M., Jaque, D., Martín, J. I., Montero, M. I., Schuller, I. K. and Vicent, J. L. (2002). Vortex lattice channeling effects in Nb films induced by anisotropic arrays of mesoscopic pinning centers, *Phys. Rev. B* **65**, 10, p. 104511.

[486] Verbanck, G., Potter, C. D., Metlushko, V., Schad, R., Moshchalkov, V. V. and Bruynseraede, Y. (1998). Coupling phenomena in superconducting Nb/Fe multilayers, *Phys. Rev. B* **57**, 10, pp. 6029–6035.

[487] Villegas, J. E., Montero, M. I., Li, C.-P. and Schuller, I. K. (2006). Correlation Length of Quasiperiodic Vortex Lattices, *Phys. Rev. Lett.* **97**, 2, p. 027002.

[488] Villegas, J. E., Savel'ev, S., Nori, F., Gonzalez, E. M., Anguita, J. V., Garca, R. and Vicent, J. L. (2003). A Superconducting Reversible Rectifier That Controls the Motion of Magnetic Flux Quanta, *Science* **302**, pp. 1188–1191.

[489] Vinnikov, L. Y., Karpinski, J., Kazakov, S. M., Jun, J., Anderegg, J., Bud'ko, S. L. and Canfield, P. C. (2003). Vortex structure in MgB_2 single crystals observed by the Bitter decoration technique, *Phys. Rev. B* **67**, 9, p. 092512.

[490] Vodolazov, D. Y. and Peeters, F. M. (2005). Superconducting rectifier based on the asymmetric surface barrier effect, *Phys. Rev. B* **72**, 17, p. 172508.

[491] Vollhardt, D. and Wlfle, P. (1990). *The Superfluid Phases of Helium 3* (Taylor & Francis, London).

[492] Volovik, G. E. (1992). *Exotic Properties of Superfluid ^3He* (World Scientific, Singapore).

[493] Wagner, P., Gordon, I., Trappeniers, L., Vanacken, J., Herlach, F., Moshchalkov, V. V. and Bruynseraede, Y. (1998). Spin Dependent Hopping and Colossal Negative Magnetoresistance in Epitaxial $Nd_{0.52}Sr_{0.48}MnO_3$ Films in Fields up to 50 T, *Phys. Rev. Lett.* **81**, 18, pp. 3980–3983.

[494] Wahl, A., Hardy, V., Provost, J., Simon, C. and Buzdin, A. (1995). Unusual field dependence of the reversible magnetization in heavy ions irradiated thallium-based single crystals, *Physica C* **250**, 1–2, pp. 163–169.

[495] Wambaugh, J. F., Reichhardt, C., Olson, C. J., Marchesoni, F. and Nori, F. (1999). Superconducting Fluxon Pumps and Lenses, *Phys. Rev. Lett.* **83**, 24, pp. 5106–5109.

[496] Wang, J. and Ma, Z.-S. (1995). Aharonov–Bohm effect induced by mutual inductance for an array of mesoscopic rings, *Phys. Rev. B* **52**, 20, pp. 14829–14833.

[497] Watson, G. I. (1997). Repulsive particles on a two-dimensional lattice, *Physica A* **246**, 1–2, pp. 253–274.

[498] Watson, G. I. and Canright, G. S. (1993). Frustration-induced disorder of flux-line structures in layered superconductors, *Phys. Rev. B* **48**, 21, pp. 15950–15956.

[499] Welker, H. and Bayer, S. B. (1938). Über ein elektronen-theoretisches Modell des Supraleiters, *Akad. Wiss.* **14**, p. 115.

[500] Welp, U., Xiao, Z. L., Jiang, J. S., Vlasko-Vlasov, V. K., Bader, S. D., Crabtree, G. W., Liang, J., Chik, H. and Xu, J. M. (2002). Superconducting transition and vortex pinning in nb films patterned with nanoscale hole arrays, *Phys. Rev. B* **66**, 21, p. 212507.

[501] Wilks, C. W., Bojko, R. and Chaikin, P. M. (1991). Field dependence of the resistive transition for a square wire network, *Phys. Rev. B* **43**, 4, pp. 2721–2725.

[502] Wolf, S. A., Fuller, W. W., Huang, C. Y., Harrison, D. W., Luo, H. L. and Maekawa, S. (1982). Magnetic-field-induced superconductivity, *Phys. Rev. B* **25**, 3, pp. 1990–1992.

[503] Wong, H. K., Jin, B. Y., Yang, H. Q., Ketterson, J. B. and Hilliard, J. E. (1986). Superconducting properties of V/Fe superlattices, *J. Low Temp. Phys.* **63**, 9, pp. 307–315.

[504] Wördenweber, R., Dymashevski, P. and Misko, V. R. (2004). Guidance of vortices and the vortex ratchet effect in high-Tc superconducting thin films obtained by arrangement of antidots, *Phys. Rev. B* **69**, 18, p. 184504.

[505] Wördenweber, R., Hollmann, E., Schubert, J., Kutzner, R. and Ghosh, A. K. (2009). Pattern induced phase transition of vortex motion in high-T_c films, *Appl. Phys. Lett.* **94**, 20, p. 202501.

[506] Xiong, P., Xiao, G. and Laibowitz, R. B. (1993). Subgap and above-gap differential resistance anomalies in superconductor-normal-metal microjunctions, *Phys. Rev. Lett.* **71**, 12, pp. 1907–1910.

[507] Yamanouchi, M., Chiba, D., Matsukura, F. and Ohno, H. (2004). Current-induced domain-wall switching in a ferromagnetic semiconductor structure, *Nature* **428**, pp. 539–542.

[508] Yang, Z., Lange, M., Volodin, A., Szymczak, R. and Moshchalkov, V. V. (2004). Domain-wall superconductivity in superconductor-ferromagnet hybrids, *Nature Mater.* **3**, pp. 793–798.

[509] Zapata, I., Bartussek, R., Sols, F. and Hänggi, P. (1996). Voltage Rectification by a SQUID Ratchet, *Phys. Rev. Lett.* **77**, 11, pp. 2292–2295.

[510] Zhang, T., Cheng, P., Li, W.-J., Sun, Y.-J., Wang, G., Zhu, X.-G., He, K., Wang, L., Ma, X., Chen, X., Wang, Y., Liu, Y., Lin, H.-Q., Jia, J.-F. and Xue, Q.-K. (2010). Superconductivity in one-atomic-layer metal films grown on Si(111), *Nature Physics* **6**, pp. 104–108.

[511] Zhang, X. and Price, J. C. (1997). Susceptibility of a mesoscopic superconducting ring, *Phys. Rev. B* **55**, pp. 3128–3140.

[512] Zhang, X. X., Hernàndez, J. M., Tejada, J., Solé, R. and Ruiz, X. (1996). Magnetic properties and domain-wall motion in single-crystal $BaFe_{10.2}Sn_{0.74}Co_{0.66}O_{19}$, *Phys. Rev. B* **53**, 6, pp. 3336–3340.

[513] Zhao, H., Fomin, V. M., Devreese, J. T. and Moshchalkov, V. V. (2003). A new vortex state with non-uniform vorticity in superconducting mesoscopic rings, *Solid State Communications* **125**, 1, pp. 59–63.

[514] Zharkov, G. F. (2001). Paramagnetic Meissner effect in superconductors from self-consistent solution of Ginzburg–Landau equations, *Phys. Rev. B* **63**, 21, p. 214502.

[515] Zhitomirsky, M. E. and Dao, V.-H. (2004). Ginzburg–Landau theory of vortices in a multigap superconductor, *Phys. Rev. B* **69**, 5, p. 054508.

[516] Zhu, B. Y., Look, L. V., Moshchalkov, V. V., Zhao, B. R. and Zhao, Z. X. (2001). Vortex dynamics in regular arrays of asymmetric pinning centers, *Phys. Rev. B* **64**, 1, p. 012504.

[517] Zhu, B. Y., Marchesoni, F., Moshchalkov, V. V. and Nori, F. (2003). Controllable step motors and rectifiers of magnetic flux quanta using periodic arrays of asymmetric pinning defects, *Phys. Rev. B* **68**, 1, p. 014514.

[518] Zhu, B. Y., Marchesoni, F. and Nori, F. (2004). Controlling the motion of magnetic flux quanta, *Phys. Rev. Lett.* **92**, 18, p. 180602.

[519] Zhu, J.-X., Wang, D. Z. and Wang, Q. (1996). Flux State in a System of Two Magnetically Coupled Mesoscopic Rings, *J. Phys. Soc. Japan* **65**, 8, pp. 2602–2605.

Index

www.ingramcontent.com/pod-product-compliance
Lightning Source LLC
Chambersburg PA
CBHW050541190326
41458CB00007B/1863